Principles of Highway Engineering and Traffic Analysis

Third Edition

Fred L. Mannering
Purdue University

Walter P. Kilareski
The Pennsylvania State University

Scott S. Washburn
University of Florida

John Wiley & Sons, Inc.

ACQUISITIONS EDITOR   Jenny Welter

SENIOR PRODUCTION EDITOR   Norine M. Pigliucci

SENIOR MARKETING MANAGER   Jenny Powers

NEW MEDIA EDITOR   Thomas Kulesa

SENIOR DESIGNER   Dawn Stanley

PRODUCTION MANAGEMENT SERVICES   Publication Services

This book was set in Times Roman by Publication Services and printed and bound by Malloy, Inc. The cover was printed by Phoenix Color.

This book is printed on acid-free paper. ∞

Copyright © 2005, by John Wiley & Sons, Inc. All rights reserved.

No part of this publication may be reproduced, stored in a retrieval system or transmitted in any form or by any means, electronic, mechanical, photocopying, recording, scanning or otherwise, except as permitted under Sections 107 or 108 of the 1976 United States Copyright Act, without either the prior written permission of the Publisher, or authorization through payment of the appropriate per-copy fee to the Copyright Clearance Center, 222 Rosewood Drive, Danvers, MA 01923, (508) 750-8400, fax (508) 750-4470. Requests to the Publisher for permission should be addressed to the Permissions Department, John Wiley & Sons, Inc., 111 River Street, Hoboken, NJ 07030, (201) 748-6011, fax (201) 748-6008, E-Mail: PERMREQ@WILEY.COM.
To order books or for customer service please call 1-800-CALL-WILEY (225-5945).

Library of Congress Cataloging-in-Publication Data
Mannering, Fred L.
  Principles of highway engineering and traffic analysis / Fred L. Mannering, Walter P. Kilareski, Scott S. Washburn -- 3rd ed.
     p. cm.
   ISBN 0-471-47256-5 (cloth)
   ISBN 0-471-66156-2 (WIE)
     1. Highway engineering. 2. Traffic engineering. I. Kilareski, Walter P. II. Washburn, Scott S. III. Title.
TE147.M28 2004
625.7—dc22
                                                                    2004042225
Printed in the United States of America
10 9 8 7 6 5 4 3 2 1

Preface

INTRODUCTION

The first two editions of *Principles of Highway Engineering and Traffic Analysis* sought to redefine how entry-level transportation engineering courses are taught. When the first edition was published, we saw the need for an entry-level transportation engineering book that focused on highway transportation and provided (1) the depth of coverage needed to serve as a basis for future transportation courses and (2) the material needed to answer questions likely to appear on the Fundamentals of Engineering (FE) or Principles and Practice of Engineering (PE) exams in civil engineering. The subsequent use of our book at some of the largest and most prestigious schools in the United States suggests that our vision of a highly focused, well-written, entry-level book was shared by many other educators.

APPROACH

In this, the third edition of *Principles of Highway Engineering and Traffic Analysis,* we continue the spirit of the first two editions by again focusing exclusively on highway transportation and providing the depth of coverage necessary to solve the highway-related problems that are most likely to be encountered in engineering practice. The focus on highway transportation seems a natural one in light of the dominance of the highway mode in the United States. Offering depth of coverage is a more risky proposition because, in order to keep the book at a manageable length, one must carefully choose the topics that most deserve coverage. Using the first two editions as a basis, along with the comments of other instructors and students who have used earlier versions of the book, we have carefully selected the topics that are fundamental to highway engineering and traffic analysis. This chosen material serves as a core upon which instructors can expand, if they wish, with material that they personally feel deserves additional attention. However, the core of material provided in the book is important for effective teaching because it allows instructors to be confident that students are learning the fundamentals necessary for them to undertake upper-level transportation courses, to enter transportation employment with a basic knowledge of highway engineering and traffic analysis, and to answer transportation-related questions on the civil engineering FE and PE exams.

Within the basic philosophical approach described above, this book addresses the complaint of some students that highway transportation is not as mathematically challenging or rigorous as other fields of study. This is not easily done because there is a dichotomy with regard to mathematical rigor, with relatively simple mathematics used in practice-oriented material and complex mathematics used in research. Thus it is common for instructors to either insult students' mathematical proficiency or vastly exceed it. This book makes an effort to find that elusive middle ground of mathematical rigor that matches junior and senior engineering students' mathematical abilities.

iv Preface

This third edition of *Principles of Highway Engineering and Traffic Analysis* has evolved from well over a decade of teaching introductory transportation engineering classes at the University of Washington, the University of Florida, Purdue University, and the Pennsylvania State University; feedback from users of the first two editions, and experiences in teaching engineering licensure exam review courses. The book's material and presentation style (which is characterized by the liberal use of example problems) are largely responsible for transforming much-maligned introductory transportation engineering courses into courses that are consistently rated by students as among the best civil engineering courses.

CHAPTER TOPICS AND ORGANIZATION

The book begins with a short introductory chapter that stresses the significance of highway transportation to the social and economic underpinnings of society. This chapter provides students with a basic overview of the problems facing the field of highway engineering and traffic analysis. The chapters that follow are arranged in sequences that focus on highway engineering (Chapters 2, 3, and 4) and traffic analysis (Chapters 5, 6, 7, and 8).

Chapter 2 introduces the basic elements of road vehicle performance. This chapter represents a major departure from the vehicle performance material presented in all other transportation and highway engineering books, in that it is far more involved and detailed. The additional level of detail is justified on two grounds. First, because students own and drive automobiles, they have a basic interest that can be linked to their freshman and sophomore coursework in physics, statics, and dynamics. The absence of such a link has been a common criticism of introductory transportation and highway engineering courses. Second, it is important that engineering students understand the principles involved with vehicle performance and the effect that continuing advances in vehicle technologies will have on engineering practice.

Chapter 3 presents current theory and design practices for the geometric alignment of highways. This chapter provides a high level of detail on vertical-curve design and the fundamental elements of horizontal-curve design. Basic material on highway classifications, turning radii, and so on is not included due to the great potential of alienating introductory-level students with too many design-oriented details.

Chapter 4 provides an overview of the current theory and practice of pavement design. The pavement-related material covers both flexible and rigid pavements in a thorough and consistent manner. The material in this chapter also links well with the geotechnical and materials courses that are likely to be part of the student's curriculum.

Chapter 5 presents the fundamentals of traffic flow and queuing theory, which provide the basic tools of traffic analysis. Relationships and models of basic traffic stream parameters are introduced, as well as queuing analysis models for deterministic and stochastic processes. Considerable effort was expended to make the material in this chapter accessible to junior and senior engineering students.

Chapter 6 presents some of the current methods used to assess highway levels of performance. Fundamentals and concepts are discussed along with the complexities involved in measuring and calculating highway level of service.

Chapter 7 introduces the basic elements of traffic control at a signalized intersection. It also builds upon the traffic analysis methods introduced in Chapter 5 and applies these to the analysis of signalized intersections. Using pretimed, isolated signals as the basis, this chapter covers both theoretical and practical elements associated with traffic signal phasing and timing.

Chapter 8, the final chapter, provides an overview of travel demand and traffic forecasting. This chapter uses a theoretically and mathematically consistent approach to travel demand and traffic forecasting. This chapter provides the student with an important understanding of the current state of travel demand and traffic forecasting, and some critical insight into the deficiencies of forecasting methods currently used.

NEW TO THIS EDITION

Dual Units

In this edition, an effort has been made to provide dual units (U.S. Customary and metric) throughout, with the primary units being U.S. Customary. This has been accomplished for all chapters except Chapter 4. Because the current pavement design guide is based only on U.S. Customary units, Chapter 4 provides only U.S. Customary units in a number of contexts where metric conversions are cumbersome.

Presentation Enhancements

Variable definitions for equations and figures are now presented in a tabular format to provide for quick accessibility and identification within the text.

New and Revised Problems

Over 50% of the problems are either new or have been revised. Users of the book will find the end-of-chapter problems to be extremely valuable in supporting the material presented in the book. Our intent has been to make these problems precise and challenging, a combination rarely found in transportation/highway engineering books.

Appendices for Problems in Metric

For those wishing to stress metric units, we have provided an appendix that contains the in-chapter example problems worked in metric units and an appendix that contains the end-of-chapter problems with metric units. It should be noted that these appendices include only problems that use units that can be converted to metric. For example, problems that deal only with units of vehicles or time are not repeated in these appendices. It should also be noted that some conversions are inexact in order to allow for consistent use of design tables, regardless of the units being used.

Some Chapter-Specific Enhancements

Chapter 1 The highway safety discussion has been updated to reflect current trends and has been integrated into the introductory section to better emphasize its significance to the entire field of transportation engineering.

Chapter 2 Section 2.9.5, Practical Stopping Distance, has been completely revised to be consistent with new design standards.

Chapter 3 Section 3.3.6, Underpass Sight Distance, has been added since this is often an important consideration in sag curve design. Numerous other revisions have been made to accommodate new design standards.

Chapter 5 Section 5.2, Traffic Stream Parameters, has been moderately expanded and revised to the provide the reader with a more thorough understanding of these concepts.

Chapter 6

- Section 6.3, Level-of-Service Determination, has been completely revised to provide a general overview of the process for determining level of service for all uninterrupted-flow highway segments.
- Sections 6.4–6.6, Basic Freeway Segments, Multilane Highways, and Two-Lane Highways, have been significantly revised in response to new design standards.

Chapter 7

- Section 7.2, Intersection and Signal Control Characteristics, has been added to introduce the reader to the basic concepts of the physical control of traffic signals.
- Section 7.3, Analysis of Traffic at Signalized Intersections, includes expanded coverage of interrupted traffic flow theory and concepts.
- Sections 7.3.3 and 7.6, Signalized Intersection Analysis for Level of Service and Level-of-Service Determination, provide explicit coverage of signal analysis for level of service.
- Section 7.5, Development of a Traffic Signal Phasing and Timing Plan, has been completely rewritten to be consistent with the latest guidelines and design standards.

Chapter 8

- Section 8.4.2, Trip Generation with Count Data Models, has been added to provide the reader with insight on a more mathematically consistent method for trip generation estimation.
- Appendix 8B, Maximum-Likelihood Estimation, has been added to introduce the reader to another method for estimating travel demand and traffic forecasting model coefficients.

Solutions Manual

The Solutions Manual includes solutions both in U.S. Customary and metric units. The Solutions Manual is available only to instructors. Instructors who are using this book for their course should visit the book's Web site (www.wiley.com/college/mannering), and go to the Instructor Companion Site portion of the Web site to register for a password and download the Solutions Manual.

Text Figures in Electronic Format

The figures from the text are available to instructors in electronic format, for easy creation of lecture slides. Visit the Instructor Companion Site portion of the book's Web site (www.wiley.com/college/mannering) to register for a password and download the text figures.

ACKNOWLEDGMENTS

We would like to thank Peter G. Furth, Aemal Khattak, Peter T. Martin, Ram M. Pendyala, Lee Robinson, Karen S. Schurr, William J. Sproule, James W. Stoner, and Michael Strub for offering many valuable suggestions for the third edition. We would also like to thank David Kirschner, Jessica Morriss, and Jason Starr for assisting with manuscript proofreading and the preparation of problem solutions.

Fred L. Mannering
Walter P. Kilareski
Scott S. Washburn

Contents

Chapter 1 Introduction to Highway Engineering and Traffic Analysis   1

    1.1 Introduction   1

    1.2 Technology   2
        1.2.1 Infrastructure Technologies   3
        1.2.2 Vehicle Technologies   3
        1.2.3 Traffic Control Technologies   4

    1.3 Human Behavior   4
        1.3.1 Dominance of Single-Occupant Private Vehicles   5
        1.3.2 Demographic Trends   5

    1.4 Scope of Study   6

Chapter 2 Road Vehicle Performance   7

    2.1 Introduction   7

    2.2 Tractive Effort and Resistance   7

    2.3 Aerodynamic Resistance   9

    2.4 Rolling Resistance   12

    2.5 Grade Resistance   14

    2.6 Available Tractive Effort   15
        2.6.1 Maximum Tractive Effort   15
        2.6.2 Engine-Generated Tractive Effort   18

    2.7 Vehicle Acceleration   21

    2.8 Fuel Efficiency   25

    2.9 Principles of Braking   26
        2.9.1 Braking Forces   26
        2.9.2 Braking Force Ratio and Efficiency   28
        2.9.3 Antilock Braking Systems   32
        2.9.4 Theoretical Stopping Distance   32
        2.9.5 Practical Stopping Distance   35
        2.9.6 Distance Traveled during Driver Perception/Reaction   39

Chapter 3 Geometric Design of Highways   45

    3.1 Introduction   45

    3.2 Principles of Highway Alignment   46

    3.3 Vertical Alignment   47
        3.3.1 Vertical Curve Fundamentals   48
        3.3.2 Stopping Sight Distance   57
        3.3.3 Stopping Sight Distance and Crest Vertical Curve Design   58
        3.3.4 Stopping Sight Distance and Sag Vertical Curve Design   63

3.3.5 Passing Sight Distance and Crest Vertical Curve Design   70
3.3.6 Underpass Sight Distance and Sag Vertical Curve Design   72

3.4 Horizontal Alignment   75
3.4.1 Vehicle Cornering   76
3.4.2 Horizontal Curve Fundamentals   80
3.4.3 Stopping Sight Distance and Horizontal Curve Design   82

Chapter 4 Pavement Design   91

4.1 Introduction   91

4.2 Pavement Types   91
4.2.1 Flexible Pavements   92
4.2.2 Rigid Pavements   93

4.3 Pavement System Design: Principles for Flexible Pavements   93
4.3.1 Calculation of Flexible Pavement Stresses and Deflections   94

4.4 The AASHTO Flexible-Pavement Design Procedure   103
4.4.1 Serviceability Concept   104
4.4.2 Flexible-Pavement Design Equation   104
4.4.3 Structural Number   112

4.5 Pavement System Design: Principles for Rigid Pavements   115
4.5.1 Calculation of Rigid-Pavement Stresses and Deflections   116

4.6 The AASHTO Rigid-Pavement Design Procedure   119

Chapter 5 Fundamentals of Traffic Flow and Queuing Theory   135

5.1 Introduction   135

5.2 Traffic Stream Parameters   135
5.2.1 Traffic Flow, Speed, and Density   136

5.3 Basic Traffic Stream Models   141
5.3.1 Speed-Density Model   141
5.3.2 Flow-Density Model   143
5.3.3 Speed-Flow Model   144

5.4 Models Of Traffic Flow   146
5.4.1 Poisson Model   146
5.4.2 Limitations of the Poisson Model   151

5.5 Queuing Theory and Traffic Flow Analysis   151
5.5.1 Dimensions of Queuing Models   152
5.5.2 *D/D/*1 Queuing   153
5.5.3 *M/D/*1 Queuing   156
5.5.4 *M/M/*1 Queuing   158
5.5.5 *M/M/N* Queuing   160

5.6 Traffic Analysis at Highway Bottlenecks   163

Chapter 6 Highway Capacity and Level of Service Analysis 170

    6.1 Introduction 170

    6.2 Level-of-Service Concept 171

    6.3 Level-of-Service Determination 174
        6.3.1 Base Conditions and Capacity 174
        6.3.2 Determine Free-Flow Speed 174
        6.3.3 Determine Analysis Flow Rate 175
        6.3.4 Calculate Service Measure(s) and Determine LOS 175

    6.4 Basic Freeway Segments 175
        6.4.1 Base Conditions and Capacity 176
        6.4.2 Service Measure 179
        6.4.3 Determining Free-Flow Speed 179
        6.4.4 Determining Analysis Flow Rate 182
        6.4.5 Calculating Density and Determining LOS 188

    6.5 Multilane Highways 191
        6.5.1 Base Conditions and Capacity 193
        6.5.2 Service Measure 194
        6.5.3 Determining Free-Flow Speed 194
        6.5.4 Determining Analysis Flow Rate 197
        6.5.5 Calculating Density and Determining LOS 197

    6.6 Two-Lane Highways 200
        6.6.1 Base Conditions and Capacity 201
        6.6.2 Service Measures 201
        6.6.3 Determining Free-Flow Speed 202
        6.6.4 Determining Analysis Flow Rate 203
        6.6.5 Calculate Service Measures 205
        6.6.6 Determine LOS 208

    6.7 Design Traffic Volumes 211

Chapter 7 Traffic Control and Analysis at Signalized Intersections 220

    7.1 Introduction 220

    7.2 Intersection and Signal Control Characteristics 221

    7.3 Analysis of Traffic at Signalized Intersections 226
        7.3.1 Concepts and Definitions 226
        7.3.2 Signalized Intersection Analysis with $D/D/1$ Queuing 230
        7.3.3 Signalized Intersection Analysis for Level of Service 236

    7.4 Optimal Traffic Signal Timing 241

    7.5 Development of a Traffic Signal Phasing and Timing Plan 243
        7.5.1 Select Signal Phasing 243
        7.5.2 Establish Analysis Lane Groups 247

xii Contents

       7.5.3 Calculate Analysis Flow Rates and Adjusted Saturation Flow Rates   249
       7.5.4 Determine Critical Lane Groups and Total Cycle Lost Time   250
       7.5.5 Calculate Cycle Length   253
       7.5.6 Allocate Green Time   255
       7.5.7 Calculate Change and Clearance Intervals   256
       7.5.8 Check Pedestrian Crossing Time   259

   7.6 Level-of-Service Determination   260

Chapter 8   Travel Demand and Traffic Forecasting   270

   8.1 Introduction   270

   8.2 Traveler Decisions   271

   8.3 Scope of the Travel Demand and Traffic Forecasting Problem   272

   8.4 Trip Generation   275
       8.4.1 Typical Trip Generation Models   276
       8.4.2 Trip Generation with Count Data Models   279

   8.5 Mode and Destination Choice   281
       8.5.1 Methodological Approach   281
       8.5.2 Logit Model Applications   283

   8.6 Highway Route Choice   289
       8.6.1 Highway Performance Functions   289
       8.6.2 User Equilibrium   290
       8.6.3 Mathematical Programming Approach to User Equilibrium   296
       8.6.4 System Optimization   297

   8.7 The State of Travel Demand and to Traffic Forecasting in Practice   301

      Appendix 8A Least Squares Estimation   302

      Appendix 8B Maximum-Likelihood Estimation   304

Appendix A   **Metric Example Problems**   311

Appendix B   **Metric End-of-Chapter Problems**   352

Appendix C   **Unit Conversions**   363

Index   367

Chapter 1

Introduction to Highway Engineering and Traffic Analysis

1.1 INTRODUCTION

The importance of highway transportation to the industrial and technological complex of the United States and other industrialized nations cannot be overstated. Virtually every aspect of modern economies, and the ways of life they support, can be tied directly or indirectly to highway transportation. From the movement of freight and people to the impact on residential, commercial, and industrial locations, highways have had, and continue to have, a profound effect on the world economy and societal development. In the United States, the manner in which highways have come to dominate the transportation system has been studied for decades as a cultural, political, and economic phenomenon. Without a doubt, the demand for unrestricted mobility and unlimited access to resources has played an important role and helped to quickly move highway transportation to its dominant position from the middle of the 20th century onward. The construction of the U.S. interstate highway system remains to this day the largest infrastructure project in human history. At the time, it underscored the nation's commitment to the unrestricted mobility of its populace and to the economic opportunities that such a system would provide its industrial and service industries. Today, additional highway expansion and maintenance of existing highway systems continue to represent an enormous investment in public infrastructure—an investment with an immeasurable impact on society in terms of mobility, economic opportunities, and environmental implications, including consumption of resources and pollution.

The mobility and opportunities that highway infrastructure provides also has a human cost. Although safety has always been a primary consideration in highway design and operation, highways continue to exact a terrible toll in terms of loss of life, injuries, property damage, and reduced productivity as a result of vehicle accidents. To be sure, the elements of highway safety are complex. They involve technical and behavioral components and the complexities of the human-machine interface. Because of the high costs of highway accidents, efforts to improve highway safety have intensified dramatically in the last decade. This has resulted in the implementation of new highway design guidelines and countermeasures (some technical and some behavioral) aimed at reducing the frequency and severity of highway accidents. These efforts sought to reverse an upward trend in U.S. highway fatalities and injuries that saw such fatalities exceed 50,000 per year in the 1970s. Fortunately, efforts aimed at improving highway design (such as more stringent design guidelines and breakaway signs), vehicle occupant protection (safety belts, padded dashboards, collapsible steering columns, driver- and passenger-side airbags, and improved bumper design), and vehicle accident avoidance (antilock braking and traction control systems), combined with accident countermeasures (campaigns to reduce drunk driving), have reduced annual U.S. highway fatalities to around 40,000 in the 1990s. However, in spite of continuing efforts and unprecedented advancements in vehicle safety technologies, the number of fatalities in recent years has begun to rise again (although fatalities per mile driven continue to decline).

To explain this, evidence points to a number of potential factors: an increase in the overall level of aggressive driving; increasing levels of disrespect for traffic control devices (running of red lights and stop signs being two of the more notable examples); in-vehicle driving distractions, such as cell phones; and poor driving skills in the younger and older driving populations. Two other phenomena are being observed that may be contributing to the increased number of fatalities. One is that some people drive more aggressively in vehicles with advanced safety features, thus leveraging some of the benefits of new safety technologies against the desire for increased mobility (speed). The other is that many people are more influenced by style and function than by safety features when making vehicle purchase decisions. This is evidenced by the growing popularity of vehicles such as sport utility vehicles (SUVs), minivans, and pickup trucks, despite their consistently overall lower rankings in certain safety categories, such as rollover probability, relative to sedan-style passenger cars. These issues underscore the importance of the highway engineer's need to consider the complex interaction between human behavior/factors and technology in highway design.

It is against this backdrop that engineers must strive to provide high levels of mobility (minimizing travel times and delays) and high levels of safety. These two goals are not only often contradictory (higher speeds minimize travel time but may also decrease safety), but must be achieved while giving full consideration to the complex impacts that are likely to result from highway-related projects. Such impacts can be broadly classified as economic (the cost and economic impact of these projects), political (the community-related impacts of projects), and environmental (the impact of projects on the environment measured in terms of air, water, land, noise, and quality of life).

Although attempting to provide higher levels of both mobility and safety may seem an impossible task at times, it is a task that engineers have an ethical responsibility to undertake. In so doing, they must use all of the technologies at their disposal while giving full consideration to the associated effects on human behavior.

1.2 TECHNOLOGY

As with all fields of engineering, technological advances offer the promise of solving complex problems and achieving lofty goals. For highways, technologies can be classified into those impacting infrastructure, vehicles, and traffic control.

1.2.1 Infrastructure Technologies

Investments in highway infrastructure have been made continuously throughout the twentieth and twenty-first centuries. Such investments have understandably varied over the years in response to need, and political and national priorities. For example, in the United States, an extraordinary capital investment in highways during the 1960s and 1970s was undertaken in constructing the interstate highway system and upgrading and constructing many other highways. The economic and political climate that permitted such an ambitious construction program was unprecedented and has not been replicated since. It is difficult to imagine, in today's economic and political environment, that a project of the magnitude of the interstate highway system would ever be seriously considered. This is because of the prohibitive costs that are associated with land acquisition and construction and the community and environmental impacts that would result.

It is also important to realize that highways are durable, long-lasting investments that require maintenance and rehabilitation at regular intervals. The legacy of a major capital investment in highway infrastructure is the proportionate maintenance and rehabilitation schedule that will follow. For example, most of the U.S. interstate system was designed with pavements that had a finite lifetime before major rehabilitation became necessary. Thus an unfortunate consequence of the extensive interstate construction program is the maintenance and rehabilitation programs that are needed today. Although there are sometimes compelling reasons to defer maintenance and rehabilitation (including the associated construction costs and the impact of the reconstruction on traffic), such deferral can result in unacceptable losses in mobility and safety.

Engineers must now deal with new construction and infrastructure maintenance and rehabilitation by developing and applying new technologies to extend the life of new facilities and to economically combat aging infrastructure. Included in these technologies are the extensive development and application of new sensing technologies in the emerging field of structural health monitoring. There are also opportunities to extend the life expectancy of new infrastructure with the ongoing nanotechnology advances in material science. Such technological advances are essential elements in the future of highway infrastructure.

1.2.2 Vehicle Technologies

Until the 1970s, vehicle technologies evolved slowly and often in response to mild trends in the vehicle market as opposed to an underlying trend toward technological development. Beginning in the 1970s, three factors began a cycle of unparalleled advances in vehicle technology that continues to this day: (1) government regulations on air quality, fuel efficiency, and vehicle occupant safety; (2) energy shortages and fuel price increases; and (3) intense competition from foreign vehicle manufacturers. The aggregate effect of these factors has been vehicle consumers demanding new technology. Vehicle manufacturers have found it necessary to reallocate resources and to restructure manufacturing and inventory control processes to meet this demand. In recent years, consumer demand and competition among vehicle manufacturers has resulted in the widespread implementation of new technologies including supplemental restraint systems, antilock brake systems, traction control systems, and a host of other applications to improve safety and comfort in highway vehicles. There is little doubt that the combination of consumer demand and intense competition in the vehicle industry will continue to spur technological innovations.

Although the development of new vehicle technologies usually lies in the domain of disciplines such as mechanical and electrical engineering, the influence of such technologies on highway design and traffic analysis is an important concern for highway engineers. This is because highway engineers must be able to account for new technologies in the design and rehabilitation of highways in their ongoing effort to provide the highest possible levels of mobility and safety.

1.2.3 Traffic Control Technologies

Intersection traffic signals are a familiar traffic control technology. At signalized intersections, the trade-off between mobility and safety is brought into sharp focus. Procedures for developing traffic signal control plans (allocating green time to conflicting traffic movements) have made significant advances over the years. Today, signals at critical intersections respond quickly to prevailing traffic flows, groups of signals are sequenced to allow for a smooth through-flow of traffic, and in some cases, computers control entire networks of signals. In addition to signal controls, numerous safety, navigational, and congestion mitigation technologies are now reaching the market under the broad heading of Intelligent Transportation Systems (ITS). Such technological efforts offer the potential to significantly reduce traffic congestion and improve safety on highways by providing an unprecedented level of traffic control. There are, however, many obstacles associated with ITS implementation, including system reliability, human response, and the human-machine interface. It is therefore important not only that highway engineers participate in the development and implementation of new traffic control technologies, but that they also recognize the limitations associated with these technologies.

1.3 HUMAN BEHAVIOR

The volume and speed of highway traffic determines both highway mobility and safety and is a behavioral phenomenon that is an outgrowth of people's travel-related choices. Over the years, these choices have resulted in crippling congestion in many urban areas and millions of highway injuries and deaths, as well as property and productivity losses that have easily exceeded a trillion dollars. Despite congestion and safety problems, the trend in traffic growth continues upward. This is rather surprising because one would expect the supply-demand effect (a congested highway infrastructure should restrain the growth of private-vehicle use) to be an effective factor in limiting urban traffic congestion. However, it has been found that people are willing to tolerate high amounts of congestion just to continue to use private vehicles, most of which only have a single occupant. The continued growth in traffic congestion is a serious obstacle to improving mobility.

Addressing traffic growth is a perplexing problem. Current economic and political environments do not favor large-scale construction programs aimed at increasing highway capacity. Technological innovations in traffic control have the potential to offer some relief, but it is questionable whether such innovations can keep pace with the growth in traffic volume. This points toward the more fundamental issue of the behavioral processes and preferences that generate highway traffic.

1.3.1 Dominance of Single-Occupant Private Vehicles

Of the available urban transportation modes (bus, commuter train, subway, and private vehicle), private vehicles (and single-occupant private vehicles in particular) offer a level of mobility that is unequaled. The single-occupant private vehicle is such an overwhelmingly dominant choice that travelers are willing to pay substantial capital and operating costs, confront high levels of congestion, and struggle with parking-related problems, just to have the flexibility in travel departure time and destination choices that is uniquely provided by private vehicles. The dominance of the private-vehicle mode is supported by historical data that shows continuing trends toward its use. For example, in the last 40 years, the percentage of trips taken in private vehicles has risen from about 69% to over 90% (public transit, walking, and other modes make up the balance). Over this same period, average private-vehicle occupancy has dropped from 1.22 to 1.09 persons per vehicle, reflecting the fact that the single-occupant vehicle is becoming the increasingly dominant mode of travel.

Dealing with extensive private-vehicle use and low vehicle occupancy presents engineers with a classic mobility dilemma. On one hand, programs that encourage travelers to take alternative modes of transportation (bus fare incentives and increases in private-vehicle parking fees) or to increase vehicle occupancy (high-occupancy vehicle lanes and employer-based ride sharing programs) have the potential to provide some traffic congestion relief on highways and provide remaining highway users with higher mobility. However, such programs have the adverse effect of directing people toward travel modes that inherently provide lower levels of mobility because no other mode offers the departure time and destination choice flexibility provided by private, single-occupant vehicles. With this mobility dilemma in mind, it is clear that controversial congestion-related compromises must be reached.

1.3.2 Demographic Trends

Travelers' commuting patterns are inextricably intertwined with their socioeconomic characteristics, such as age, income, household size, education, and job type, as well as the distribution of residential, commercial, and industrial developments within the region. Many American metropolitan areas have experienced population declines in central cities accompanied by a growth in suburban areas. Many have argued that the population shift from the central cities to the suburbs was made possible by the increased mobility provided by the major highway projects undertaken during the 1960s and 1970s. This mobility enabled people to improve their quality of life by gaining access to affordable housing and land, while still being able to access jobs in the central city with acceptable travel times. Conventional wisdom suggested that as overall metropolitan traffic congestion grew (making the suburb-to-city commuting pattern much less attractive), commuters would seek to avoid traffic congestion by reverting back to public transport modes and/or once again choosing to reside in the central city. However, a different trend has emerged. Employment centers have developed in the suburbs and now provide a viable alternative to the suburb-to-city commute (the suburb-to-suburb commute). The result is a continuing tendency toward low-density, private vehicle–based development as people seek to retain the high quality of life associated with such development.

Ongoing demographic trends also present engineers with an ever-moving target that further complicates the problem of providing mobility and safety. An example is the rising average age of the U.S. population as a result of population cohorts (the baby boom following the Second World War) and advances in medical technology that prolong life. Because older people tend to have slower reaction times, taking longer to respond to driving situations that require action, engineers must confront the possibility of changing highway design guidelines and practices to accommodate slower reaction times and the potentially higher variance in reaction times among highway users.

1.4 SCOPE OF STUDY

As can be readily determined from the preceding discussions, the social, political, environmental, and technological impacts of highway infrastructures are exceedingly complex and impossible to quantify. Therefore, the remaining chapters in this book make no attempt to do so. Instead, the material in Chapters 2 through 8 is carefully chosen to provide the reader with the fundamental elements and methodological approaches that are currently used in the design of highways and the assessment of their performance. This material constitutes the fundamental principles of highway engineering and traffic analysis that are needed to begin to grasp the complex social, political, and economic environment in which highways are constructed and function.

Chapter 2

Road Vehicle Performance

2.1 INTRODUCTION

The performance of road vehicles forms the basis for highway design guidelines and traffic analysis. For example, in highway design, the determination of the length of freeway acceleration and deceleration lanes, maximum highway grades, stopping sight distances, passing sight distances, and numerous accident prevention devices all rely on a basic understanding of vehicle performance. Similarly, vehicle performance is a major consideration in the selection and design of traffic control devices, the determination of speed limits, and the timing and control of traffic signal systems.

Studying vehicle performance serves two important functions. First, it provides insight into highway design and traffic operations and the compromises that are necessary to accommodate the wide variety of vehicles (from high-powered sports cars to heavily laden trucks) that use highways. Second, it forms a basis from which the impact of advancing vehicle technologies on existing highway design guidelines can be assessed. This second function is particularly important in light of the ongoing, unprecedented advances in vehicle technology. Such advances will necessitate more frequent updating of highway design guidelines as well as engineers that have a better understanding of the fundamental principles underlying vehicle performance.

The objective of this chapter is to introduce the basic principles of road vehicle performance. Primary attention will be given to the straight-line performance of vehicles (acceleration, deceleration, top speed, and the ability to ascend grades). Cornering performance of vehicles is overviewed in Chapter 3, but detailed presentations of this material are better suited to more specialized sources [Campbell 1978; Wong 1978; Brewer and Rice 1983].

2.2 TRACTIVE EFFORT AND RESISTANCE

Tractive effort and resistance are the two primary opposing forces that determine the straight-line performance of road vehicles. Tractive effort is simply the force available, at the roadway surface, to perform work and is expressed in lb (N). Resistance, also expressed in lb (N), is defined as the force impeding vehicle motion. The three major sources of vehicle resistance are (1) aerodynamic resistance, (2) rolling resistance (which originates from the roadway surface–tire interface), and (3) grade or gravitational resistance. To illustrate these forces, consider the vehicle force diagram shown in Fig. 2.1.

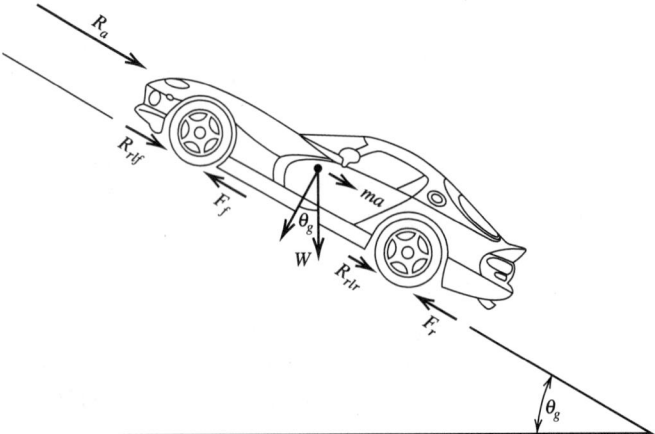

Figure 2.1 Forces acting on a road vehicle.

In this figure,

R_a = aerodynamic resistance in lb (N),
R_{rlf} = rolling resistance of the front tires in lb (N),
R_{rlr} = rolling resistance of the rear tires in lb (N),
F_f = available tractive effort of the front tires in lb (N),
F_r = available tractive effort of the rear tires in lb (N),
W = total vehicle weight in lb (N),
θ_g = angle of the grade in degrees,
m = vehicle mass in slugs (kg), and
a = acceleration in ft/s² (m/s²).

Summing the forces along the vehicle's longitudinal axis provides the basic equation of vehicle motion:

$$F_f + F_r = ma + R_a + R_{rlf} + R_{rlr} + R_g \tag{2.1}$$

where R_g is the grade resistance and is equal to $W \sin \theta_g$. For exposition purposes it is convenient to let F be the sum of available tractive effort delivered by the front and rear tires ($F_f + F_r$) and similarly to let R_{rl} be the sum of rolling resistance ($R_{rlf} + R_{rlr}$). This notation allows Eq. 2.1 to be written as

$$F = ma + R_a + R_{rl} + R_g \tag{2.2}$$

Sections 2.3 to 2.8 present a thorough discussion of the components and implications of Eq. 2.2.

2.3 AERODYNAMIC RESISTANCE

Aerodynamic resistance is a resistive force that can have significant impacts on vehicle performance. At high speeds, where this component of resistance can become overwhelming, proper vehicle aerodynamic design is essential. Attention to aerodynamic efficiency in design has long been the rule in racing and sports cars. More recently, concerns over fuel efficiency and overall vehicle performance have resulted in more efficient aerodynamic designs in common passenger cars, although not necessarily in pickup trucks or sport utility vehicles (SUVs).

Aerodynamic resistance originates from a number of sources. The primary source (typically accounting for over 85% of total aerodynamic resistance) is the turbulent flow of air around the vehicle body. This turbulence is a function of the shape of the vehicle, particularly the rear portion, which has been shown to be a major source of air turbulence. To a much lesser extent (on the order of 12% of total aerodynamic resistance), the friction of the air passing over the body of the vehicle contributes to resistance. Finally, approximately 3% of the total aerodynamic resistance can be attributed to air flow through vehicle components such as radiators and air vents.

Based on these sources, the equation for determining aerodynamic resistance is

$$R_a = \frac{\rho}{2} C_D A_f V^2 \qquad (2.3)$$

where

R_a = aerodynamic resistance in lb (N),
ρ = air density in slugs/ft^3 (kg/m^3),
C_D = coefficient of drag (unitless),
A_f = frontal area of the vehicle (projected area of the vehicle in the direction of travel) in ft^2 (m^2), and
V = speed of the vehicle in ft/s (m/s).

To be truly accurate, for aerodynamic resistance computations, V is actually the speed of the vehicle relative to the prevailing wind speed. To simplify the exposition of concepts soon to be presented, the wind speed is assumed to be equal to zero for all problems and derivations in this book.

Air density is a function of both elevation and temperature, as indicated in Table 2.1. Equation 2.3 indicates that as the air becomes more dense, total aerodynamic resistance increases. The drag coefficient (C_D) is a term that implicitly accounts for all three of the aerodynamic resistance sources discussed above. The drag coefficient is measured from empirical data either from wind tunnel experiments or actual field tests in which the vehicle is allowed to decelerate from a known speed with other sources of resistance (rolling and grade) accounted for. Table 2.2 gives some approximation of the range of drag coefficients for different types of road vehicles. Table 2.3 presents drag coefficients for specific automobiles covering the last 35+ years.

Table 2.1 Typical Values of Air Density under Specified Atmospheric Conditions

Altitude		Temperature		Pressure		Air density	
(ft)	(m)	(°F)	(°C)	(lb/in²)	(kPa)	(slugs/ft³)	(kg/m³)
0	0	59.0	15.0	14.7	101.4	0.002378	1.2256
5,000	1500	41.2	9.7	12.2	84.4	0.002045	1.0567
10,000	3000	23.4	−4.5	10.1	70.1	0.001755	0.9096

Table 2.2 Ranges of Drag Coefficients for Typical Road Vehicles

Vehicle type	Drag coefficient (C_D)
Automobile	0.25–0.55
Bus	0.5–0.7
Tractor-Trailer	0.6–1.3
Motorcycle	0.27–1.8

Table 2.3 Drag Coefficients of Selected Automobiles

Vehicle	Drag coefficient (C_D)
1967 Chevrolet Corvette	0.50
1967 Volkswagen Beetle	0.46
1977 Triumph TR7	0.40
1977 Jaguar XJS	0.36
1987 Acura Integra	0.34
1987 Ford Taurus	0.32
1993 Acura Integra	0.32
1993 Ford Probe GT	0.31
1993 Ford Ranger (truck)	0.45
1997 Lexus LS400	0.29
1997 Infiniti Q45	0.29
2000 Honda Insight (hybrid)	0.25
2002 Acura NSX	0.30
2002 Lexus LS430	0.25
2003 Dodge Caravan (minivan)	0.35
2003 Ford Explorer (SUV)	0.41
2003 Dodge Ram (truck)	0.53
2004 Mercedes Benz C240	0.27

The general trend toward lower drag coefficients over this period of time reflects the continuing efforts of the automotive industry to improve overall vehicle efficiency by minimizing resistance forces. The table also includes some larger personal vehicles, such as pickup trucks and sport utility vehicles, which generally represent the upper range of drag coefficients for automobiles.

Figure 2.2 illustrates the effect of automobile operating conditions on drag coefficients. As indicated, even minor factors, such as the opening of windows, can have a significant effect on a vehicle's drag coefficient and thus its total aerodynamic resistance. Projected frontal area (approximated as the height of the vehicle multiplied by its width) typically ranges from 10 ft² to 30 ft² (1.0 m² to 2.5 m²) for passenger cars and is also a major factor in determining aerodynamic resistance.

Because aerodynamic resistance is proportional to the square of the vehicle's speed, it is clear that such resistance will increase rapidly at higher speeds. The magnitude of this increase can be underscored by considering an expression for the power (hp_{R_a} or P_{R_a}) required to overcome aerodynamic resistance. With power being the product of force and speed, the multiplication of Eq. 2.3 by speed gives

U.S. Customary   **Metric**

$$hp_{R_a} = \frac{\rho C_D A_f V^3}{1100} \qquad P_{R_a} = \frac{\rho}{2} C_D A_f V^3 \qquad (2.4)$$

where

hp_{R_a} = horsepower required to overcome aerodynamic resistance (1 horsepower equals 550 ft-lb/s),

P_{R_a} = power required to overcome aerodynamic resistance in N-m/s (watts), and

Other terms are as defined previously.

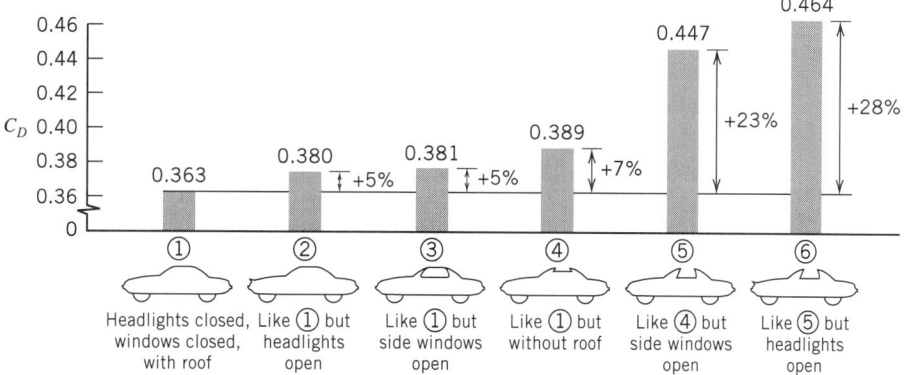

Figure 2.2 Effect of operational factors on the drag coefficient of an automobile.
Reproduced by permission from L. J. Janssen and W. H. Hucho, "The Effect of Various Parametres on the Aerodynamic Drag of Passenger Cars," *Advances in Road Vehicle Aerodynamics 1973*, BMRA Fluid Engineering, Cranfield, England.

Thus, the amount of power required to overcome aerodynamic resistance increases with the cube of speed, indicating, for example, that eight times as much power is required to overcome aerodynamic resistance if the vehicle speed is doubled.

2.4 ROLLING RESISTANCE

Rolling resistance refers to the resistance generated from a vehicle's internal mechanical friction and from pneumatic tires and their interaction with the roadway surface. The primary source of this resistance is the deformation of the tire as it passes over the roadway surface. The force needed to overcome this deformation accounts for approximately 90% of the total rolling resistance. Depending on the vehicle's weight and the material composition of the roadway surface, the penetration of the tire into the surface and the corresponding surface compression can also be a significant source of rolling resistance. However, for typical vehicle weights and pavement types, penetration and compression constitute only around 4% of the total rolling resistance. Finally, frictional motion due to the slippage of the tire on the roadway surface and, to a lesser extent, air circulation around the tire and wheel (the fanning effect) are sources accounting for roughly 6% of the total rolling resistance [Taborek 1957].

Considering the sources of rolling resistance, three factors are worthy of note. First, the rigidity of the tire and the roadway surface will influence the degree of tire penetration, surface compression, and tire deformation. Hard, smooth, and dry roadway surfaces provide the lowest rolling resistance. Second, tire conditions, including inflation pressure and temperature, can have a substantial impact on rolling resistance. High tire inflation decreases rolling resistance on hard paved surfaces as a result of reduced friction but increases rolling resistance on soft unpaved surfaces due to additional surface penetration. Also, higher tire temperatures make the tire body more flexible, and thus less resistance is encountered during tire deformation. The third and final factor is the vehicle's operating speed, which affects tire deformation. Increasing speed results in additional tire flexing and vibration and thus a higher rolling resistance.

Due to the wide range of factors that determine rolling resistance, a simplifying approximation is used. Studies have shown that overall rolling resistance can be approximated as the product of a friction term (coefficient of rolling resistance) and the weight of the vehicle acting normal to the roadway surface. The coefficient of rolling resistance for road vehicles operating on paved surfaces is approximated as

U.S. Customary Metric

$$f_{rl} = 0.01\left(1 + \frac{V}{147}\right) \qquad f_{rl} = 0.01\left(1 + \frac{V}{44.73}\right) \tag{2.5}$$

where

f_{rl} = coefficient of rolling resistance (unitless), and
V = vehicle speed in ft/s (m/s).

2.4 Rolling Resistance

By inspection of Fig. 2.1, the rolling resistance, in lb (N), will simply be the coefficient of rolling resistance multiplied by $W \cos \theta_g$, the vehicle weight acting normal to the roadway surface. For most highway applications θ_g is quite small, so it can be assumed that $\cos \theta_g = 1$, giving the equation for rolling resistance (R_{rl}) as

$$R_{rl} = f_{rl} W \tag{2.6}$$

From this, the amount of power required to overcome rolling resistance is

U.S. Customary $\qquad$ Metric

$$\text{hp}_{R_{rl}} = \frac{f_{rl} W V}{550} \qquad P_{R_{rl}} = f_{rl} W V \tag{2.7}$$

where

$\text{hp}_{R_{rl}}$ = horsepower required to overcome rolling resistance
(1 horsepower equals 550 ft-lb/s),

$P_{R_{rl}}$ = power required to overcome rolling resistance in N-m/s (watts), and

W = total vehicle weight in lb (N).

EXAMPLE 2.1

A 2500-lb (11.1-kN) car is driven at sea level ($\rho = 0.002378$ slugs/ft^3 or 1.2256 kg/m^3) on a level paved surface. The car has $C_D = 0.38$ and 20 ft^2 (1.86 m^2) of frontal area. It is known that at maximum speed, 50 hp (37.3 kW) is being expended to overcome rolling and aerodynamic resistance. Determine the car's maximum speed.

SOLUTION

It is known that at maximum speed (V_m),

$$\text{available horsepower} = R_a V_m + R_{rl} V_m$$

or

$$\text{available hp} = \frac{\frac{\rho}{2} C_D A_f V_m^3 + f_{rl} W V_m}{550}$$

Substituting,

$$50 = \frac{\frac{0.002378}{2}(0.38)(20) V_m^3 + 0.01\left(1 + \frac{V_m}{147}\right)(2500) V_m}{550}$$

or

$$27{,}500 = 0.00904 V_m^3 + 0.17 V_m^2 + 25 V_m$$

Solving for V_m gives

$$V_m = \underline{133 \text{ ft/s}} \quad \text{or} \quad \underline{90 \text{ mi/h}}$$

2.5 GRADE RESISTANCE

Grade resistance is simply the gravitational force (the component parallel to the roadway) acting on the vehicle. As suggested in Fig. 2.1, the expression for grade resistance (R_g) is

$$R_g = W \sin \theta_g \tag{2.8}$$

As was the case in the development of the rolling resistance formula (Eq. 2.6), highway grades are usually very small, so $\theta_g \cong \tan \theta_g$. Rewriting Eq. 2.8,

$$R_g \cong W \tan \theta_g = WG \tag{2.9}$$

where

G = grade, defined as the vertical rise per some specified horizontal distance (opposite side of the force triangle, Fig. 2.1, divided by the adjacent side) in ft/ft (m/m).

Grades are generally specified as percentages for ease of understanding. Thus a roadway that rises 5 ft (m) vertically per 100 ft (m) horizontally ($G = 0.05$ and $\theta_g = 2.86°$) is said to have a 5% grade.

EXAMPLE 2.2

A 2000-lb (8.9-kN) car has $C_D = 0.40$, $A_f = 20$ ft^2 (1.86 m^2), and an available tractive effort of 255 lb (1.134 kN). If the car is traveling at an elevation of 5000 ft (1524 m) ($\rho = 0.002045$ slugs/ft^3 or 1.0567 kg/m^3) on a paved surface at a speed of 70 mi/h (112.6 km/h), what is the maximum grade that this car could ascend and still maintain the 70-mi/h (112.6-km/h) speed?

SOLUTION

To maintain the speed, the available tractive effort will be exactly equal to the summation of resistances. Thus no tractive effort will remain for vehicle acceleration ($ma = 0$). Therefore, Eq. 2.2 can be written as

$$F = R_a + R_{rl} + R_g$$

For grade resistance (using Eq. 2.9),

$$R_g = WG = 2000G$$

for aerodynamic resistance (using Eq. 2.3),

$$R_a = \frac{\rho}{2} C_D A_f V^2$$
$$= \frac{0.002045}{2}(0.4)(20)(70 \times 5280/3600)^2$$
$$= 86.22 \text{ lb}$$

and for rolling resistance (using Eq. 2.6),

$$R_{rl} = f_{rl} W$$
$$= 0.01\left(1 + \frac{70 \times 5280/3600}{147}\right) \times 2000$$
$$= 33.97 \text{ lb}$$

Therefore,

$$F = 255 = 86.22 + 33.97 + 2000G$$
$$G = \underline{0.0674} \quad \text{or a} \quad \underline{6.74\%} \text{ grade}$$

2.6 AVAILABLE TRACTIVE EFFORT

With the resistance terms in the basic equation of vehicle motion (Eq. 2.2) discussed, attention can now be directed toward available tractive effort (F) as used in Example 2.2. The tractive effort available to overcome resistance and/or to accelerate the vehicle is determined either by the force generated by the vehicle's engine or by some maximum value that will be a function of the vehicle's weight distribution and the characteristics of the roadway surface–tire interface. The basic concepts underlying these two determinants of available tractive effort are presented here.

2.6.1 Maximum Tractive Effort

No matter how much force a vehicle's engine makes available at the roadway surface, there is a point beyond which additional force merely results in the spinning of tires and does not overcome resistance or accelerate the vehicle. To explain what determines this point of maximum tractive effort (the limiting value beyond which tire spinning begins), a force and moment-generating diagram is provided in Fig. 2.3.

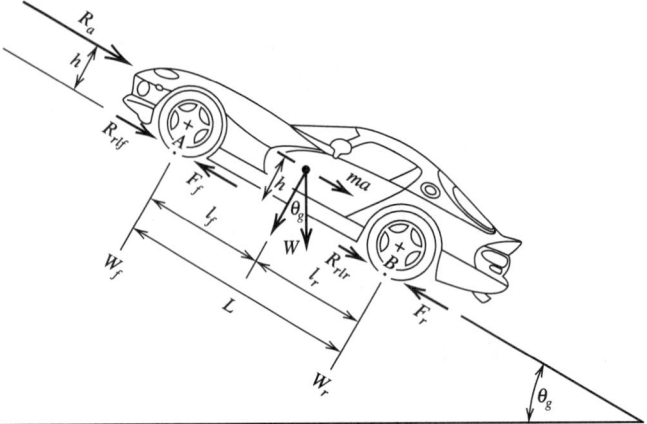

Figure 2.3 Vehicle forces and moment-generating distances.

In this figure,

R_a = aerodynamic resistance in lb (N),
R_{rlf} = rolling resistance of the front tires in lb (N),
R_{rlr} = rolling resistance of the rear tires in lb (N),
F_f = available tractive effort of the front tires in lb (N),
F_r = available tractive effort of the rear tires in lb (N),
W = total vehicle weight in lb (N),
W_f = weight of the vehicle on the front axle in lb (N),
W_r = weight of the vehicle on the rear axle in lb (N),
θ_g = angle of the grade in degrees,
m = vehicle mass in slugs (kg),
a = acceleration in ft/s² (m/s²),
L = length of wheelbase,
h = height of the center of gravity above the roadway surface,
l_f = distance from the front axle to the center of gravity, and
l_r = distance from the rear axle to the center of gravity.

To determine the maximum tractive effort that the roadway surface–tire contact can support, it is necessary to examine the normal loads on the axles. The normal load on the rear axle (W_r) is given by summing the moments about point A (see Fig. 2.3):

$$W_r = \frac{R_a h + W l_f \cos \theta_g + mah \pm Wh \sin \theta_g}{L} \qquad (2.10)$$

In this equation the grade moment ($Wh \sin \theta_g$) is positive for an upward slope and negative for a downward slope. Rearranging terms (assuming $\cos \theta_g = 1$ for the small grades encountered in highway applications) and substituting into Eq. 2.2 gives

$$W_r = \frac{l_f}{L}W + \frac{h}{L}(F - R_{rl}) \tag{2.11}$$

From basic physics, the maximum tractive effort as determined by the roadway surface–tire interaction will be the normal force multiplied by the coefficient of road adhesion (μ), so for a rear-wheel–drive car

$$F_{max} = \mu W_r \tag{2.12}$$

and substituting Eq. 2.11 into Eq. 2.12,

$$F_{max} = \mu \left[\frac{l_f}{L}W + \frac{h}{L}(F_{max} - R_{rl}) \right] \tag{2.13}$$

$$F_{max} = \frac{\mu W(l_f - f_{rl}h)/L}{1 - \mu h/L} \tag{2.14}$$

Similarly, by summing moments about point B (see Fig. 2.3), it can be shown that for a front-wheel–drive vehicle

$$F_{max} = \frac{\mu W(l_r + f_{rl}h)/L}{1 + \mu h/L} \tag{2.15}$$

Note that in Eqs. 2.14 and 2.15, because of canceling of units, h, l_f, l_r, and L can be in any unit of length (feet, inches, meters, etc.). However, all of these terms must be in the same chosen unit of measure. The units of F_{max} will be the same as the units for W (lb or N).

EXAMPLE 2.3

A 2500-lb (11.1-kN) car is designed with a 120-inch (3.048-m) wheelbase. The center of gravity is located 22 inches (559 mm) above the pavement and 40 inches (1.12 m) behind the front axle. If the coefficient of road adhesion is 0.6, what is the maximum tractive effort that can be developed if the car is (a) front-wheel drive and (b) rear-wheel drive?

SOLUTION

For the front-wheel–drive case Eq. 2.15 is used:

$$F_{max} = \frac{\mu W(l_r + f_{rl}h)/L}{1 + \mu h/L}$$

and, from Eq. 2.5, $f_{rl} = 0.01$ because $V = 0$ ft/s, so

$$F_{max} = \frac{[0.6 \times 2500 \times (80 + 0.01(22))]/120}{1 + (0.6 \times 22)/120}$$
$$= \underline{903.38 \text{ lb}}$$

For the rear-wheel–drive case, Eq. 2.14 is used:

$$F_{max} = \frac{[0.6 \times 2500 \times (40 - 0.01(22))]/120}{1 - (0.6 \times 22)/120}$$

$$= \underline{558.71 \text{ lb}}$$

2.6.2 Engine-Generated Tractive Effort

The amount of tractive effort generated by the vehicle's engine is a function of a variety of engine and driveline design factors. For engine design, critical factors in determining output include the shape of the combustion chamber, the quantity of air drawn into the combustion chamber during the induction phase, the type of fuel used, and fuel intake design. Although a complete description of engine design is beyond the scope of this book, an understanding of how engine output is measured and used is important to the study of vehicle performance. The two most commonly used measures of engine output are torque and power. Torque is the work generated by the engine (the twisting moment) and is expressed in foot-pounds (ft-lb) or newton-meters (N-m). Power is the rate of engine work, expressed in horsepower (hp) or kilowatts (kW), and is related to the engine's torque by the following equation:

U.S. Customary

$$hp_e = \frac{2\pi M_e n_e}{550}$$

Metric

$$P_e = \frac{2\pi M_e n_e}{1000} \qquad (2.16)$$

where

hp_e = engine-generated horsepower (1 horsepower equals 550 ft-lb/s),
P_e = engine-generated power in kW,
M_e = engine torque in ft-lb (N-m), and
n_e = engine speed in crankshaft revolutions per second.

Figure 2.4 presents a torque-power diagram for a typical gasoline-powered engine.

EXAMPLE 2.4

It is known that an experimental engine has a torque curve of the form $M_e = an_e - bn_e^2$, where M_e is engine torque in ft-lb, n_e is engine speed in revolutions per second, and a and b are unknown parameters. If the engine develops a maximum torque of 92 ft-lb (124.75 N-m) at 3200 rev/min (revolutions per minute), what is the engine's maximum power?

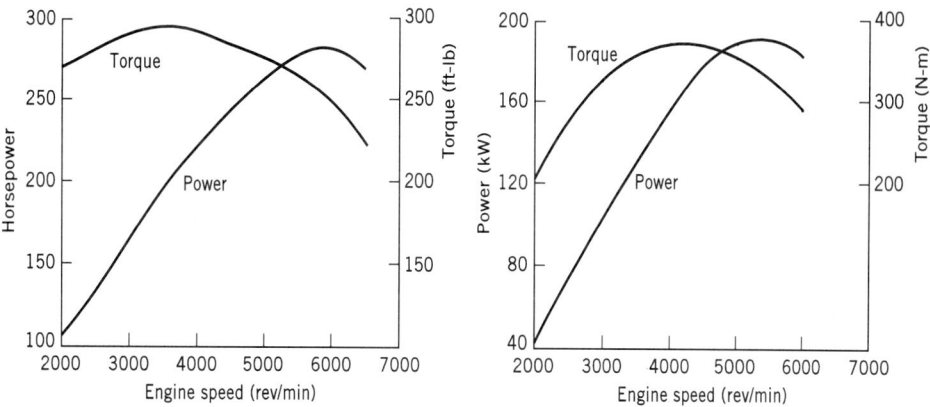

Figure 2.4 Typical torque-power curves for a gasoline-powered automobile engine.

SOLUTION

At maximum torque, $n_e = 53.33$ rev/s (3200/60) and

$$\frac{dM_e}{dn_e} = 0 = a - 2bn_e$$

$$a = 2(53.33)b = 106.67b$$

Also, at maximum torque,

$$M_e = an_e - bn_e^2$$

$$92 = a(53.33) - b(53.33)^2$$

Using these two equations to solve for the two unknowns (a and b), we find that $b = 0.032$ and $a = 3.450$. Using Eq. 2.16 and $M_e = an_e - bn_e^2$,

$$\text{hp}_e = \frac{2\pi(an_e - bn_e^2)n_e}{550}$$

$$= \frac{2\pi(3.450n_e - 0.032n_e^2)n_e}{550}$$

The first derivative of the power equation is used to solve for the engine speed at maximum power:

$$\frac{d\text{hp}_e}{dn_e} = 0 = (0.01142)(6.90n_e - 0.096n_e^2)$$

$$n_e = 71.88 \text{ rev/s}$$

so the engine's maximum power is

$$\text{hp}_e = \frac{2\pi(3.450n_e - 0.032n_e^2)n_e}{550}$$

$$= \frac{2\pi[3.450(71.88) - 0.032(71.88)^2]71.88}{550}$$

$$= \underline{67.87 \text{ hp}}$$

Given the output measures of a vehicle's engine, focus can be directed toward the relationship between engine-generated torque and the tractive effort ultimately delivered to the driving wheels. Unfortunately, the tractive effort needed for acceptable vehicle performance (to provide adequate acceleration characteristics) is greater at lower vehicle speeds, and because maximum engine torque is developed at fairly high engine speeds (crankshaft revolutions), the use of gasoline-powered engines requires some form of gear reduction, as illustrated in Fig. 2.5. This gear reduction provides the mechanical advantage necessary for acceptable vehicle acceleration.

With gear reductions there are two factors that determine the amount of tractive effort reaching the drive wheels. First, the mechanical efficiency of the driveline (the gear reduction devices, including the transmission and differential) must be considered. Typically, 5% to 25% of the tractive effort generated by the engine is lost in

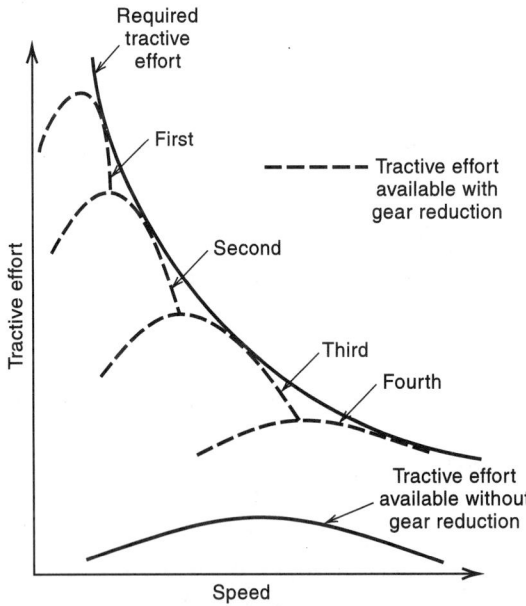

Figure 2.5 Tractive effort requirements and tractive effort generated by a typical gasoline-powered vehicle.

gear reduction devices, which corresponds to a mechanical efficiency of the driveline (η_d) of 0.75 to 0.95. Second, the overall gear reduction ratio (ε_0), which includes the gear reductions of the transmission and differential, plays a key role in the determination of tractive effort. By definition, the overall gear reduction ratio refers to the relationship between the revolutions of the engine's crankshaft and the revolutions of the drive wheels. For example, an overall gear reduction ratio of 4 to 1 ($\varepsilon_0 = 4$) means that the engine's crankshaft turns four revolutions for every one revolution of the drive wheels.

With these terms defined, the engine-generated tractive effort reaching the drive wheels is given as

$$F_e = \frac{M_e \varepsilon_0 \eta_d}{r} \tag{2.17}$$

where

F_e = engine-generated tractive effort reaching the drive wheels in lb (N),
M_e = engine torque in ft-lb (N-m),
ε_0 = overall gear reduction ratio,
η_d = mechanical efficiency of the driveline, and
r = radius of the drive wheels in ft (m).

It follows that the relationship between vehicle speed and engine speed is

$$V = \frac{2\pi r n_e (1 - i)}{\varepsilon_0} \tag{2.18}$$

where

V = vehicle speed in ft/s (m/s),
n_e = engine speed in crankshaft revolutions per second,
i = slippage of the driveline, generally taken as 2 to 5 percent ($i = 0.02$ to 0.05) for passenger vehicles, and
Other terms are as defined previously.

To summarize this section, the available tractive effort (F in Eq. 2.2) at any given speed is the lesser of the maximum tractive effort (F_{max}) and the engine-generated tractive effort (F_e).

2.7 VEHICLE ACCELERATION

As defined in the previous section, available tractive effort (F) can be used to determine a number of vehicle performance characteristics including vehicle acceleration and top speed. For determining vehicle acceleration, Eq. 2.2 can be applied with an additional term to account for the inertia of the vehicle's rotating parts that must be overcome during acceleration. This term is referred to as the mass factor (γ_m) and is introduced in Eq. 2.2 as

$$F - \sum R = \gamma_m m a \qquad (2.19)$$

where the mass factor is approximated as

$$\gamma_m = 1.04 + 0.0025\, \varepsilon_0^2 \qquad (2.20)$$

Two measures of vehicle acceleration are worthy of note: the time to accelerate and the distance to accelerate. For both, the force available to accelerate is $F_{net} = F - \Sigma R$. The basic relationship between the force available to accelerate, F_{net}, the available tractive effort, F (the lesser of F_{max} and F_e), and the summation of resistances is illustrated in Fig. 2.6. In this figure, F_{net} is the vertical distance between the lesser of the F_{max} and the F_e curves and the total resistance curve. So, referring to Fig. 2.6, at speed V', F_{net} will be $F_{max} - \Sigma R$, and at V'', F_{net} will be $F_e - \Sigma R$. It follows that when $F_{net} = 0$, the vehicle cannot accelerate and is at its maximum speed for specified conditions (grade, air density, engine torque, and so on). Such was the case for the vehicle described in Example 2.2. When F_{net} is greater than zero (the vehicle is traveling at a speed less than its maximum speed), Eq. 2.19 can be written in differential form,

$$F_{net} = \gamma_m m \frac{dV}{dt} \quad \text{or} \quad dt = \frac{\gamma_m m\, dV}{F_{net}}$$

and because F_{net} is itself a function of vehicle speed [$F_{net} = f(V)$], integration gives the time to accelerate as

$$t = \gamma_m m \int_{V_1}^{V_2} \frac{dV}{f(V)} \qquad (2.21)$$

where V_1 is the initial vehicle speed and V_2 is the final vehicle speed. Similarly, it can be shown that the distance to accelerate is

$$d_a = \gamma_m m \int_{V_1}^{V_2} \frac{V\, dV}{f(V)} \qquad (2.22)$$

To solve Eqs. 2.21 and 2.22, numerical integration is necessary because the functional forms of these equations do not lend themselves to closed-form solutions. Such numerical integration is straightforward but requires a computer. Consequently, we will not provide an example of solving these equations.

2.7 Vehicle Acceleration

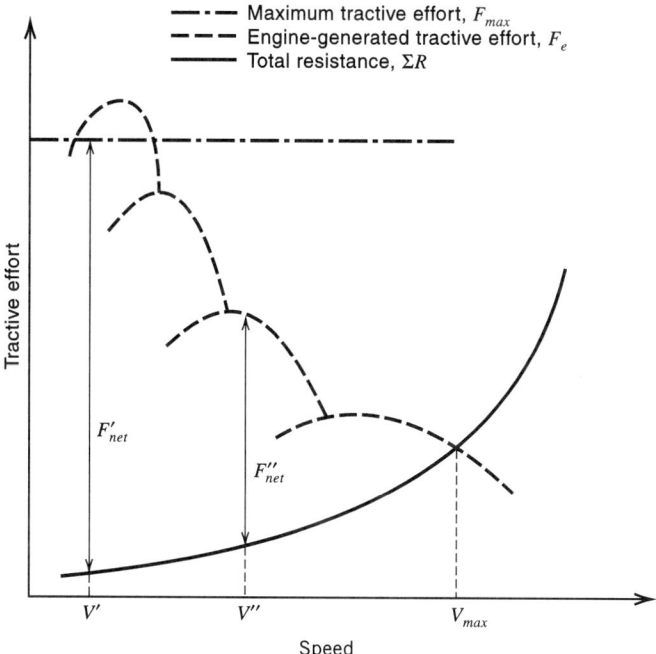

Figure 2.6 Relationship among the forces available to accelerate, available tractive effort, and total vehicle resistance.

EXAMPLE 2.5

A car is traveling at 10 mi/h (16.1 km/h) on a roadway covered with hard-packed snow (coefficient of road adhesion of 0.20). The car has $C_D = 0.30$, $A_f = 20$ ft^2 (1.86 m^2), and $W = 3000$ lb (13.34 kN). The wheelbase is 120 inches (3.048 m), and the center of gravity is 20 inches (508 mm) above the roadway surface and 50 inches (1.27 m) behind the front axle. The air density is 0.002045 slugs/ft^3 (1.054 kg/m^3). The car's engine is producing 95 ft-lb (128.8 N-m) of torque and is in a gear that gives an overall gear reduction ratio of 4.5 to 1, the wheel radius is 14 inches (356 mm), and the mechanical efficiency of the driveline is 80%. If the driver needs to accelerate quickly to avoid an accident, what would the acceleration be if the car is (a) front-wheel drive and (b) rear-wheel drive?

SOLUTION

We begin by computing the resistances, tractive effort generated by the engine, and mass factor because all of these factors will be the same for both front- and rear-wheel drive.

The aerodynamic resistance is (from Eq. 2.3)

$$R_a = \frac{\rho}{2} C_D A_f V^2$$

$$= \frac{0.002045}{2}(0.3)(20)(10 \times 5280/3600)^2$$

$$= 1.32 \text{ lb}$$

The rolling resistance is (from Eq. 2.6)

$$R_{rl} = f_{rl} W$$

$$= 0.01\left(1 + \frac{10 \times 5280/3600}{147}\right) \times 3000$$

$$= 32.99 \text{ lb}$$

The engine-generated tractive effort is (from Eq. 2.17)

$$F_e = \frac{M_e \varepsilon_0 \eta_d}{r}$$

$$= \frac{95(4.5)(0.8)}{\frac{14}{12}}$$

$$= 293.14 \text{ lb}$$

The mass factor is (from Eq. 2.20)

$$\gamma_m = 1.04 + 0.0025\,\varepsilon_0^2$$

$$= 1.04 + 0.0025(4.5)^2$$

$$= 1.091$$

Recall that to determine acceleration, we need the resistances (already computed) and the available tractive effort, F, which is the lesser of F_e and F_{max}. For the case of the front-wheel–drive car, Eq. 2.15 can be applied to determine F_{max}:

$$F_{max} = \frac{\mu W(l_r + f_{rl}h)/L}{1 + \mu h/L}$$

$$= \frac{[0.2 \times 3000 \times (70 + 0.011(20))]/120}{1 + (0.2 \times 20)/120}$$

$$= 339.77 \text{ lb}$$

Thus for a front-wheel–drive car $F = 293.14$ lb (the lesser of 293.14 and 339.77), and the acceleration is (from Eq. 2.19)

$$F - \sum R = \gamma_m ma$$

$$a = \frac{F - \sum R}{\gamma_m m} = \frac{293.14 - 34.31}{1.09(3000/32.2)} = \underline{2.546 \text{ ft/s}^2}$$

For the case of the rear-wheel–drive car, Eq. 2.14 can be applied to determine F_{max}:

$$F_{max} = \frac{[0.2 \times 3000 \times (50 - 0.011(20))]/120}{1 - (0.2 \times 20)/120} = 257.48 \text{ lb}$$

Thus for a rear-wheel–drive car $F = 257.48$ lb (the lesser of 293.14 and 257.48), and the acceleration is (from Eq. 2.19)

$$a = \frac{F - \sum R}{\gamma_m m} = \frac{257.48 - 34.31}{1.09(3000/32.2)} = \underline{2.196 \text{ ft/s}^2}$$

Typical values for the coefficient of road adhesion (μ) are shown later, in Table 2.4. However, in determining acceleration (and braking and cornering, as will be shown later in this book) two points are worthy of note. First, it is possible that the coefficient of road adhesion can exceed 1.0. This is because there is a micro-interaction at the tire-pavement interface that results in a "cog-type" effect that, for some high-performance tires that use softer compounds, can increase μ above 1.0. This explains why many race cars, particularly drag-racing cars, have initial acceleration rates well in excess of 1 g (32.2 ft/s^2 or 9.807 m/s^2). Second, at high speed, vehicle aerodynamics can create downward forces that effectively increase W in the preceding equations, and this allows for greater acceleration. Drag-racing cars and some open-wheel race cars (such as Formula One–style cars) are examples of aerodynamic designs that use air deflectors designed to generate significant downward forces to enhance acceleration, braking, and cornering.

2.8 FUEL EFFICIENCY

Given the factors discussed in the preceding sections of this chapter, it is clear what elements determine a vehicle's fuel efficiency. One of the most critical determinants relates to engine design (how the engine-generated tractive effort is produced). Engine designs that increase the quantity of air entering the combustion chamber, improve fuel delivery to the combustion chamber, and decrease internal engine friction lead to improved fuel efficiency. Improvements to other mechanical components, such as decreasing slippage and improving the mechanical efficiency of the driveline, also increase the overall fuel efficiency.

In terms of resistance-reducing options, decreasing overall vehicle weight (W) will lower grade and rolling resistances, thus reducing fuel consumption (all other factors held constant). Similarly, aerodynamic improvements such as a lower drag coefficient (C_D) and a reduced frontal area (A_f) can produce significant fuel savings. Finally, improved tire designs with lower rolling resistance can improve overall fuel efficiency.

2.9 PRINCIPLES OF BRAKING

In terms of highway design and traffic analysis, the braking characteristics of road vehicles is arguably the single most important aspect of vehicle performance. The braking behavior of road vehicles is critical in the determination of stopping sight distance, roadway surface design, and accident avoidance systems. Moreover, ongoing advances in braking technology make it essential that transportation engineers have a basic comprehension of the underlying principles involved.

2.9.1 Braking Forces

To begin the discussion of braking principles, consider the force and moment-generating diagram in Fig. 2.7. During vehicle braking there is a load transfer from the rear to the front axle. To illustrate this, expressions for the normal loads on the front and rear axles can be written by summing the moments about roadway surface–tire contact points A and B (as was done in deriving Eqs. 2.14 and 2.15, with $\cos \theta_g$ assumed to be equal to 1 because of the small grades encountered in highway applications):

$$W_f = \frac{1}{L}[Wl_r + h(ma - R_a \pm W \sin \theta_g)] \qquad (2.23)$$

and

$$W_r = \frac{1}{L}[Wl_f - h(ma - R_a \pm W \sin \theta_g)] \qquad (2.24)$$

where, in this case, the contribution of grade resistance ($W \sin \theta_g$) is negative for uphill grades and positive for downhill grades.

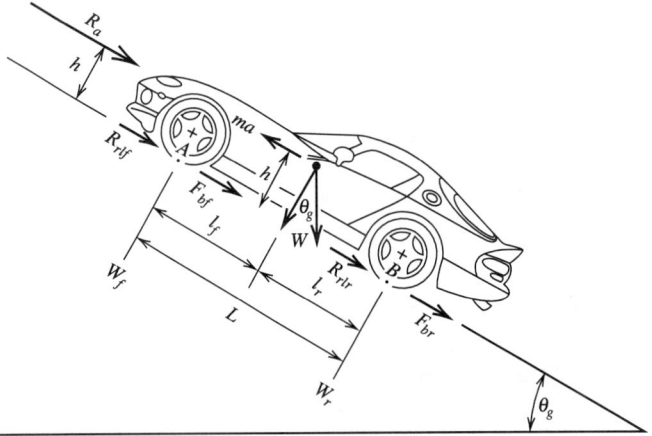

Figure 2.7 Forces acting on a vehicle during braking, with driveline resistance ignored.

2.9 Principles of Braking

In this figure,

R_a = aerodynamic resistance in lb (N),
R_{rlf} = rolling resistance of the front tires in lb (N),
R_{rlr} = rolling resistance of the rear tires in lb (N),
F_{bf} = braking force on the front tires in lb (N),
F_{br} = braking force on the rear tires in lb (N),
W = total vehicle weight in lb (N),
W_f = weight of the vehicle on the front axle in lb (N),
W_r = weight of the vehicle on the rear axle in lb (N),
θ_g = angle of the grade in degrees,
m = vehicle mass in slugs (kg),
a = acceleration in ft/s² (m/s²),
L = length of wheelbase,
h = height of the center of gravity above the roadway surface,
l_f = distance from the front axle to the center of gravity, and
l_r = distance from the rear axle to the center of gravity.

From the summation of forces along the vehicle's longitudinal axis,

$$F_b + f_{rl} W = ma - R_a \pm W \sin \theta_g \tag{2.25}$$

with the rolling resistance equal to the coefficient of rolling resistance multiplied by the vehicle weight (from Eq. 2.6, $R_{rl} = f_{rl} W$) and $F_b = F_{bf} + F_{br}$. Substituting Eq. 2.25 into Eqs. 2.23 and 2.24 gives

$$W_f = \frac{1}{L} [Wl_r + h(F_b + f_{rl} W)] \tag{2.26}$$

and

$$W_r = \frac{1}{L} [Wl_f - h(F_b + f_{rl} W)] \tag{2.27}$$

Because the maximum vehicle braking force ($F_{b\,max}$) is equal to the coefficient of road adhesion (μ), multiplied by the vehicle weights normal to the roadway surface,

$$\begin{aligned} F_{bf\,max} &= \mu W_f \\ &= \frac{\mu W}{L} [l_r + h(\mu + f_{rl})] \end{aligned} \tag{2.28}$$

and

$$\begin{aligned} F_{br\,max} &= \mu W_r \\ &= \frac{\mu W}{L} [l_f - h(\mu + f_{rl})] \end{aligned} \tag{2.29}$$

To develop maximum braking forces, the tires should be at the point of an impending slide. If the tires begin to slide (the brakes lock), a significant reduction in

Table 2.4 Typical Values of Coefficients of Road Adhesion

Pavement	Coefficient of road adhesion	
	Maximum	Slide
Good, dry	1.00*	0.80
Good, wet	0.90	0.60
Poor, dry	0.80	0.55
Poor, wet	0.60	0.30
Packed snow or ice	0.25	0.10

*In some instances, the coefficient of road adhesion values can exceed 1.0. See discussion at the end of Section 2.7.
Source: S. G. Shadle, L. H. Emery, and H. K. Brewer, "Vehicle Braking, Stability, and Control," *SAE Transactions,* vol. 92, paper 830562, 1983.

road adhesion will result. An indication of the extent of the reduction in road adhesion as the result of tire slide, under various pavement and weather conditions, is presented in Table 2.4. It is clear from this table that the braking forces will decline dramatically when the wheels are locked (resulting in tire slide). Avoiding this locked condition is the function of antilock braking systems in cars. Such systems will be discussed later in this chapter.

2.9.2 Braking Force Ratio and Efficiency

On a given roadway surface, the maximum attainable vehicle deceleration (using the vehicle's braking system) is equal to μg, where μ is the coefficient of road adhesion and g is the gravitational constant (32.2 ft/s^2 or 9.807 m/s^2). To approach this maximum vehicle deceleration, vehicle braking systems must correctly distribute braking forces between the vehicle's front and rear brakes. This is typically done by the allocation of hydraulic pressures within the braking system. This front/rear proportioning of braking forces (within the vehicle's braking system) will be optimal (achieving a deceleration rate equal to μg) when it is in the exact same proportion as the ratio of the maximum braking forces on the front and rear axles ($F_{bf\,max}/F_{br\,max}$). Thus maximum braking forces (with the tires at the point of impending slide) will be developed when the brake force ratio (front force over rear force) is

$$BFR_{f/r\,max} = \frac{l_r + h(\mu + f_{rl})}{l_f - h(\mu + f_{rl})} \tag{2.30}$$

where

$BFR_{f/r\,max}$ = the brake force ratio, allocated by the vehicle's braking system, that results in maximum (optimal) braking forces, and

Other terms are as defined previously.

It follows that the percentage of braking force that the braking system should allocate to the front axle (PBF_f) for maximum braking is

$$PBF_f = 100 - \frac{100}{1 + BFR_{f/r\ max}} \qquad (2.31)$$

and the percentage of braking force that the braking system should allocate to the rear axle (PBF_r) for maximum braking is

$$PBF_r = \frac{100}{1 + BFR_{f/r\ max}} \qquad (2.32)$$

EXAMPLE 2.6

A car has a wheelbase of 100 inches (2.54 m) and a center of gravity that is 40 inches (1.016 m) behind the front axle at a height of 24 inches (609 mm). If the car is traveling at 80 mi/h (128.7 km/h) on a road with poor pavement that is wet, determine the percentages of braking force that should be allocated to the front and rear brakes (by the vehicle's braking system) to ensure that maximum braking forces are developed.

SOLUTION

The coefficient of rolling resistance is

$$f_{rl} = 0.01\left(1 + \frac{80 \times 5280/3600}{147}\right) = 0.018$$

and $\mu = 0.6$ from Table 2.4 (maximum because we want the tires to be at the point of impending slide). Applying Eq. 2.30 gives

$$BFR_{f/r\ max} = \frac{l_r + h(\mu + f_{rl})}{l_f - h(\mu + f_{rl})}$$

$$= \frac{60 + 24(0.6 + 0.018)}{40 - 24(0.6 + 0.018)}$$

$$= 2.973$$

Using Eq. 2.31, the percentage of the force allocated to the front brakes should be

$$PBF_f = 100 - \frac{100}{1 + BFR_{f/r\ max}}$$

$$= 100 - \frac{100}{1 + 2.973}$$

$$= \underline{74.83\%}$$

and using Eq. 2.32 (or simply $100 - PBF_f$), the percentage of the force allocated to the rear brakes should be

$$PBF_r = \frac{100}{1 + BFR_{f/r\ max}}$$

$$= \frac{100}{1 + 2.973}$$

$$= \underline{25.17\%}$$

It is clear from Eq. 2.30 that the design of a vehicle's braking system is not an easy task because the optimal brake force proportioning changes with both vehicle and road conditions. For example, the addition of vehicle cargo and/or passengers will not only change the weight of the vehicle (which affects f_{rl} in Eq. 2.30), but will also change the distribution of the weight, shifting the height of the center of gravity and its location along the vehicle's longitudinal axis. This will change the optimal brake force proportioning ($BFR_{f/r\ max}$). Similarly, changes in road conditions will produce different coefficients of adhesion, again changing optimal brake force proportioning.

Figure 2.8 illustrates the effects of vehicle loadings on the optimal distribution of braking forces on a road with a coefficient of adhesion equal to 0.85. Figure 2.8a shows the effect of loading conditions on the braking performance of a light truck. This figure's curves define the points of impending wheel lockup, so, for example, with 20% of the braking force allocated to the front brakes by the vehicle's braking system, a 0.6-g (19.32 ft/s², 5.88 m/s²) deceleration will be achieved before the brakes lock and the coefficient of adhesion drops to slide values. Point A defines the optimal brake force proportioning under loaded conditions as roughly 42% of the total braking force allocated to the front brakes (as would be computed by Eq. 2.31). At this point a maximum deceleration of 0.85 g's (23.37 ft/s², 8.34 m/s²) will be achieved as limited by the coefficient of road adhesion ($\mu = 0.85$). Note that when only 20% of the braking force is allocated to the front brakes, the deceleration (19.32 ft/s², 5.88 m/s²) is substantially less than the maximum (23.37 ft/s², 8.34 m/s²). The point A' indicates that optimal brake force proportioning under unloaded conditions has about 72% of the braking force allocated to the front brakes. This disparity in optimal brake force allocation between loaded and unloaded conditions (points A and A') presents a dilemma that necessitates a compromised braking system design (the compromised brake proportioning represented by point 1). This compromise gives suboptimal braking force allocations under both loaded and unloaded conditions. Figure 2.8b shows that the disparity of optimal braking forces between loaded and unloaded conditions is less pronounced for passenger cars but still necessitates brake system compromises.

It is important to note that studies have indicated that if wheel lockup is to occur, it is preferable to have the front wheels lock first because having the rear wheels lock first can result in uncontrollable vehicle spin. Front-wheel lockup will result in the loss of steering control, but the vehicle will at least continue to brake in a straight line. Technological advancements in braking systems since the late 1970s have

2.9 Principles of Braking

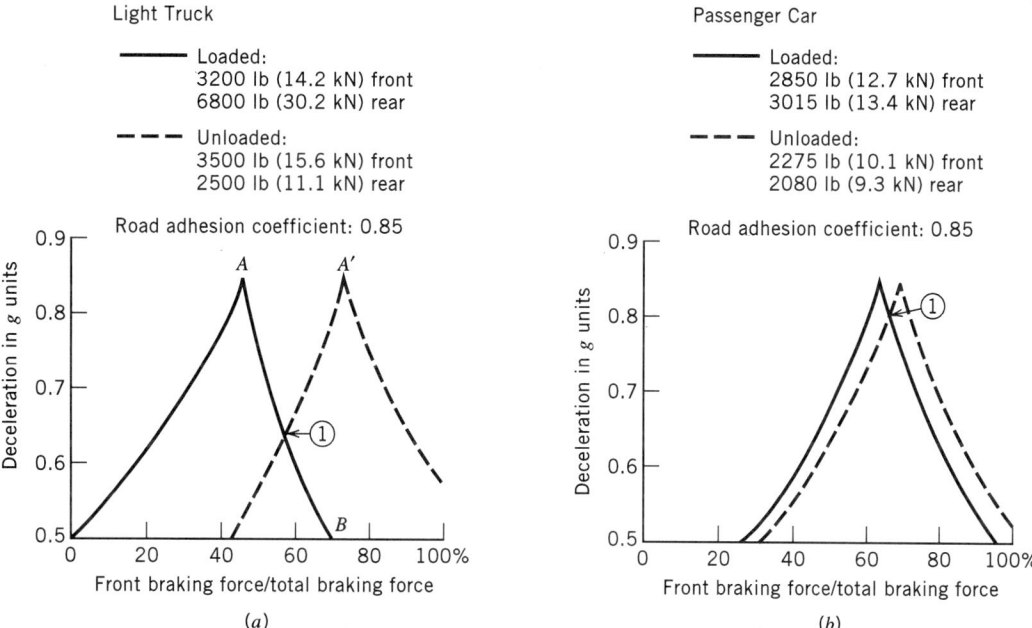

Figure 2.8 Effect of brake force proportioning on the braking performance of a light truck and a passenger car.
Reproduced by permission of the Society of Automotive Engineers from D. J. Bickerstaff and G. Hartley, "Light Truck Tire Traction Properties and Their Effects on Braking Performance," *SAE Transactions*, Vol. 83, paper 741137, 1974.

resulted in vehicles that are increasingly capable of proportioning brake forces in a manner that is closer to optimal and avoiding the dangerous rear-wheel–first lockup due to front-wheel underbraking.

Because true optimal brake force proportioning is seldom achieved in standard non-antilock braking systems, it is useful to define a braking efficiency term that reflects the degree to which the braking system is operating below optimal. Simply stated, braking efficiency is defined as the ratio of the maximum rate of deceleration, expressed in g's (g_{max}), achievable prior to any wheel lockup to the coefficient of road adhesion.

$$\eta_b = \frac{g_{max}}{\mu} \qquad (2.33)$$

where

η_b = braking efficiency,
g_{max} = maximum deceleration in g units (with the absolute maximum = μ), and
μ = coefficient of road adhesion.

As an example of the application of this equation, suppose we want to find the braking efficiency of the loaded truck in Fig. 2.8a when 40% of the available braking force is allocated to the front brakes. The figure shows that the deceleration rate is

0.75 g's, so the application of Eq. 2.33 gives a braking efficiency of 0.8824, or 88.24% ($\eta_b = 0.75/0.85 = 0.8824$).

2.9.3 Antilock Braking Systems

Many modern cars have braking systems that are designed to prevent the wheels from locking during braking applications (antilock braking systems). In theory, antilock braking systems serve two purposes. First, they prevent the coefficient of road adhesion from dropping to slide values (see Table 2.4). Second, they have the potential to raise the braking efficiency to 100%. In practice, designing an antilock braking system that avoids slide coefficients of adhesion and achieves 100% braking efficiency ($\eta_b = 1.0$) is a difficult task. This is because most antilock braking system technologies detect which wheels have locked and release them momentarily before reapplying the brake on the locking wheel. The wheel lock detection speed, speed of brake force reallocation, and braking system design (the amount of braking forces that can be accommodated by the vehicle's front and rear brake discs and calipers) all impact the overall effectiveness of the antilock braking system. Early antilock braking systems often fell short of achieving 100% braking efficiency, and in many cases, an expert driver operating a non-antilock braking car could modulate the brakes to achieve shorter stopping distances than cars equipped with antilock brakes. However, advances in antilock braking system technology continue to bring us closer to 100% braking efficiency.

2.9.4 Theoretical Stopping Distance

With a basic understanding of brake force proportioning and the resulting braking efficiency, attention can now be directed toward developing expressions for minimum stopping distances. By inspection of Fig. 2.7, it can be seen that the relationship between the stopping distance, braking force, vehicle mass, and vehicle speed is

$$a \, ds = \left[\frac{F_b + \sum R}{\gamma_b m} \right] ds \qquad (2.34)$$
$$= V \, dV$$

where

γ_b = mass factor accounting for moments of inertia during braking, which is given the value of 1.04 for automobiles [Wong 1978], and
Other terms are as defined in Fig. 2.7.

Integrating to determine stopping distance (S) gives

2.9 Principles of Braking

$$S = \int_{V_1}^{V_2} \gamma_b m \frac{V\, dV}{F_b + \sum R} \qquad (2.35)$$

Substituting in the resistances (see Fig. 2.7), we obtain

$$S = \gamma_b m \int_{V_1}^{V_2} \frac{V\, dV}{F_b + R_a + f_{rl}W \pm W \sin \theta_g} \qquad (2.36)$$

where

V_1 = initial vehicle speed in ft/s (m/s),
V_2 = final vehicle speed in ft/s (m/s),
$f_{rl}W$ = rolling resistance,
$W \sin \theta_g$ = grade resistance (positive for uphill slopes and negative for downhill slopes), and
Other terms are as defined previously.

To simplify notation, let

$$K_a = \frac{\rho}{2} C_D A_f \qquad (2.37)$$

so that Eq. 2.3 is

$$R_a = K_a V^2 \qquad (2.38)$$

Continuing, assume that the effect of speed on the coefficient of rolling resistance, f_{rl}, is constant and can be approximated by using the average of initial (V_1) and final (V_2) speeds in Eq. 2.5 [$V = (V_1 + V_2)/2$]. With this assumption (which will introduce only a very small amount of error), and letting $m = W/g$ and $F_b = \mu W$, integration of Eq. 2.36 gives

$$S = \frac{\gamma_b W}{2gK_a} \ln \left[\frac{\mu W + K_a V_1^2 + f_{rl} W \pm W \sin \theta_g}{\mu W + K_a V_2^2 + f_{rl} W \pm W \sin \theta_g} \right] \qquad (2.39)$$

If the vehicle is assumed to stop ($V_2 = 0$),

$$S = \frac{\gamma_b W}{2gK_a} \ln \left[1 + \frac{K_a V_1^2}{\mu W + f_{rl} W \pm W \sin \theta_g} \right] \qquad (2.40)$$

With braking efficiency considered, the actual braking force is

$$F_b = \eta_b \mu W \qquad (2.41)$$

Therefore, by substitution into Eq. 2.40, the theoretical stopping distance is

$$S = \frac{\gamma_b W}{2gK_a}\ln\left[1 + \frac{K_a V_1^2}{\eta_b \mu W + f_{rl} W \pm W \sin\theta_g}\right] \qquad (2.42)$$

Similarly, Eq. 2.39 can be written to include braking efficiency. Finally, if aerodynamic resistance is ignored (due to its comparatively small contribution to braking), integration of Eq. 2.35 gives the theoretical stopping distance as

$$S = \frac{\gamma_b(V_1^2 - V_2^2)}{2g(\eta_b \mu + f_{rl} \pm \sin\theta_g)} \qquad (2.43)$$

EXAMPLE 2.7

A new experimental 2500-lb (11.12-kN) car, with $C_D = 0.25$ and $A_f = 18$ ft² (1.67 m²), is traveling at 90 mi/h (144.8 km/h) down a 10% grade. The coefficient of road adhesion is 0.7 and the air density is 0.0024 slugs/ft³ (1.24 kg/m³). The car has an advanced antilock braking system that gives it a braking efficiency of 100%. Determine the theoretical minimum stopping distance for the case where aerodynamic resistance is considered and the case where aerodynamic resistance is ignored.

SOLUTION

With aerodynamic resistance considered, Eq. 2.42 can be applied with $\gamma_b = 1.04$, $\theta_g = 5.71°$, and

$$f_{rl} = 0.01\left(1 + \frac{90 \times 5280/3600 + 0}{147}\right) = 0.0145$$

$$K_a = \frac{0.0024}{2}(0.25)(18) = 0.0054$$

Then

$$S = \frac{1.04(2500)}{2(32.2)(0.0054)}\ln\left[1 + \frac{0.0054(90 \times 5280/3600)^2}{(1.0)(0.7)(2500) + (0.0145)(2500) - 2500\sin(5.71°)}\right]$$

$$= \underline{444.07 \text{ ft}}$$

With aerodynamic resistance excluded, Eq. 2.43 is used:

$$S = \frac{1.04(90 \times 5280/3600)^2}{2(32.2)(0.7 + 0.0145 - \sin(5.71°))}$$

$$= \underline{457.53 \text{ ft}}$$

EXAMPLE 2.8

A car is traveling at 80 mi/h (128.7 km/h) and has a braking efficiency of 80%. The brakes are applied to miss an object that is 150 ft (45.72 m) from the point of brake application, and the coefficient of road adhesion is 0.85. Ignoring aerodynamic resistance and assuming theoretical minimum stopping distance, estimate how fast the car will be going when it strikes the object if (a) the surface is level and (b) the surface is on a 5% upgrade.

SOLUTION

In both cases rolling resistance will be approximated as

$$f_{rl} = 0.01\left(1 + \frac{\frac{80 \times 5280/3600 + V_2}{2}}{147}\right) = 0.014 + 0.000034 V_2$$

Applying Eq. 2.43 for the level grade with $\gamma_b = 1.04$, $\theta_g = 0°$,

$$S = \frac{\gamma_b(V_1^2 - V_2^2)}{2g(\eta_b\mu + f_{rl} \pm \sin\theta_g)}$$

$$150 = \frac{1.04[(80 \times 5280/3600)^2 - V_2^2]}{2(32.2)[0.8(0.85) + (0.014 + 0.000034 V_2) \pm 0]}$$

$V_2 = \underline{85.40 \text{ ft/s}}$ or $\underline{58.23 \text{ mi/h}}$

On a 5% grade with $\theta_g = 2.86°$,

$$150 = \frac{1.04[(80 \times 5280/3600)^2 - V_2^2]}{2(32.2)[0.8(0.85) + (0.014 + 0.000034 V_2) \pm 0.05]}$$

$V_2 = \underline{82.64 \text{ ft/s}}$ or $\underline{56.35 \text{ mi/h}}$

2.9.5 Practical Stopping Distance

As mentioned earlier, one of the most critical concerns in the design of a highway is the provision of adequate driver sight distance to permit a safe stop. The theoretical assessment of vehicle stopping distance presented in the previous section provided the principles of braking for an individual vehicle under specified roadway surface conditions. However, highway engineers face a more complex problem because they must design for a variety of driver skill levels (which can affect whether or not the brakes lock and reduce the coefficient of road adhesion to slide values), vehicle types (with varying aerodynamics, weight distributions, and brake efficiencies), and weather conditions (which change the roadway's coefficient of adhesion). As a result of the wide variability inherent in the determination of braking distance, an equation is required that provides an estimate of typical observed braking distances, and is more simplistic and usable than Eq. 2.42.

The basic physics equation on rectilinear motion, assuming constant deceleration, is chosen as the basis of a practical equation for stopping distance:

$$V_2^2 = V_1^2 + 2ad \tag{2.44}$$

where

V_2 = final vehicle speed in ft/s (m/s),
V_1 = initial vehicle speed in ft/s (m/s),
a = acceleration (negative for deceleration) in ft/s^2 (m/s^2), and
d = deceleration distance (practical stopping distance) in ft (m).

Rearranging Eq. 2.44 and assuming a will be negative for deceleration gives

$$d = \frac{V_1^2 - V_2^2}{2a} \tag{2.45}$$

If $V_2 = 0$ (the vehicle comes to a complete stop), the practical-stopping-distance equation is

$$d = \frac{V_1^2}{2a} \tag{2.46}$$

To make this equation generally applicable for design purposes, a deceleration rate, a, must be chosen that is representative of appropriately conservative braking behavior. AASHTO [2001] recommends a deceleration rate of 11.2 ft/s^2 (3.4 m/s^2). Empirical studies [Fambro et al. 1997] have shown that most drivers decelerate at rates greater than this, and that this deceleration rate is well within a driver's capability to maintain steering control during a braking maneuver on wet surfaces. Additionally, empirical studies [Fambro et al. 1997] have confirmed that most vehicle braking systems and tire-pavement friction levels are capable of supporting this deceleration rate, even under wet conditions.

To account for the effect of grade, Eq. 2.46 is modified as follows:

$$d = \frac{V_1^2}{2g\left(\dfrac{a}{g} \pm G\right)} \qquad (2.47)$$

where

g = gravitational constant, 32.2 ft/s² (9.807 m/s²),
G = roadway grade (+ for uphill, − for downhill) in percent/100, and
Other terms are as defined previously.

It is important to note that Eq. 2.47 is consistent with Eq. 2.43 (the theoretical stopping distance ignoring aerodynamic resistance). Rewriting Eq. 2.43 with the assumption that the vehicle comes to a stop ($V_2 = 0$), that $\sin\theta_g = \tan\theta_g = G$ (for small grades), and that γ_b and f_{rl} can be ignored due to their small and essentially offsetting effects, we have

$$S = \frac{V_1^2}{2g(\eta_b\mu \pm G)} \qquad (2.48)$$

Recall that $\eta_b\mu = g_{max}$ (Eq. 2.33). However, rather than determining the maximum deceleration rate (in g's) for a specific vehicle braking efficiency and specific coefficient of road adhesion, the AASHTO recommended maximum deceleration rate (again, an appropriately conservative value for the overall driver and vehicle population) is used. Thus, a maximum deceleration of 0.35 g's (11.2/32.2 or 3.4/9.807) is used for Eq. 2.47.

The recommended deceleration rate as determined empirically already accounts for the effects of aerodynamic resistance, braking efficiency, coefficient of road adhesion, and inertia during braking (the braking mass factor). This value is reflective of current vehicle technologies and driving behavior. It is important to recognize that as vehicle braking technology and other vehicle characteristics change, as well as possibly driver behavior, the recommended value of a should be reviewed to determine if it is still applicable for highway design purposes. The relationship between changing vehicle characteristics and changing highway design guidelines is one that must always be kept in the design engineer's mind.

EXAMPLE 2.9

A car [W = 2200 lb (9.79 kN), C_D = 0.25, A_f = 21.5 ft² (2 m²)] has an antilock braking system that gives it a braking efficiency of 100%. The car's stopping distance is tested on a level roadway with poor, wet pavement (with tires at the point of impending skid), and $\rho = 0.00238$ slugs/ft³ (1.227 kg/m³). How inaccurate will the stopping distance predicted by the practical-stopping-distance equation be compared with the theoretical stopping distance, assuming the car is initially traveling 60 mi/h (96.5 km/h)? How inaccurate will the practical-stopping-distance equation be if the same car has a braking efficiency of 85%?

SOLUTION

First, to calculate the theoretical minimum stopping distance, Eq. 2.42 is applied with $\gamma_b = 1.04$, $\theta_g = 0°$, $\mu = 0.60$ (maximum for poor, wet pavement, from Table 2.4), and

$$f_{rl} = 0.01\left(1 + \frac{\frac{60 \times 5280/3600 + 0}{2}}{147}\right) = 0.013$$

$$K_a = \frac{0.00238}{2}(0.25)(21.5) = 0.0064$$

so from Eq. 2.42,

$$S = \frac{1.04(2200)}{2(32.2)(0.0064)} \ln\left[1 + \frac{0.0064(60 \times 5280/3600)^2}{(1.0)(0.60)(2200) + (0.013)(2200) \pm 0}\right]$$

$$= 200.35 \text{ ft}$$

For the same conditions but with a vehicle braking efficiency of 85%, Eq. 2.42 gives

$$S = \frac{1.04(2200)}{2(32.2)(0.0064)} \ln\left[1 + \frac{0.0064(60 \times 5280/3600)^2}{(0.85)(0.60)(2200) + (0.013)(2200) \pm 0}\right]$$

$$= 234.11 \text{ ft}$$

Now applying Eq. 2.46 (since $G = 0$) for the practical stopping distance, we find

$$d = \frac{(60 \times 5280/3600)^2}{2(11.2)} = 345.71 \text{ ft}$$

In the first case, the error is $\underline{145.36 \text{ ft}}$. In the case of 85% braking efficiency, the error is $\underline{111.60 \text{ ft}}$. Rearranging Eq. 2.46 to solve for a, we find that the stopping distances of 200.35 ft and 234.11 ft correspond to deceleration rates of 19.33 ft/s² and 16.54 ft/s², respectively. Studies [Fambro et al. 1997] have shown that most drivers decelerate at rates of 18.4 ft/s² (5.6 m/s²) or greater for emergency stopping situations. Thus, this range of theoretical values is consistent with observed distances for situations in which minimum stopping distances are trying to be achieved. Comparing these theoretical values to the AASHTO recommended deceleration rate of 11.2 ft/s²,

2.9.6 Distance Traveled during Driver Perception/Reaction

Until now the focus has been directed toward the distance required to stop the vehicle from the point of brake application. However, in providing sufficient sight distance for a driver to stop safely, it is also necessary to consider the distance traveled during the time the driver is perceiving and reacting to the need to stop. The distance traveled during perception/reaction (d_r) is given by

$$d_r = V_1 \times t_r \tag{2.49}$$

where

V_1 = initial vehicle speed in ft/s (m/s), and
t_r = time required to perceive and react to the need to stop, in s.

The perception/reaction time of a driver is a function of a number of factors, including the driver's age, physical condition, and emotional state, as well as the complexity of the situation and the strength of the stimuli requiring a stopping action. For highway design, a conservative perception/reaction time has been determined to be 2.5 seconds [AASHTO 2001]. For comparison, average drivers have perception/reaction times of approximately 1.0 to 1.5 seconds.

Thus, the total required stopping distance is a combination of the braking distance, either theoretical (Eq. 2.42 or 2.43) or practical (Eq. 2.47), and the distance traveled during perception/reaction (Eq. 2.49), as shown in Eq. 2.50.

$$d_s = d + d_r \tag{2.50}$$

where

d_s = total stopping distance (including perception/reaction) in ft (m),
d = distance traveled during braking in ft (m), and
d_r = distance traveled during perception/reaction in ft (m).

The combination of practical stopping distance and the distance traveled during perception/reaction is a primary consideration in highway design, as will be discussed in detail in Chapter 3.

EXAMPLE 2.10

Two drivers each have a reaction time of 2.5 seconds. One is obeying a 55-mi/h (88.5-km/h) speed limit, and the other is traveling illegally at 70 mi/h (112.6 km/h). How much distance will each of the drivers cover while perceiving/reacting to the need to stop, and what will the total stopping distance be for each driver (using practical stopping distance and assuming $G = -2.5\%$)?

SOLUTION

The distances traveled by each driver during perception/reaction will be calculated first, using Eq. 2.49. For the driver traveling at 55 mi/h,

$$d_r = V_1 \times t_r = (55 \times 5280/3600)(2.5) = \underline{201.67 \text{ ft}}$$

For the driver traveling at 70 mi/h,

$$d_r = V_1 \times t_r = (70 \times 5280/3600)(2.5) = \underline{256.67 \text{ ft}}$$

Therefore, driving at 70 mi/h increases the distance traveled during perception/reaction by 55.0 ft.

Next, the distance traveled during braking will be calculated for each driver, using the equation for practical stopping distance (Eq. 2.47). For the driver traveling at 55 mi/h,

$$d = \frac{(55 \times 5280/3600)^2}{2(32.2)\left(\frac{11.2}{32.2} - 0.025\right)} = \frac{6507.11}{20.79} = 312.99 \text{ ft}$$

For the driver traveling at 70 mi/h,

$$d = \frac{(70 \times 5280/3600)^2}{2(32.2)\left(\frac{11.2}{32.2} - 0.025\right)} = \frac{10{,}540.44}{20.79} = 507.00 \text{ ft}$$

The total stopping distance for each driver is now calculated with Eq. 2.50. For the driver traveling at 55 mi/h,

$$d_s = d + d_r = 312.99 + 201.67 = \underline{514.66 \text{ ft}}$$

For the driver traveling at 70 mi/h,

$$d_s = d + d_r = 507.00 + 256.67 = \underline{763.67 \text{ ft}}$$

Therefore, driving at 70 mi/h increases the total stopping distance by a very substantial 249.01 ft.

NOMENCLATURE FOR CHAPTER 2

a	acceleration (deceleration if negative)
A_f	frontal area of vehicle
$BFR_{f/r\ max}$	brake force ratio (front over rear) for maximum braking force
C_D	coefficient of aerodynamic drag
d	practical stopping distance
d_a	distance to accelerate
d_r	distance traveled during driver perception/reaction
d_s	total stopping distance (vehicle braking distance plus perception/reaction distance)
F	total available tractive effort
F_b	total braking force
F_{bf}	front-axle braking force
$F_{bf\ max}$	maximum front-axle braking force
F_{br}	rear-axle braking force
$F_{br\ max}$	maximum rear-axle braking force
F_e	engine-generated tractive effort
F_f	available tractive effort at the front axle
F_{max}	maximum tractive effort
F_r	available tractive effort at the rear axle
f_{rl}	coefficient of rolling resistance
G	roadway grade in ft/ft (m/m) (percent grade divided by 100; $G = 0.05$ is a 5% grade)
g	gravitational constant (32.2 ft/s^2, 9.807 m/s^2)
g_{max}	maximum deceleration achieved before wheel lockup
h	height of vehicle's center of gravity above the roadway surface
hp_e	engine-generated power, measured in horsepower
i	driveline slippage
K_a	elements of aerodynamic resistance that are not a function of speed
L	vehicle wheelbase
l_f	distance from vehicle's center of gravity to front axle
l_r	distance from vehicle's center of gravity to rear axle
M_e	engine torque
m	mass
n_e	engine speed in crankshaft revolutions per second
P_e	engine-generated power, measured in kW
PBF_f	optimal percent of braking force on the front axle
PBF_r	optimal percent of braking force on the rear axle
R_a	aerodynamic resistance
R_g	grade resistance
R_{rl}	rolling resistance
r	radius of vehicle drive wheels
S	minimum theoretical stopping distance
t_r	driver perception/reaction time
V	vehicle speed
W	total vehicle weight
W_f	vehicle weight acting normal to the roadway surface on the front axle
W_r	vehicle weight acting normal to the roadway surface on the rear axle
γ_b	braking mass factor
γ_m	acceleration mass factor
ε_0	gear reduction ratio
η_b	braking efficiency
η_d	driveline efficiency
θ_g	angle of grade
μ	coefficient of road adhesion
ρ	air density

REFERENCES

AASHTO (American Association of State Highway and Transportation Officials). *A Policy on Geometric Design of Highways and Streets,* 4th ed., Washington, DC: AASHTO, 2001.

Brewer, H. K., and R. S. Rice. "Tires: Stability and Control." *SAE Transactions*, Vol. 92, paper 830561, 1983.

Campbell, C. *The Sports Car: Its Design and Performance.* Cambridge, MA: Robert Bently, 1978.

Fambro, D. B., K. Fitzpatrick, and R. J. Koppa. *Determination of Stopping Sight Distances*. NCHRP Report 400. Transportation Research Board, Washington, DC, 1997.

Taborek, J. J. "Mechanics of Vehicles." *Machine Design*, 1957.

Wong, J. H. *Theory of Ground Vehicles*. New York: John Wiley & Sons, 1978.

PROBLEMS

2.1 A new sports car has a drag coefficient of 0.29 and a frontal area of 20 ft^2, and is traveling at 100 mi/h. How much power is required to overcome aerodynamic drag if $\rho = 0.002378$ slugs/ft^3?

2.2 A vehicle manufacturer is considering an engine for a new sedan ($C_D = 0.30$, $A_f = 21$ ft^2). The car is being designed to achieve a top speed of 100 mi/h on a paved surface at sea level ($\rho = 0.002378$ slugs/ft^3). The car currently weighs 2100 lb, but the designer initially selected an underpowered engine because he did not account for aerodynamic and rolling resistances. If 2 lb of additional vehicle weight is added for each unit of horsepower needed to overcome the neglected resistance, what will be the final weight of the car if it is to achieve the 100-mi/h top speed?

2.3 For Example 2.3, how far back from the front axle would the center of gravity have to be to ensure that the maximum tractive effort developed for front- and rear-wheel–drive options is equal (assume that all other variables are unchanged)?

2.4 A rear-wheel–drive 3000-lb drag race car has a 200-inch wheelbase and a center of gravity 20 inches above the pavement and 140 inches behind the front axle. The owners wish to achieve an initial acceleration from rest of 15 ft/s^2 on a level paved surface. What is the minimum coefficient of road adhesion needed to achieve this acceleration? (Assume $\gamma_m = 1.00$.)

2.5 If the race car in Problem 2.4 has a center of gravity 36 inches above the roadway and is run on a pavement with a coefficient of adhesion of 1.0, how far back from the front axle would the center of gravity have to be to develop a maximum acceleration from rest of 1.0 g (32.2 ft/s^2)? (Assume $\gamma_m = 1.00$.)

2.6 A rear-wheel–drive car weighs 2700 lb, has 14-inch–radius wheels, a driveline efficiency of 95%, and an engine that develops 540 ft-lb of torque. Its wheelbase is 8.2 ft, and the center of gravity is 18 inches above the road surface and 3.3 ft behind the front axle. What is the lowest gear reduction ratio that would allow this car to achieve the highest possible acceleration from rest on good, dry pavement?

2.7 A newly designed car has a 9.2-ft wheelbase, is rear-wheel drive, and has a center of gravity 18 inches above the road and 4.3 ft behind the front axle. The car weighs 2450 lb, the mechanical efficiency of the driveline is 90%, and the wheel radius is 14 inches. The base engine develops 185 ft-lb of torque, and a modified version of the engine develops 215 ft-lb of torque. If the overall gear reduction ratio is 9 to 1, what is the maximum acceleration from rest for the car with the 185 ft-lb engine and for the car with the 215 ft-lb engine? (It is on good, dry, and level pavement.)

2.8 A 3000-lb car is traveling on a paved road with $C_D = 0.35$, $A_f = 21$ ft^2, and $\rho = 0.002378$ slugs/ft^3. Its engine is running at 3000 rev/min and is producing 250 ft-lb of torque. The car's gear reduction ratio is 3.5 to 1, driveline efficiency is 90%, driveline slippage is 3.5%, and the road-wheel radius is 15 inches. What will the car's maximum acceleration rate be under these conditions on a level road? (Assume that the available tractive effort is the engine-generated tractive effort.)

2.9 A 2150-lb car is traveling at sea level at a constant speed. Its engine is running at 4500 rev/min and is producing 150 ft-lb of torque. It has a driveline efficiency of 90%, a driveline slippage of 2%, 15-inch–radius wheels, and an overall gear reduction ratio of 3 to 1. If the car's frontal area is 19.4 ft^2, what is the drag coefficient?

2.10 A 2500-lb car has a maximum speed (at sea level, and on a level, paved surface) of 150 mi/h with 14-inch–radius wheels, a gear reduction of 3 to 1, and a driveline efficiency of 90%. It is known that at the car's top speed the engine is producing 200 ft-lb of torque. If the car's frontal area is 25 ft^2, what is its drag coefficient?

2.11 A 2500-lb car ($C_D = 0.35$, $A_f = 25$ ft^2, and $\rho = 0.002378$ slugs/ft^3) has 14-inch–radius wheels, a driveline efficiency of 90%, an overall gear reduction ratio of 3.2 to 1, and driveline slippage of 3.5%. The engine develops a maximum torque of 200 ft-lb at 3500 rev/min. What is the maximum grade this vehicle could ascend, on a paved surface, while the engine is developing maximum torque? (Assume that the available tractive effort is the engine-generated tractive effort.)

2.12 A 2700-lb car is traveling in third gear (overall gear reduction ratio of 2.5 to 1) on a level road at its top speed of 124 mi/h. The air density is 0.00206 slugs/ft^3. The car has a frontal area of 19.4 ft^2, a drag coefficient of 0.28, a wheel radius of 12.6 inches, a driveline slippage of 0.03, and a driveline efficiency of 90%. At this vehicle speed, what torque is the engine producing and what is the engine speed (in revolutions per minute)?

2.13 A rear-wheel–drive car weighs 2500 lb, has an 80-inch wheelbase, a center of gravity 20 inches above the roadway surface and 30 inches behind the front axle, a driveline efficiency of 75%, 14-inch–radius wheels, and an overall gear reduction of 11 to 1. The car's torque/engine speed curve is given by

$$M_e = 6n_e - 0.045n_e^2$$

If the car is on a paved, level roadway surface with a coefficient of adhesion of 0.75, determine its maximum acceleration from rest.

2.14 Consider the car in Problem 2.13. If it is known that the car achieves maximum speed at an overall gear reduction ratio of 2 to 1 with a driveline slippage of 3.5%, how fast would the car be going if it could achieve its maximum speed when its engine is producing maximum power?

2.15 Consider the situation described in Example 2.5. If the vehicle is redesigned with 13-inch–radius wheels (assume that the mass factor is unchanged) and a center of gravity located at the same height but at the midpoint of the wheelbase, determine the acceleration for front- and rear-wheel–drive options.

2.16 An engineer designs a rear-wheel–drive car (without an engine) that weighs 2000 lb and has a 100-inch wheelbase, driveline efficiency of 80%, 14-inch–radius wheels, an overall gear reduction ratio of 10 to 1, and a center of gravity (without engine) that is 22 inches above the roadway surface and 55 inches behind the front axle. An engine that weighs 3 lb for each ft-lb of developed torque is to be placed in the front portion of the car. Calculations show that for every 20 lb of engine weight added, the car's center of gravity moves 1 inch closer to the front axle (but stays at the same height above the roadway surface). If the car is starting from rest on a level paved roadway with a coefficient of adhesion of 0.8, select an engine size (weight and associated torque) that will result in the highest possible available tractive effort.

2.17 If the car in Example 2.7 had $C_D = 0.45$ and $A_f = 25$ ft^2, what would have been the difference in minimum theoretical stopping distances with and without aerodynamic resistance considered (all other factors the same as in Example 2.7)?

2.18 A 3500-lb vehicle ($C_D = 0.40$, $A_f = 28$ ft^2, $\rho = 0.002378$ slugs/ft^3) is driven on a surface with a coefficient of adhesion of 0.5, and the coefficient of rolling friction is approximated as 0.015 for all speeds. Assuming minimum theoretical stopping distances, if the vehicle comes to a stop 250 ft after brake application on a level surface and has a braking efficiency of 0.78, what was its initial speed (a) if aerodynamic resistance is considered and (b) if aerodynamic resistance is ignored?

2.19 A level test track has a coefficient of road adhesion of 0.75, and a car being tested has a coefficient of rolling friction that is approximated as 0.018 for all speeds. The vehicle is tested unloaded and achieves the theoretical minimum stop in 200 ft (from brake application). The initial speed was 60 mi/h. Ignoring aerodynamic resistance, what is the unloaded braking efficiency?

2.20 A driver is traveling at 110 mi/h down a 3% grade on good, wet pavement. An accident investigation team noted that braking skid marks started 590 ft before a parked car was hit at an estimated 55 mi/h. Ignoring air resistance, and using theoretical stopping distance, what was the braking efficiency of the car?

2.21 A small truck is to be driven down a 4% grade at 75 mi/h. The coefficient of road adhesion is 0.95, and it is known that the braking efficiency is 80% when the truck is empty and it decreases by one percentage point for every 100 lb of cargo added. Ignoring aerodynamic resistance, if the driver wants the truck to be able to achieve a minimum theoretical stopping distance of 300 ft from the point of brake application, what is the maximum amount of cargo (in pounds) that can be carried?

2.22 Consider the conditions in Example 2.8. The car has $W = 3500$ lb, $C_D = 0.5$, $A_f = 25$ ft^2, $\rho = 0.002378$ slugs/ft^3, and a coefficient of rolling friction approximated as 0.018 for all speed conditions. If aerodynamic resistance is considered in stopping, estimate how fast the car will be going when it strikes the object on a level and a +5% grade [all other conditions (speed, etc.) as described in Example 2.8].

2.23 A car is traveling at 75 mi/h down a 3% grade on poor, wet pavement. The car's braking efficiency is 90%. The brakes were applied 300 ft before impacting an object. The car had an antilock braking system, but the system failed 200 ft after the brakes had been applied (wheels locked). What speed was the car traveling at just before it impacted the object? (Assume theoretical stopping distance, ignore air resistance, and let $f_{rl} = 0.015$.)

2.24 A driver traveling down a 4% grade collides with a roadside object in rainy conditions, and is issued a ticket for driving too fast for conditions. The posted speed limit is 65 mi/h. The accident investigation team determined the following: The vehicle was traveling 40 mi/h when it struck the object, braking skid marks started 200 ft before the struck object, the pavement is in good condition, and the braking efficiency of the vehicle was 0.95. Using theoretical stopping distance, assuming aerodynamic resistance is negligible, and with the coefficient rolling resistance approximated as 0.015, should the driver appeal the ticket? Why or why not?

2.25 A driver is traveling 70 mi/h on a road with a negative 3% grade. There is a stalled car on the road 1000 ft ahead of the driver. The driver's vehicle has a braking efficiency of 90%, and it has antilock brakes. The road is in good condition and is initially dry, but it becomes wet 150 ft before the stalled car (and stays wet until the car is reached). What is the minimum distance from the stalled car at which the driver could apply the brakes and still stop before hitting it? (Assume theoretical stopping distance, ignore air resistance, and let $f_{rl} = 0.013$.)

2.26 Two cars are traveling on level terrain at 60 mi/h on a road with a coefficient of adhesion of 0.8. The driver of car 1 has a 2.5-s perception/reaction time and the driver of car 2 has a 2.0-s perception/reaction time. Both cars are traveling side by side and the drivers are able to stop their respective cars in the same amount of distance from first seeing a roadway obstacle (perception/reaction plus vehicle stopping distance). If the braking efficiency of car 2 is 0.75, determine the braking efficiency of car 1. (Assume minimum theoretical stopping distance and ignore aerodynamic resistance.)

2.27 An engineering student is driving on a level roadway and sees a construction sign 600 ft ahead in the middle of the roadway. The student strikes the sign at a speed of 35 mi/h. If the student was traveling at 55 mi/h when the sign was first spotted, what was the student's associated perception/reaction time (use practical stopping distance)?

2.28 An engineering student claims that a country road can be safely negotiated at 70 mi/h in rainy weather. Because of the winding nature of the road, one stretch of level pavement has a sight distance of only 590 ft. Assuming practical stopping distance, comment on the student's claim.

2.29 A driver is traveling at 55 mi/h on a wet road. An object is spotted on the road 450 ft ahead and the driver is able to come to a stop just before hitting the object. Assuming standard perception/reaction time and practical stopping distance, determine the grade of the road.

2.30 A test of a driver's perception/reaction time is being conducted on a special testing track with wet pavement and a driving speed of 55 mi/h. When the driver is sober, a stop can be made just in time to avoid hitting an object that is first visible 520 ft ahead. After a few drinks under the exact same conditions, the driver fails to stop in time and strikes the object at a speed of 35 mi/h. Determine the driver's perception/reaction time before and after drinking. (Assume practical stopping distance.)

Chapter 3

Geometric Design of Highways

3.1 INTRODUCTION

With the understanding of vehicle performance provided in Chapter 2, attention can now be directed toward highway design. The design of highways necessitates the determination of specific design elements, which include the number of lanes, lane width, median type (if any) and width, length of acceleration and deceleration lanes for on- and off-ramps, need for truck climbing lanes for steep grades, curve radii required for vehicle turning, and the alignment required to provide adequate stopping and passing sight distances. Many of these design elements are influenced by the performance characteristics of vehicles. For example, vehicle acceleration and deceleration characteristics have a direct impact on the design of acceleration and deceleration lanes (the length needed to provide a safe and orderly flow of traffic) and the highway alignment needed to provide adequate passing and stopping sight distances. Furthermore, vehicle performance characteristics determine the need for truck climbing lanes on steep grades (where the poor performance of large trucks necessitates a separate lane) as well as the number of lanes required because the observed spacing between vehicles in traffic is directly related to vehicle performance characteristics (this will be discussed further in Chapter 5). In addition, the physical dimensions of vehicles affect a number of design elements, such as the radii required for low-speed turning, height of highway overpasses, and lane widths.

When one considers the diversity of vehicles' performance and physical dimensions, and the interaction of these characteristics with the many elements constituting highway design, it is clear that proper design is a complex procedure that requires numerous compromises. Moreover, it is important that design guidelines evolve over time in response to changes in vehicle performance and dimensions, and in response to evidence collected as to the effectiveness of existing highway design practices, such as the relationship between crash rates and various roadway design characteristics. Current guidelines of highway design are presented in detail in *A Policy on Geometric Design of Highways and Streets*, published by the American Association of State Highway and Transportation Officials [AASHTO 2001].

Because of the sheer number of geometric elements involved in highway design, a detailed discussion of each design element is beyond the scope of this book, and the reader is referred to [AASHTO 2001] for a complete discussion of current design practices. Instead, this book focuses exclusively on the key elements of highway alignment, which are arguably the most important components of geometric design. As will be shown, the alignment topic is particularly well suited for demonstrating the effect of vehicle performance (specifically braking performance) and vehicle dimensions (such as driver's eye height, headlight height, and taillight height) on the design of highways. By concentrating on the specifics of the highway alignment problem, the reader will develop an understanding of the procedures and compromises inherent in the design of all highway-related geometric elements.

3.2 PRINCIPLES OF HIGHWAY ALIGNMENT

The alignment of a highway is a three-dimensional problem measured in x, y, and z coordinates. This is illustrated, from a driver's perspective, in Fig. 3.1. However, in highway design practice, three-dimensional design computations are cumbersome, and, what is perhaps more important, the actual implementation and construction of a design based on three-dimensional coordinates has historically been prohibitively difficult. As a consequence, the three-dimensional highway alignment problem is reduced to 2 two-dimensional alignment problems, as illustrated in Fig. 3.2. One of the alignment problems in this figure corresponds roughly to x and z coordinates and is referred to as horizontal alignment. The other corresponds to highway length (measured along some constant elevation) and y coordinates (elevation) and is referred to as vertical alignment. Referring to Fig. 3.2, note that the horizontal alignment of a highway is referred to as the plan view, which is roughly equivalent to the perspective of an aerial photo of the highway. The vertical alignment is represented in a profile view, which gives the elevation of all points measured along the length of the highway (again, with length measured along a constant elevation reference).

Aside from considering the alignment problem as 2 two-dimensional problems, one further simplification is made. That is, instead of using x and z coordinates, highway positioning and length are defined as the distance along the highway (usually measured along the centerline of the highway, on a horizontal, constant-elevation plane) from a specified point. This distance is measured in terms of stations, with each station consisting of 100 ft (1000 m for metric) of highway alignment distance. The notation

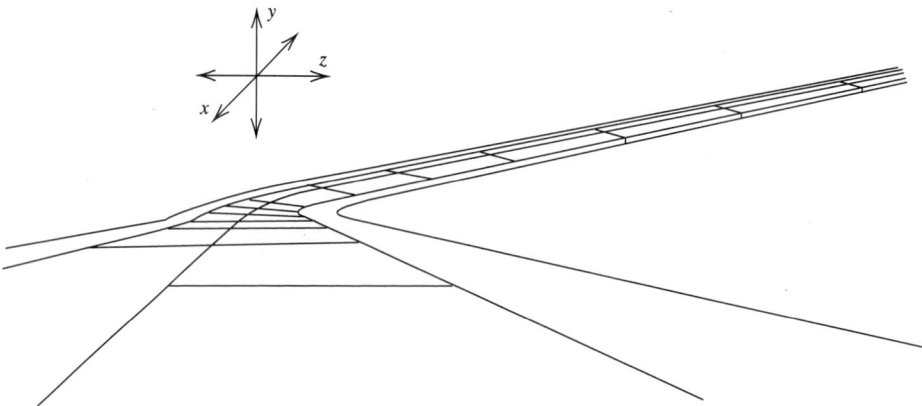

Figure 3.1 Highway alignment in three dimensions.

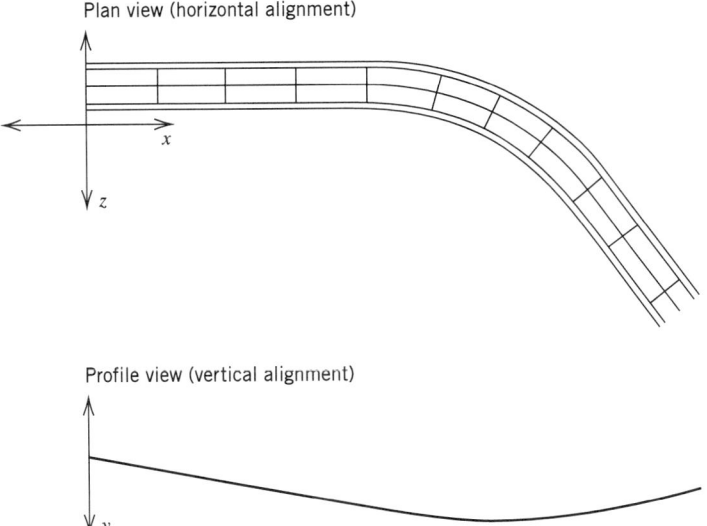

Figure 3.2 Highway alignment in two-dimensional views.

for stationing distance is such that a point on a highway 4250 ft (1295.3 m) from a specified point of origin is said to be at station 42 + 50 (1 + 295.300), that is, 42 stations and 50 ft (1 station and 295.300 m), with the point of origin being at station 0 + 00 (0 + 000 for metric). This stationing concept, combined with the highway's alignment direction given in the plan view (horizontal alignment) and the elevation corresponding to stations given in the profile view (vertical alignment), gives a unique identification of all highway points in a manner that is virtually equivalent to using true x, y, and z coordinates.

3.3 VERTICAL ALIGNMENT

Vertical alignment specifies the elevation of points along a roadway. The elevation of these roadway points are usually determined by the need to provide an acceptable level of driver safety, driver comfort, and proper drainage (from rainfall runoff). A primary concern in vertical alignment is establishing the transition of roadway elevations between two grades. This transition is achieved by means of a vertical curve.

Vertical curves can be broadly classified into crest vertical curves and sag vertical curves, as illustrated in Fig. 3.3 [AASHTO 2001]. Note that in Fig. 3.3, the distance from the *PVC* to the *PVI* is $L/2$. This is used in this figure because in practice the vast majority of vertical curves are arranged such that half of the curve length is positioned before the *PVI* and half after. Curves that satisfy this criterion are said to be equal-tangent vertical curves.

For referencing points on a vertical curve, it is important to note that the profile views presented in Fig. 3.3 correspond to all highway points even if a horizontal curve occurs concurrently with a vertical curve (as in Fig. 3.1 and Fig 3.2). Thus, each roadway point is uniquely defined by stationing (which is measured along a horizontal plane) and elevation. This will be made clearer through forthcoming examples.

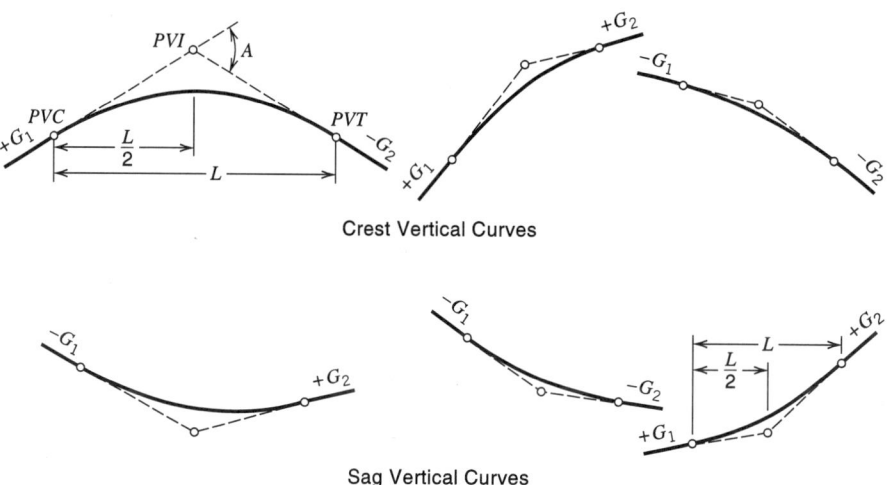

Figure 3.3 Types of vertical curves.
Reproduced by permission from American Association of State Highway and Transportation Officials, *A Policy on Geometric Design of Highways and Streets*, Washington, DC, 2001.

In this figure,

G_1 = initial roadway grade in percent or ft/ft (m/m) (this grade is also referred to as the initial tangent grade, viewing Fig. 3.3 from left to right),

G_2 = final roadway (tangent) grade in percent or ft/ft (m/m),

A = absolute value of the difference in grades (initial minus final, usually expressed in percent),

PVC = point of the vertical curve (the initial point of the curve),

PVI = point of vertical intersection (intersection of initial and final grades),

PVT = point of vertical tangent, which is the final point of the vertical curve (the point where the curve returns to the final grade or, equivalently, the final tangent), and

L = length of the curve in stations or ft (m) measured in a constant-elevation horizontal plane.

3.3.1 Vertical Curve Fundamentals

In connecting roadway grades (tangents) with an appropriate vertical curve, a mathematical relationship defining elevations at all points (or equivalently, stations) along the vertical curve is needed. A parabolic function has been found suitable in this regard because, among other things, it provides a constant rate of change of slope and implies equal curve tangents. The general form of the parabolic equation, as applied to vertical curves, is

$$y = ax^2 + bx + c \tag{3.1}$$

where

y = roadway elevation at distance x from the beginning of the vertical curve (the *PVC*) in ft (m),
x = distance from the beginning of the vertical curve in stations or ft (m),
a, b = coefficients defined below, and
c = elevation of the *PVC* (because $x = 0$ corresponds to the *PVC*) in ft (m).

In defining a and b, note that the first derivative of Eq. 3.1 gives the slope and is

$$\frac{dy}{dx} = 2ax + b \tag{3.2}$$

At the *PVC*, $x = 0$, so, using Eq. 3.2,

$$b = \frac{dy}{dx} = G_1 \tag{3.3}$$

where G_1 is the initial slope in ft/ft (m/m), as defined in Fig. 3.3. Also note that the second derivative of Eq. 3.1 is the rate of change of slope and is

$$\frac{d^2y}{dx^2} = 2a \tag{3.4}$$

However, the average rate of change of slope, by observation of Fig. 3.3, can also be written as

$$\frac{d^2y}{dx^2} = \frac{G_2 - G_1}{L} \tag{3.5}$$

Equating Eqs. 3.4 and 3.5 gives

$$a = \frac{G_2 - G_1}{2L} \tag{3.6}$$

with all terms as defined previously (see Fig. 3.3). Please note that the selection of units for coefficients a and b in Eqs. 3.3 and 3.6 must be such that they provide ft (m) when multiplied by x^2 and x, respectively. The preceding equations define all of the terms in the parabolic vertical curve equation (Eq. 3.1). The following example gives a typical application of this equation.

EXAMPLE 3.1

A 600-ft (182.880-m) equal-tangent sag vertical curve has the *PVC* at station 170 + 00 (5 + 181.600) and elevation 1000 ft (304.800 m). The initial grade is −3.5% and the final grade is +0.5%. Determine the stationing and elevation of the *PVI*, the *PVT*, and the lowest point on the curve.

SOLUTION

Since the curve is equal tangent, the *PVI* will be 300 ft or three stations (measured in a horizontal plane) from the *PVC*, and the *PVT* will be 600 ft or six stations from the *PVC*. Therefore, the stationing of the *PVI* and *PVT* are 173 + 00 and 176 + 00, respectively. For the elevations of the *PVI* and *PVT*, it is known that a −3.5% grade can be equivalently written as −3.5 ft/station (a 3.5 ft drop per 100 ft of horizontal distance). Since the *PVI* is three stations from the *PVC*, which is known to be at elevation 1000 ft, the elevation of the *PVI* is

$$1000 - 3.5 \text{ ft/station} \times (3 \text{ stations}) = \underline{989.5 \text{ ft}}$$

Similarly, with the *PVI* at elevation 989.5 ft, the elevation of the *PVT* is

$$989.5 + 0.5 \text{ ft/station} \times (3 \text{ stations}) = \underline{991.0 \text{ ft}}$$

It is clear from the values of the initial and final grades that the lowest point on the vertical curve will occur when the first derivative of the parabolic function (Eq. 3.1) is zero because the initial and final grades are opposite in sign. When initial and final grades are not opposite in sign, the low (or high) point on the curve will not be where the first derivative is zero because the slope along the curve will never be zero. For example, a sag curve with an initial grade of −2.0% and a final grade of −1.0% will have its lowest elevation at the *PVT*, and the first derivative of Eq. 3.1 will not be zero at any point along the curve. However, in our example problem the derivative will be equal to zero at some point, so the low point will occur when

$$\frac{dy}{dx} = 2ax + b = 0$$

From Eq. 3.3 we have

$$b = G_1 = -3.5$$

with G_1 in percent. From Eq. 3.6 (with *L* in stations and G_1 and G_2 in percent),

$$a = \frac{0.5 - (-3.5)}{2(6)} = 0.33333$$

Substituting for *a* and *b* gives

$$\frac{dy}{dx} = 2(0.33333)x + (-3.5) = 0$$

$$x = 5.25 \text{ stations}$$

This gives the stationing of the low point at $175 + 25$ ($5 + 25$ stations from the *PVC*). For the elevation of the lowest point on the vertical curve, the values of a, b, c (elevation of the *PVC*), and x are substituted into Eq. 3.1, giving

$$y = 0.33333(5.25)^2 + (-3.5)(5.25) + 1000$$
$$= \underline{990.81 \text{ ft}}$$

Note that the preceding equations can also be solved with grades expressed as the decimal equivalent of percent (for example, 0.02 ft/ft for 2%) if x is expressed in feet instead of stations. Care must be taken not to mix units. A dimensional analysis of Eq. 3.1 must ensure that each right-side element of the equation has resulting units of feet.

Another interesting vertical curve problem that is sometimes encountered is one in which the curve must be designed so that the elevation of a specific location is met. An example might be to have the roadway connect with another (at the same elevation) or to have the roadway at some specified elevation to pass under another roadway. This type of problem is referred to as a curve-through-a-point problem and is demonstrated by the following example.

EXAMPLE 3.2

An equal-tangent vertical curve is to be constructed between grades of -2.0% (initial) and $+1.0\%$ (final). The *PVI* is at station $110 + 00$ ($3 + 352.800$) and at elevation 420 ft (128.016 m). Due to a street crossing the roadway, the elevation of the roadway at station $112 + 00$ ($3 + 413.760$) must be at 424.5 ft (129.388 m). Design the curve.

SOLUTION

The design problem is one of determining the length of the curve required to ensure that station $112 + 00$ is at elevation 424.5 ft. To begin, we use Eq. 3.1:

$$y = ax^2 + bx + c$$

From Eq. 3.3,

$$b = G_1 = -2.0$$

and from Eq. 3.6,

$$a = \frac{G_2 - G_1}{2L}$$

Substituting $G_1 = -2.0$ and $G_2 = 1.0$, we have

$$a = \frac{G_2 - G_1}{2L} = \frac{1.0 - (-2.0)}{2L} = \frac{1.5}{L}$$

Now note that c (the elevation of the PVC) in Eq. 3.1 will be equal to the elevation of the PVI plus $G_1 \times 0.5L$ (this is simply using the slope of the initial grade to determine the elevation difference between the PVI and PVC). With G_1 in percent (which is ft/station) and the curve length L in stations, we have

$$c = 420 + 2.0(0.5L) = 420 + L$$

Finally, the value of x to be used in Eq. 3.1 will be $0.5L + 2$ because the point of interest (station $112 + 00$) is two stations from the PVI (which is at station $110 + 00$). Substituting $b = -2.0$, the expressions for a, c, and x, and $y = 424.5$ ft (the given elevation) into Eq. 3.1 gives

$$424.5 = (1.5/L)(0.5L + 2)^2 + (-2.0)(0.5L + 2) + (420 + L)$$
$$4.5 = 0.375L + 3 + 6/L - 4$$
$$0 = -0.375L^2 + 5.5L - 6$$

Solving this quadratic equation gives $L = 1.187$ stations (which is not feasible because we know that the point of interest is 2.00 stations beyond the PVI, so the curve length must be longer than 1.187 stations) or $L = 13.466$ stations (which is the only feasible solution). This means that the curve must be 1346.6 ft long. Using this value of L,

elevation of $PVC = c = 420 + L = 420 + 13.466 = 433.47$ ft

stationing of $PVC = 110 + 00 - (13 + 46.6)/2 = 103 + 26.7$

elevation of PVT = elevation of $PVI + (0.5L)G_2 = 420 + [0.5(13.466)](1.0) = 426.73$ ft

stationing of $PVT = 110 + 00 + (13 + 46.6)/2 = 116 + 73.3$

and

$$x = 0.5L + 2.0 = 6.733 + 2.0 = 8.733 \text{ stations from the } PVC$$

To check the elevation of the curve at station $112 + 00$, we apply Eq. 3.1 with $x = 8.733$:

$$y = ax^2 + bx + c$$
$$= \frac{3}{2(13.466)}(8.733)^2 + (-2.0)(8.733) + 433.47$$
$$= 424.5 \text{ ft}$$

Therefore, all calculations are correct.

Some additional properties of vertical curves can now be formalized. For example, offsets, which are vertical distances from the initial tangent to the curve, as illustrated in Fig. 3.4, are extremely important in vertical curve design and construction.

3.3 Vertical Alignment

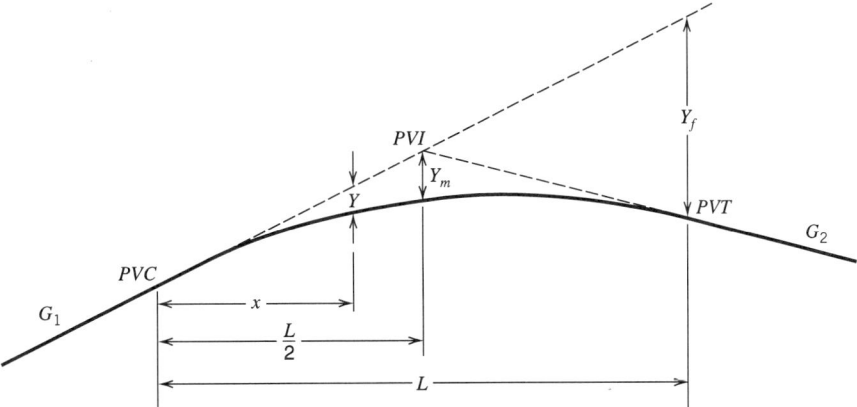

Figure 3.4 Offsets for equal-tangent vertical curves.

In this figure,

- G_1 = initial roadway grade in percent or ft/ft (m/m) (this grade is also referred to as the initial tangent grade, viewing Fig. 3.4 from left to right),
- G_2 = final roadway grade in percent or ft/ft (m/m),
- PVC = point of the vertical curve (the initial point of the curve),
- PVI = point of vertical intersection (intersection of initial and final grades),
- PVT = point of vertical tangent, which is the final point of the vertical curve (the point where the curve returns to the final grade or, equivalently, the final tangent),
- L = length of the curve in stations or ft (m) measured in a constant-elevation horizontal plane,
- x = distance from the PVC in ft (m),
- Y = offset at any distance x from the PVC in ft (m),
- Y_m = midcurve offset in ft (m), and
- Y_f = offset at the end of the vertical curve in ft (m).

Referring to the elements shown in Fig. 3.4, the properties of an equal-tangent parabola can be used to give

$$Y = \frac{A}{200L}x^2 \qquad (3.7)$$

where

A = absolute value of the difference in grades ($|G_1 - G_2|$) expressed in percent, and Other terms are as defined in Fig. 3.4.

Note that in this equation, 200 is used in the denominator instead of 2 because A is expressed in percent instead of ft/ft (m/m) (this division by 100 also applies to Eqs. 3.8 and 3.9 below). It follows from Fig. 3.4 that

$$Y_m = \frac{AL}{800} \tag{3.8}$$

and

$$Y_f = \frac{AL}{200} \tag{3.9}$$

Another useful vertical curve property is one that gives the length of curve required to effect a 1% change in slope. Because the parabolic equation used for roadway elevations (Eq. 3.1) gives a constant rate of change of slope, it can be shown that the horizontal distance required to change the slope by 1% is

$$K = \frac{L}{A} \tag{3.10}$$

where

K = horizontal distance, in ft (m), required to effect a 1% change in the slope of the vertical curve,

L = length of curve in ft (m), and

A = absolute value of the difference in grades ($|G_1 - G_2|$), expressed as a percentage.

This K-value can also be used to compute the high and low point locations of crest and sag vertical curves, respectively (provided the high or low point does not occur at the *PVC* or *PVT*). As shown in Example 3.1, setting $dy/dx = 0$ in Eq. 3.2 and solving for x gives the distance from the *PVC* to the high/low point. If Eq. 3.6 is used to substitute for a in Eq. 3.2 (with $L = KA$), it can be shown that setting $dy/dx = 0$ in Eq. 3.2 gives

$$x_{hl} = K \times |G_1| \tag{3.11}$$

where

x_{hl} = distance from the *PVC* to the high/low point in ft (m), and
Other terms are as defined previously.

In addition to high/low point computations, K-values have an important application in the design of vertical curves, as will be demonstrated in Sections 3.3.3 and 3.3.4.

EXAMPLE 3.3

A curve has initial and final grades of +3% and −4%, respectively, and is 700 ft (213.360 m) long. The *PVC* is at elevation 100 ft. Graph the vertical curve elevations and the slope of the curve against the length of curve. Compute the K-value and use it to locate the high point of the curve (distance from the PVC).

SOLUTION

Recall that to find the slope at any point on the curve, we take the derivative of Eq. 3.1, which gives Eq. 3.2. To apply this equation, a and b need to be determined. From Eq. 3.6,

$$a = \frac{-4.0 - (3.0)}{2(7)} = -0.5$$

and from Eq. 3.3,

$$b = G_1 = 3$$

The results of applying Eq. 3.2 and solving for the slope at all points along the curve, as well as a profile view of the curve itself (by application of Eq. 3.1), are shown graphically in Fig. 3.5 (exaggerating the vertical scale). Figure 3.5 shows the constant rate of change of the slope along the length of the curve. The circular points on the slope-of-curve line correspond to changes in grade of 1%, and these points occur at equal intervals of 100 ft.

To show that this is consistent with the K-value, Eq. 3.10 gives

$$K = \frac{L}{A} = \frac{700}{|3 - (-4)|} = \underline{100 \text{ ft}}$$

This indicates that there should be a change in grade of 1% for every 100 ft of curve length (measured in the horizontal plane), and this is consistent with Fig. 3.5. Applying Eq. 3.11 with the K-value of 100 ft gives the high point at $\underline{300 \text{ ft}}$ from the beginning of the curve ($x_{hl} = 100 \times 3 = 300$ ft). This is shown in Fig. 3.5, where the slope of the curve at 300 ft is zero (the same result obtained by setting the derivative of Eq. 3.2 equal to zero and solving for x). This result can also be explained conceptually based on the definition of the K-value. The K-value gives the horizontal distance required to

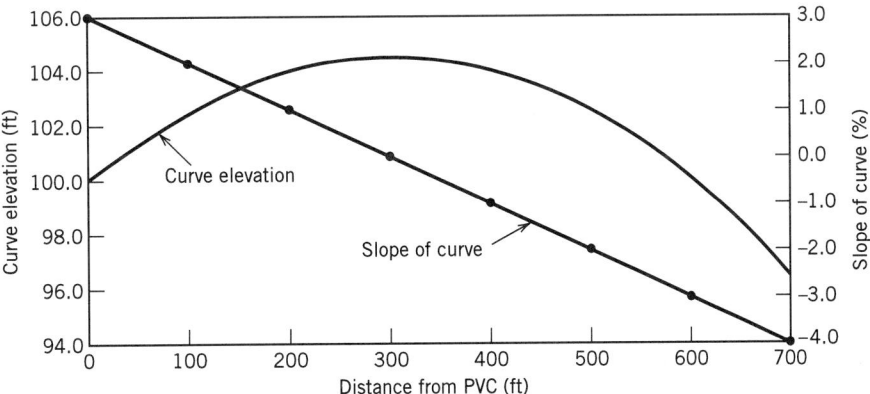

Figure 3.5 Profile view of vertical curve for Example 3.3 with the graph of the slope at all points along the curve overlayed.

EXAMPLE 3.4

A vertical curve crosses a 4-ft (1.219-m) diameter pipe at right angles. The pipe is located at station 110 + 85 (3 + 378.708) and its centerline is at elevation 1091.60 ft (332.720 m). The *PVI* of the vertical curve is at station 110 + 00 (3 + 352.800) and elevation 1098.4 ft (334.792 m). The vertical curve is equal tangent, 600 ft (182.880 m) long, and connects an initial grade of +1.20% and a final grade of −1.08%. Using offsets, determine the depth, below the surface of the curve, of the top of the pipe and determine the station of the highest point on the curve.

SOLUTION

The *PVC* is at station 107 + 00 (110 + 00 minus 3 + 00, which is half of the curve length), so the pipe is 385 ft (110 + 85 minus 107 + 00) from the beginning of the curve (*PVC*). The elevation of the *PVC* will be the elevation of the *PVI* minus the drop in grade over one-half the curve length,

$$1098.4 - (3 \text{ stations} \times 1.2 \text{ ft/station}) = 1094.8 \text{ ft}$$

Using this, the elevation of the initial tangent above the pipe is

$$1094.8 + (3.85 \text{ stations} \times 1.2 \text{ ft/station}) = 1099.42 \text{ ft}$$

Using Eq. 3.7 to determine the offset above the pipe at $x = 385$ ft (the distance of the pipe from the *PVC*), we have

$$Y = \frac{A}{200L}x^2$$

$$Y = \frac{|1.2-(-1.08)|}{200(600)}(385)^2 = 2.82 \text{ ft}$$

Thus the elevation of the curve above the pipe is 1096.6 ft (1099.42 − 2.82). The elevation of the top of the pipe is 1093.60 ft (elevation of the centerline plus one-half of the pipe's diameter), so the pipe is <u>3.0 ft</u> below the surface of the curve (1096.6 − 1093.6).

To determine the location of the highest point on the curve, we find K from Eq. 3.10 as

$$K = \frac{600}{|1.2-(-1.08)|} = 263.16$$

and the distance from the *PVC* to the highest point is (from Eq. 3.11)

$$x_{hl} = K \times |G_1| = 263.16 \times 1.2 = 315.79 \text{ ft}$$

This gives the station of the highest point at 110 + 15.79 (107 + 00 plus 3 + 15.79). Note that this example could also be solved by applying Eq. 3.1, setting Eq. 3.2 equal to zero (for determining the location of the highest point on the curve), and following the procedure used in Example 3.1.

3.3.2 Stopping Sight Distance

Construction of a vertical curve is generally a costly operation requiring the movement of significant amounts of earthen material. Thus one of the primary challenges facing highway designers is to minimize construction costs (usually by making the vertical curve as short as possible) while still providing an adequate level of safety. An appropriate level of safety is usually defined as that level of safety that provides drivers with sufficient sight distance to allow them to safely stop their vehicles to avoid collisions with objects obstructing their forward motion. The provision of adequate roadway drainage is sometimes an important concern as well, but is not discussed in terms of vertical curves in this book (see [AASHTO 2001]). Referring back to the vehicle braking performance concepts discussed in Chapter 2, we can compute this necessary stopping sight distance (SSD) as the summation of vehicle practical stopping distance (Eq. 2.47) and the distance traveled during driver perception/reaction time (Eq. 2.49). That is,

$$\text{SSD} = \frac{V_1^2}{2g\left(\frac{a}{g} \pm G\right)} + V_1 \times t_r \qquad (3.12)$$

where

SSD = stopping sight distance in ft (m),
V_1 = initial vehicle speed in ft/s (m/s),
g = gravitational constant, 32.2 ft/s² (9.807 m/s²),
a = deceleration rate in ft/s² (m/s²),
G = roadway grade (+ for uphill and − for downhill) in percent/100, and
t_r = perception/reaction time in s.

Recall from Sections 2.9.5 and 2.9.6 that a value of 11.2 ft/s² (3.4 m/s²) for a and a value of 2.5 s for t_r were recommended for roadway design purposes. The design speed of the highway is defined as the maximum safe speed at which a highway can be negotiated assuming near worst-case conditions (wet-weather conditions). The application of Eq. 3.12 (assuming $G = 0$) produces the stopping sight distances presented in Table 3.1.

Table 3.1 Stopping Sight Distance

U.S. Customary					Metric				
Design speed (mi/h)	Brake reaction distance (ft)	Braking distance on level (ft)	Stopping sight distance		Design speed (km/h)	Brake reaction distance (m)	Braking distance on level (m)	Stopping sight distance	
			Calculated (ft)	Design (ft)				Calculated (m)	Design (m)
15	55.1	21.6	76.7	80	20	13.9	4.6	18.5	20
20	73.5	38.4	111.9	115	30	20.9	10.3	31.2	35
25	91.9	60.0	151.9	155	40	27.8	18.4	46.2	50
30	110.3	86.4	196.7	200	50	34.8	28.7	63.5	65
35	128.6	117.6	246.2	250	60	41.7	41.3	83.0	85
40	147.0	153.6	300.6	305	70	48.7	56.2	104.9	105
45	165.4	194.4	359.8	360	80	55.6	73.4	129.0	130
50	183.8	240.0	423.8	425	90	62.6	92.9	155.5	160
55	202.1	290.3	492.4	495	100	69.5	114.7	184.2	185
60	220.5	345.5	566.0	570	110	76.5	138.8	215.3	220
65	238.9	405.5	644.4	645	120	83.4	165.2	248.6	250
70	257.3	470.3	727.6	730	130	90.4	193.8	284.2	285
75	275.6	539.9	815.5	820					
80	294.0	614.3	908.3	910					

Note: Brake reaction distance is based on a time of 2.5 s; a deceleration rate of 11.2 ft/s^2 (3.4 m/s^2) is used to determine calculated stopping sight distance.
Source: American Association of State Highway and Transportation Officials, *A Policy on Geometric Design of Highways and Streets*, Washington, DC, 2001.

3.3.3 Stopping Sight Distance and Crest Vertical Curve Design

The length of curve (L in Fig. 3.3) is the critical element in providing sufficient SSD on a vertical curve. Longer curve lengths provide more SSD, all else being equal, but are more costly to construct. Shorter curve lengths are less expensive to construct but may not provide adequate SSD due to more rapid changes in slope. What is needed, then, is an expression for minimum curve length given a required SSD. In developing such an expression, crest and sag vertical curves are considered separately.

The case of designing a crest vertical curve for adequate stopping sight distance is illustrated in Fig. 3.6.

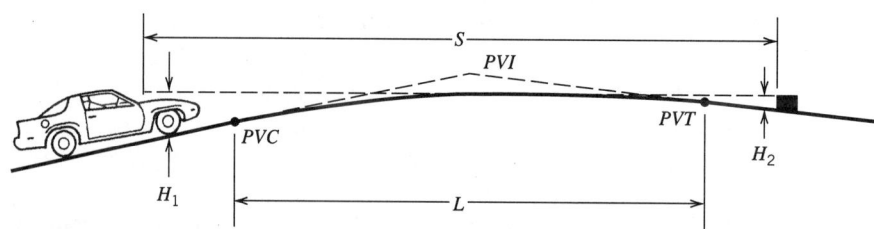

Figure 3.6 Stopping sight distance considerations for crest vertical curves.

In this figure,

S = sight distance in ft (m),
H_1 = height of driver's eye above roadway surface in ft (m),
H_2 = height of object above roadway surface in ft (m),
PVC = point of the vertical curve (the initial point of the curve),
PVI = point of vertical intersection (intersection of initial and final grades),
PVT = point of vertical tangent, which is the final point of the vertical curve (the point where the curve returns to the final grade or, equivalently, the final tangent), and
L = length of the curve in ft (m).

To determine the minimum length of curve for a required sight distance, the properties of a parabola for an equal tangent curve can be used to show that

For $S < L$

$$L_m = \frac{AS^2}{200(\sqrt{H_1} + \sqrt{H_2})^2} \tag{3.13}$$

For $S > L$

$$L_m = 2S - \frac{200(\sqrt{H_1} + \sqrt{H_2})^2}{A} \tag{3.14}$$

where

L_m = minimum length of vertical curve in ft (m),
A = absolute value of the difference in grades ($|G_1 - G_2|$), expressed as a percentage, and
Other terms are as defined in Fig. 3.6.

For the sight distance required to provide adequate SSD, current AASHTO design guidelines [2001] use a driver eye height, H_1, of 3.5 ft (1080 mm) and a roadway object height, H_2, of 2.0 ft (600 mm) (the height of an object to be avoided by stopping before a collision). In applying Eqs. 3.13 and 3.14 to determine the minimum length of curve required to provide adequate SSD, we set the sight distance, S, equal to the stopping sight distance, SSD. Substituting AASHTO guidelines for H_1 and H_2 and letting S = SSD in Eqs. 3.13 and 3.14 gives

	U.S. Customary	Metric	

For SSD < L

$$L_m = \frac{A \times SSD^2}{2158} \qquad L_m = \frac{A \times SSD^2}{658} \qquad (3.15)$$

For SSD > L

$$L_m = 2 \times SSD - \frac{2158}{A} \qquad L_m = 2 \times SSD - \frac{658}{A} \qquad (3.16)$$

where

SSD = stopping sight distance in ft (m), and
Other terms are as defined previously.

EXAMPLE 3.5

A highway is being designed to AASHTO guidelines with a 70-mi/h (113-km/h) design speed, and at one section, an equal-tangent vertical curve must be designed to connect grades of +1.0% and −2.0%. Determine the minimum length of curve necessary to meet SSD requirements.

SOLUTION

If we ignore the effect of grades ($G_s = 0$), the SSD can be read directly from Table 3.1. In this case, the SSD corresponding to a speed of 70 mi/h is 730 ft. If we assume that $L >$ SSD (an assumption that is typically made), Eq. 3.15 gives

$$L_m = \frac{A \times SSD^2}{2158} = \frac{3 \times 730^2}{2158} = \underline{740.82 \text{ ft}}$$

Since 740.82 > 730, the assumption that $L >$ SSD was correct.

The assumption that $G = 0$, made at the beginning of Example 3.5, is not really correct. If $G \neq 0$, we cannot use the SSD values in Table 3.1 and instead must apply Eq. 3.12 with the appropriate G value. In this problem, if we use the initial grade in Eq. 3.12 (+1.0%), we will underestimate the stopping sight distance because the vertical curve has a slope as steeply positive as this only at the *PVC*. If we use the final grade in Eq. 3.12 (−2.0%), we will overestimate the stopping sight distance because the vertical curve has a slope as steeply negative as this only at the *PVT*. If we knew where the vehicle began to brake, we could use the first derivative of the parabolic curve function (from Eq. 3.2) to give G in Eq. 3.12 and set up the equation to solve for SSD exactly. In practice, policies vary as to how this grade issue is handled. Fortunately, because sight distance tends to be greater on downgrades (which require longer stopping distances) than on upgrades, a self-correction for the effect of grades is generally provided. As a consequence, some design agencies ignore the effect of grades completely, while others assume G is equal to zero for grades less than 3% and use simple adjustments to the SSD, depending on the initial and final grades, for grades of 3% or more (see [AASHTO 2001]). For the remainder of this chapter, we will ignore the effect of grades ($G = 0$ will be used in Eq. 3.12). However, it must be pointed out that the use of SSD grade corrections is very easy and straightforward, and all of the equations presented herein still apply.

The use of Eqs. 3.15 and 3.16 can be simplified if the initial assumption that $L >$ SSD is made, in which case Eq. 3.15 is always used. The advantage of this assumption is that the relationship between A and L_m is linear, and Eq. 3.10 can be used to give

$$L_m = KA \qquad (3.17)$$

where

$K =$ horizontal distance, in ft (m), required to effect a 1% change in the slope (as in Eq. 3.10), defined as

U.S. Customary   **Metric**

$$K = \frac{\text{SSD}^2}{2158} \qquad K = \frac{\text{SSD}^2}{658} \qquad (3.18)$$

With known SSD for a given design speed (assuming $G = 0$), K-values can be computed for crest vertical curves as shown in Table 3.2. Thus the minimum curve length can be obtained (as shown in Eq. 3.17) simply by multiplying A by the K-value read from Table 3.2.

Some discussion about the assumption that $L >$ SSD is warranted. This assumption is made because there are two complications that could arise when SSD $> L$. First, if SSD $> L$ the relationship between A and L_m is not linear, so K-values cannot be used in the $L = KA$ formula (Eq. 3.10). Second, at low values of A, it is possible to get negative minimum curve lengths (see Eq. 3.16). As a result of these complications, the assumption that $L >$ SSD is almost always made in practice, and Eqs. 3.17 and 3.18 and the K-values presented in Table 3.2 are used. It is important to note that the assumption that $L >$ SSD (upon which Eqs. 3.17 and 3.18 are based) is a good one because in many cases, L is greater than SSD, and when it is not (SSD $> L$), use of the $L >$ SSD formula (Eq. 3.15 instead of Eq. 3.16) gives longer curve lengths and thus the error is on the conservative, safe side.

A final point relates to the smallest allowable length of curve. Very short vertical curves can be difficult to construct and may not be warranted for safety purposes. As a result, it is common practice to set minimum curve length limits that range from 100 to 325 ft (30 to 100 m) depending on individual jurisdictional guidelines. A common alternative to these limits is to set the minimum curve length limit at 3 times the design speed (with speed in mi/h and length in ft), or 0.6 times the design speed (with speed in km/h and length in m) [AASHTO 2001].

Table 3.2 Design Controls for Crest Vertical Curves Based on Stopping Sight Distance

U.S. Customary				Metric			
Design speed (mi/h)	Stopping sight distance (ft)	Rate of vertical curvature, K^*		Design speed (km/h)	Stopping sight distance (m)	Rate of vertical curvature, K^*	
		Calculated	Design			Calculated	Design
15	80	3.0	3	20	20	0.6	1
20	115	6.1	7	30	35	1.9	2
25	155	11.1	12	40	50	3.8	4
30	200	18.5	19	50	65	6.4	7
35	250	29.0	29	60	85	11.0	11
40	305	43.1	44	70	105	16.8	17
45	360	60.1	61	80	130	25.7	26
50	425	83.7	84	90	160	38.9	39
55	495	113.5	114	100	185	52.0	52
60	570	150.6	151	110	220	73.6	74
65	645	192.8	193	120	250	95.0	95
70	730	246.9	247	130	285	123.4	124
75	820	311.6	312				
80	910	383.7	384				

*Rate of vertical curvature, K, is the length of curve per percent algebraic difference in intersecting grades (A): $K = L/A$.
Source: American Association of State Highway and Transportation Officials, *A Policy on Geometric Design of Highways and Streets*, Washington, DC, 2001.

EXAMPLE 3.6

Solve Example 3.5 using the *K*-values in Table 3.2.

SOLUTION

From Example 3.5, $A = 3$. For a 70-mi/h design speed, $K = 247$ (from Table 3.2). Therefore, application of Eq. 3.17 gives

$$L_m = KA = 247(3) = \underline{741.00 \text{ ft}}$$

which is almost identical to the 740.82 ft obtained in Example 3.5. This difference is due to rounding. In this example the rounded *K* of 247 was used as opposed to the calculated *K* of 246.9. The rounded values are typically used in design for computational convenience. Note, however, that fractional calculated values are always rounded up to the nearest integer value, to be conservative.

EXAMPLE 3.7

If the grades in Example 3.5 intersect at station 100 + 00 (3 + 048.000), determine the stationing of the *PVC*, *PVT*, and curve high point for the minimum curve length based on SSD requirements.

SOLUTION

Using the curve length from Example 3.6, $L = 741$ ft. Since the curve is equal tangent (as are virtually all curves used in practice), one-half of the curve will occur before the PVI and one-half after, so that

$$PVC \text{ is at } 100 + 00 - L/2 = 100 + 00 \text{ minus } 3 + 70.5 = \underline{96 + 29.5}$$
$$PVT \text{ is at } 100 + 00 + L/2 = 100 + 00 \text{ plus } 3 + 70.5 = \underline{103 + 70.5}$$

For the stationing of the high point, Eq. 3.11 is used:

$$x_{hl} = K \times |G_1| = 247(1) = 247 \text{ ft}$$

or

$$\text{station } 96 + 29.5 \text{ plus } 2 + 47 = \underline{98 + 76.5}$$

3.3.4 Stopping Sight Distance and Sag Vertical Curve Design

Sag vertical curve design differs from crest vertical curve design in the sense that sight distance is governed by nighttime conditions because in daylight, sight distance on a sag vertical curve is unrestricted. Thus the critical concern for sag vertical curve design is the length of roadway illuminated by the vehicle headlights, which is a function of the height of the headlight above the roadway and the inclined angle of the headlight beam, relative to the horizontal plane of the car. The sag vertical curve sight distance design problem is illustrated in Fig. 3.7.

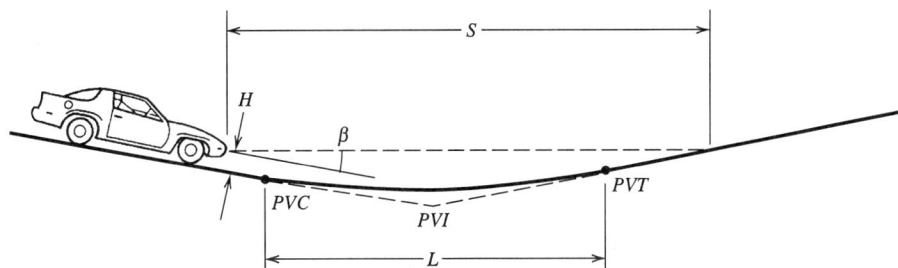

Figure 3.7 Stopping sight distance considerations for sag vertical curves.

In this figure,

S = sight distance in ft (m),
H = height of headlight in ft (m),
β = inclined angle of headlight beam in degrees,
PVC = point of the vertical curve (the initial point of the curve),
PVI = point of vertical intersection (intersection of initial and final grades),
PVT = point of vertical tangent, which is the final point of the vertical curve (the point where the curve returns to the final grade or, equivalently, the final tangent), and
L = length of the curve in ft (m).

To determine the minimum length of curve for a required sight distance, the properties of a parabola for an equal-tangent curve can be used to show that

For $S < L$

$$L_m = \frac{AS^2}{200(H + S \tan \beta)} \qquad (3.19)$$

For $S > L$

$$L_m = 2S - \frac{200(H + S \tan \beta)}{A} \qquad (3.20)$$

where

L_m = minimum length of vertical curve in ft (m),
A = absolute value of the difference in grades ($|G_1 - G_2|$), expressed as a percentage, and
Other terms are as defined in Fig. 3.7.

For the sight distance required to provide adequate SSD, current AASHTO design guidelines [2001] use a headlight height of 2.0 ft (600 mm) and an upward angle of one degree. Substituting these design guidelines and S = SSD (as was done in the crest vertical curve case) into Eqs. 3.19 and 3.20 gives

U.S. Customary   **Metric**

For SSD < L

$$L_m = \frac{A \times \text{SSD}^2}{400 + 3.5 \times \text{SSD}} \qquad L_m = \frac{A \times \text{SSD}^2}{120 + 3.5 \times \text{SSD}} \qquad (3.21)$$

For SSD > L

$$L_m = 2 \times \text{SSD} - \frac{400 + 3.5 \times \text{SSD}}{A} \qquad L_m = 2 \times \text{SSD} - \frac{120 + 3.5 \times \text{SSD}}{A}$$

$$(3.22)$$

where

SSD = stopping sight distance in ft (m), and
Other terms are as defined previously.

As was the case for crest vertical curves, K-values can be computed by assuming $L >$ SSD, which gives us the linear relationship between L_m and A as shown in Eq. 3.21. Thus for sag vertical curves (with $L_m = KA$),

U.S. Customary   **Metric**

$$K = \frac{\text{SSD}^2}{400 + 3.5 \text{ SSD}} \qquad K = \frac{\text{SSD}^2}{120 + 3.5 \text{ SSD}} \qquad (3.23)$$

where

K = horizontal distance, in ft (m), required to effect a 1% change in the slope (as in Eq. 3.10), and

SSD = stopping sight distance in ft (m).

The K-values corresponding to design speed–based SSDs are presented in Table 3.3. As was the case for crest vertical curves, some caution should be exercised in using this table because the assumption that $G = 0$ (for determining SSD) is used. Also, assume that $L >$ SSD is a safe, conservative assumption (as was the case for crest vertical curves) and the smallest allowable curve lengths for sag curves are the same as those for crest curves (see discussion in Section 3.3.3).

EXAMPLE 3.8

An existing tunnel needs to be connected to a newly constructed bridge with sag and crest vertical curves. The profile view of the tunnel and bridge is shown in Fig. 3.8. Develop a vertical alignment to connect the tunnel and bridge by determining the highest possible common design speed for the sag and crest (equal-tangent) vertical curves needed. Compute the stationing and elevations of *PVC*, *PVI*, and *PVT* curve points.

Table 3.3 Design Controls for Sag Vertical Curves Based on Stopping Sight Distance

U.S. Customary				Metric			
Design speed (mi/h)	Stopping sight distance (ft)	Rate of vertical curvature, K^*		Design speed (km/h)	Stopping sight distance (m)	Rate of vertical curvature, K^*	
		Calculated	Design			Calculated	Design
15	80	9.4	10	20	20	2.1	3
20	115	16.5	17	30	35	5.1	6
25	155	25.5	26	40	50	8.5	9
30	200	36.4	37	50	65	12.2	13
35	250	49.0	49	60	85	17.3	18
40	305	63.4	64	70	105	22.6	23
45	360	78.1	79	80	130	29.4	30
50	425	95.7	96	90	160	37.6	38
55	495	114.9	115	100	185	44.6	45
60	570	135.7	136	110	220	54.4	55
65	645	156.5	157	120	250	62.8	63
70	730	180.3	181	130	285	72.7	73
75	820	205.6	206				
80	910	231.0	231				

*Rate of vertical curvature, K, is the length of curve per percent algebraic difference in intersecting grades (A): $K = L/A$.

Source: American Association of State Highway and Transportation Officials, *A Policy on Geometric Design of Highways and Streets*, Washington, DC, 2001.

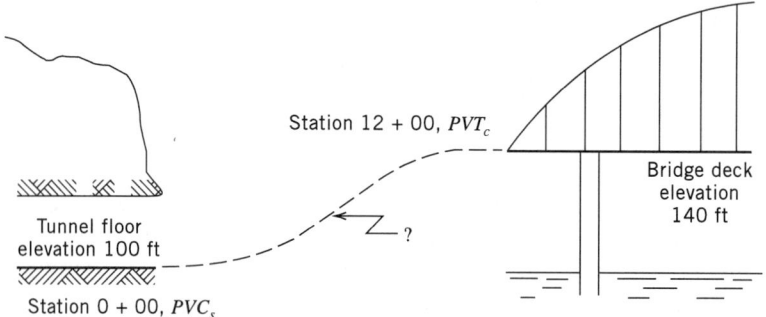

Figure 3.8 Profile view (vertical alignment diagram) for Example 3.8.

SOLUTION

From left to right (see Fig. 3.8), a sag vertical curve (with subscript s) and a crest vertical curve (with subscript c) is needed to connect the tunnel and bridge. From given information, it is known that $G_{1s} = 0\%$ (the initial slope of the sag vertical curve) and $G_{2c} = 0\%$ (the final slope of the crest vertical curve). To obtain the highest possible design speed, we want to use all of the horizontal distance available. This means we want to connect the curve such that the PVT of the sag curve (PVT_s) will be the PVC of the crest curve (PVC_c). If this is the case, $G_{2s} = G_{1c}$ and since $G_{1s} = G_{2c} = 0$, $A_s = A_c = A$, the common algebraic difference in the grades.

Since 1200 ft separates the tunnel and bridge,

$$L_s + L_c = 1200$$

Also, the summation of the end-of-curve offset for the sag curve and the beginning-of-curve offset (relative to the final grade) for the crest curve must equal 40 ft. Using the equation for the final offset, Eq. 3.9, we have

$$\frac{AL_s}{200} + \frac{AL_c}{200} = 40$$

Rearranging,

$$\frac{A}{200}(L_s + L_c) = 40$$

and since $L_s + L_c = 1200$,

$$\frac{A}{200}1200 = 40$$

Solving for A gives $A = 6.667\%$. The problem now becomes one of finding K-values that allow $L_s + L_c = 1200$. Since $L = KA$ (Eq. 3.17), we can write

$$K_s A + K_c A = 1200$$

3.3 Vertical Alignment

Substituting $A = 6.667$,

$$K_s + K_c = 180$$

To find the highest possible design speed, Tables 3.2 and 3.3 are used to arrive at K-values to solve $K_s + K_c = 180$. From Tables 3.2 and 3.3 it is apparent that the highest possible design speed is 50 mi/h, at which speed $K_c = 84$ and $K_s = 96$ (the summation of K's is 180).

To arrive at the stationing of curve points, we first determine curve lengths as

$$L_s = K_s A = 96(6.667) = 640.0 \text{ ft}$$

$$L_c = K_c A = 84(6.667) = 560.0 \text{ ft}$$

Since the station of the PVC_s is $\underline{0 + 00}$ (given), it is clear that the $PVI_s = \underline{3 + 20.0}$, $PVT_s = PVC_c = \underline{6 + 40.0}$, $PVI_c = \underline{9 + 20.0}$, and $PVT_c = \underline{12 + 00.0}$. For elevations, $PVC_s = PVI_s = \underline{100 \text{ ft}}$ and $PVI_c = PVT_c = \underline{140 \text{ ft}}$. Finally, the elevation of PVT_s and PVC_c can be computed as

$$100 + \frac{AL_s}{200} = 100 + \frac{6.667(640.0)}{200} = 121.33 \text{ ft} = \underline{\underline{121.33 \text{ ft}}}$$

EXAMPLE 3.9

Consider the conditions described in Example 3.8. Suppose a design speed of only 35 mi/h (56 km/h) is needed. Determine the lengths of curves required to connect the bridge and tunnel while keeping the connecting grade as small as possible.

SOLUTION

It is known that the 1200 ft separating the tunnel and bridge are more than enough to connect a 35-mi/h alignment because Example 3.8 showed that 50 mi/h is possible. Therefore, to connect the tunnel and bridge and keep the connecting grade as small as possible, we will place a constant-grade section between the sag and crest curves (as shown in Fig. 3.9).

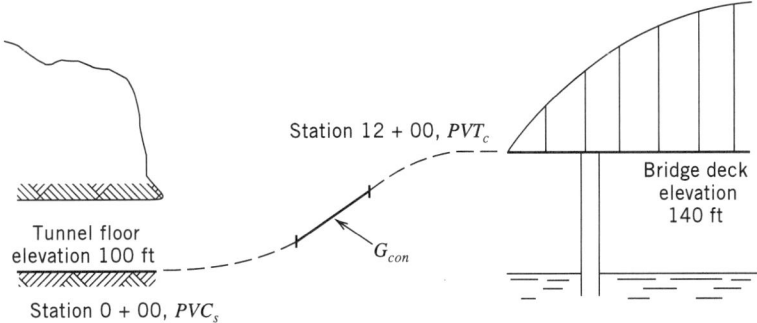

Figure 3.9 Profile view (vertical alignment diagram) for Example 3.9.

The elevation change will be the final offsets of the sag and crest curves plus the change in elevation resulting from the constant-grade section connecting the two curves. Let G_{con} be the grade of the constant-grade section. This means that $G_{2s} = G_{1c} = G_{con}$, and since $G_{1s} = G_{2c} = 0$ (as in Example 3.8), $G_{con} = A_s = A_c = A$. The equation that will solve the vertical alignment for this problem is

$$\frac{AL_s}{200} + \frac{AL_c}{200} + \frac{A(1200 - L_s - L_c)}{100} = 40$$

where the third term accounts for the elevation difference attributable to the constant-grade section connecting the sag and crest curves (the 100 in the denominator of this term converts A from percent to ft/ft). Using $L = KA$, we have

$$\frac{A^2 K_s}{200} + \frac{A^2 K_c}{200} + \frac{A(1200 - K_s A - K_c A)}{100} = 40$$

From Table 3.2, $K_c = 29$, and from Table 3.3, $K_s = 49$. Putting these values in the above equation gives

$$0.39A^2 + 12A - 0.78A^2 = 40$$

$$-0.39A^2 + 12A - 40 = 0$$

Solving this gives $A = 3.803$ and $A = 26.966$; $A = 3.803\%$ is chosen because we want to minimize the grade. For this value of A, the curve lengths are

$$L_s = K_s A = 49(3.803) = \underline{186.35 \text{ ft}}$$

$$L_c = K_c A = 29(3.803) = \underline{110.29 \text{ ft}}$$

and the length of the constant-grade section will be 903.36 ft. This means that about 34.35 ft of the elevation difference will occur in the constant-grade section, with the remainder of the elevation difference attributable to the final curve offsets.

Another variation of this type of problem is the case when the initial and final grades are not equal to zero. This makes the problem a bit more complex, as demonstrated in the following example.

EXAMPLE 3.10

Two sections of highway are separated by 1800 ft (548.640 m), as shown in Fig. 3.10. Determine the curve lengths required for a 60-mi/h (97-km/h) vertical alignment to connect these two highway segments while keeping the connecting grade as small as possible.

SOLUTION

Let Y_{fc} and Y_{fs} be the final offsets of the crest and sag curves, respectively. Let G_{con} be the slope of a constant-grade section connecting the crest and sag curves (we will assume that the horizontal distance is sufficient to connect the highway with a 60-mi/h

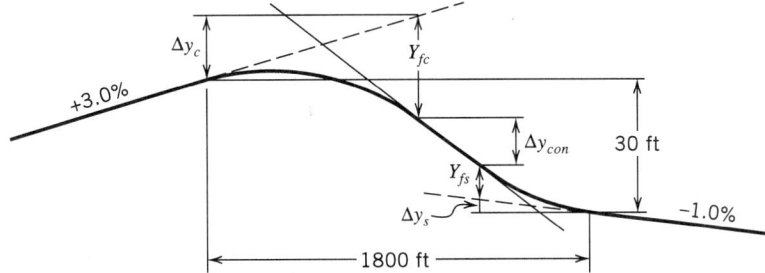

Figure 3.10 Profile view (vertical alignment diagram) for Example 3.10.

alignment; if this assumption is incorrect, the following equations will produce an obviously erroneous answer and a lower design speed will have to be chosen). Finally, let Δy_{con} be the change in elevation over the constant-grade section, and let Δy_c and Δy_s be the changes in elevation due to the extended curve tangents. The elevation equation is then (see Fig. 3.10)

$$Y_{fc} + Y_{fs} + \Delta y_{con} + \Delta y_s = 30 + \Delta y_c$$

Substituting offset equations and equations for elevation changes (with subscripts c for crest and s for sag),

$$\frac{A_c L_c}{200} + \frac{A_s L_s}{200} + \frac{G_{con}(1800 - L_c - L_s)}{100} + \frac{1.0 L_s}{100} = 30 + \frac{3.0 L_c}{100}$$

Using $L = KA$, this equation becomes

$$\frac{A_c^2 K_c}{200} + \frac{A_s^2 K_s}{200} + \frac{G_{con}(1800 - K_c A_c - K_s A_s)}{100} + \frac{1.0 K_s A_s}{100} = 30 + \frac{3.0 K_c A_c}{100}$$

From Tables 3.2 and 3.3, $K_c = 151$ and $K_s = 136$. Substituting and defining A's (and arranging the equation such that G_{con} will be positive, and assuming G_{con} will be greater than 1%) gives

$$\frac{(3 + G_{con})^2 151}{200} + \frac{(G_{con} - 1)^2 136}{200} + \frac{G_{con}[1800 - 151(3 + G_{con}) - 136(G_{con} - 1)]}{100}$$
$$+ \frac{1.0[136(G_{con} - 1)]}{100} = 30 + \frac{3.0[151(3 + G_{con})]}{100}$$

or

$$-1.435 G_{con}^2 + 14.83 G_{con} - 37.475 = 0$$

which gives $G_{con} = 4.40$ (the other possible solution is 5.93, which is rejected because we want to minimize the grade). Using $L = KA$ gives $L_c = \underline{1117.40 \text{ ft}}$ (151 × 7.40) and $L_s = \underline{462.40 \text{ ft}}$ (136 × 3.40). Accordingly, the length of the constant-grade section is $\underline{220.20 \text{ ft}}$ ($1800 - L_c - L_s$). Elevations and the locations of curve points can be readily computed with this information.

3.3.5 Passing Sight Distance and Crest Vertical Curve Design

In addition to stopping sight distance, in some instances it may be desirable to provide adequate passing sight distance, which can be an important issue in two-lane highway design (one lane in each direction). Passing sight distance is a factor only in crest vertical curve design because, for sag curves, the sight distance is unobstructed looking up or down the grade, and at night, the headlights of oncoming or opposing vehicles will be seen. In determining the sight distance required to pass on a crest vertical curve, Eqs. 3.13 and 3.14 will apply; but whereas the driver's eye height, H_1, will remain at 3.5 ft (1080 mm), H_2 will now also be set to 3.5 ft (1080 mm). This value for H_2 is the assumed value for the portion of a vehicle's height necessary to be visible such that it can be recognized as an opposing vehicle to a driver performing a passing maneuver. Using the same height for both H_1 and H_2 provides a reciprocal design relationship; that is, if the driver of the passing vehicle can see the opposing vehicle, then the opposing vehicle driver can see the passing vehicle. Substituting these H-values into Eqs. 3.13 and 3.14 and letting the sight distance S equal the passing sight distance, PSD, gives

U.S. Customary   **Metric**

For PSD < L

$$L_m = \frac{A \times \text{PSD}^2}{2800} \qquad L_m = \frac{A \times \text{PSD}^2}{864} \qquad (3.24)$$

For PSD > L

$$L_m = 2 \times \text{PSD} - \frac{2800}{A} \qquad L_m = 2 \times \text{PSD} - \frac{864}{A} \qquad (3.25)$$

where

L_m = minimum length of vertical curve in ft (m),
PSD = passing sight distance in ft (m), and
A = absolute value of the difference in grades ($|G_1 - G_2|$), expressed as a percentage.

As was the case for stopping sight distance, it is typically assumed that the length of curve is greater than the required sight distance (in this case L > PSD), so

U.S. Customary   **Metric**

$$K = \frac{\text{PSD}^2}{2800} \qquad K = \frac{\text{PSD}^2}{864} \qquad (3.26)$$

where

K = horizontal distance, in ft (m), required to effect a 1% change in the slope (as in Eq. 3.10), and
PSD = passing sight distance in ft (m).

The passing sight distance (PSD) used for design is assumed to consist of four distances: (1) the initial maneuver distance (which includes the driver's perception/reaction time and the time it takes to bring the vehicle from its trailing speed to the point of encroachment on the left lane), (2) the distance that the passing vehicle traverses while occupying the left lane, (3) the clearance length between the passing and opposing vehicles at the end of the passing maneuver, and (4) the distance traversed by the opposing vehicle during two-thirds of the time the passing vehicle occupies the left lane. The determination of these distances is undertaken using assumptions regarding the time of the initial maneuver; average vehicle acceleration; and the speeds of passing, passed, and opposing vehicles. The summation of these four distances gives the required passing sight distance. The reader is referred to [AASHTO 2001] for a complete description of the assumptions made in determining required passing sight distances.

The minimum distances needed to pass (PSDs) at various design speeds, along with the corresponding K-values as computed from Eq. 3.26, are presented in Table 3.4. Notice that the K-values in this table are much higher than those required for stopping sight distance (as given in Table 3.2). As a result, designing a crest curve to provide adequate passing sight distance is often an expensive proposition (due to the length of curve required).

Table 3.4 Design Controls for Crest Vertical Curves Based on Passing Sight Distance

U.S. Customary			Metric		
Design speed (mi/h)	Passing sight distance (ft)	Rate of vertical curvature, K^*	Design speed (km/h)	Passing sight distance (m)	Rate of vertical curvature, K^*
20	710	180	30	200	46
25	900	289	40	270	84
30	1090	424	50	345	138
35	1280	585	60	410	195
40	1470	772	70	485	272
45	1625	943	80	540	338
50	1835	1203	90	615	438
55	1985	1407	100	670	520
60	2135	1628	110	730	617
65	2285	1865	120	775	695
70	2480	2197	130	815	769
75	2580	2377			
80	2680	2565			

*Rate of vertical curvature, K, is the length of curve per percent algebraic difference in intersecting grades (A): $K = L/A$.
Source: American Association of State Highway and Transportation Officials, *A Policy on Geometric Design of Highways and Streets*, Washington, DC, 2001.

EXAMPLE 3.11

An equal-tangent crest vertical curve is 4000 ft (1219.200 m) long and connects a +2.5% and a −1.5% grade. If the design speed of the roadway is 55 mi/h (88.5 km/h), does this curve have adequate passing sight distance?

SOLUTION

To determine the length of curve required to provide adequate passing sight distance at a design speed of 55 mi/h, we use $L = KA$ with $K = 1407$ (as read from Table 3.4). This gives

$$L = 1407(4.0) = 5628 \text{ ft}$$

Since the curve is only 4000 ft long, it is not long enough to provide adequate passing sight distance. Alternatively, the K-value for the existing design can be compared with that required for a PSD-based design. The K-value for the existing design is

$$K = \frac{4000}{4} = 1000$$

Since the K-value of 1000 for the existing curve design is less than 1407, this curve does not provide adequate PSD for a 55-mi/h design speed.

3.3.6 Underpass Sight Distance and Sag Vertical Curve Design

As mentioned in Section 3.3.4, design for sag curves is based on nighttime conditions because during daytime conditions a driver can see the entire sag curve. However, in the case of a sag curve being built under an overhead structure (such as roadway or railroad crossing), a driver's line of sight may be restricted such that the entire curve length is not visible. An example of this situation is shown in Fig. 3.11.

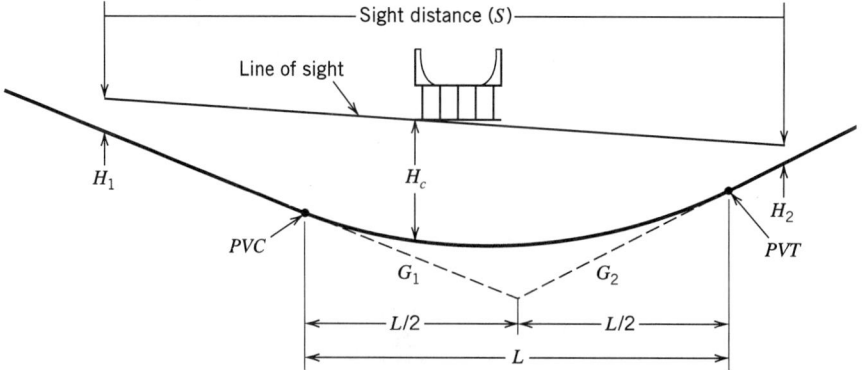

Figure 3.11 Stopping sight distance considerations for underpass sag curves.
Reproduced by permission from American Association of State Highway and Transportation Officials, *A Policy on Geometric Design of Highways and Streets*, Washington, DC, 2001.

3.3 Vertical Alignment

In this figure,

S = sight distance in ft (m),
H_1 = height of driver's eye in ft (m),
H_2 = height of object in ft (m),
H_c = clearance height of overpass structure above roadway in ft (m),
PVC = point of the vertical curve (the initial point of the curve),
PVT = point of vertical tangent (the final point of the curve),
G_1 = initial roadway grade in percent or ft/ft (m/m),
G_2 = final roadway grade in percent or ft/ft (m/m), and
L = length of the curve in ft (m).

In designing the sag curve, it is essential that the curve be long enough to provide a suitably gradual rate of curvature such that the overhead structure does not block the line of sight and allows the required stopping sight distance for the specified design speed to be maintained.

Again, by using the properties of a parabola for an equal-tangent vertical curve, it can be shown that the minimum length of sag curve for a required sight distance and clearance height is

For $S < L$

$$L_m = \frac{AS^2}{800\left(H_c - \frac{H_1 + H_2}{2}\right)} \tag{3.27}$$

For $S > L$

$$L_m = 2S - \frac{800\left(H_c - \frac{H_1 + H_2}{2}\right)}{A} \tag{3.28}$$

where

L_m = minimum length of vertical curve in ft (m),
A = absolute value of the difference in grades ($|G_1 - G_2|$), expressed as a percentage, and
Other terms are as defined in Fig. 3.11.

Current AASHTO design guidelines [2001] use a driver eye height, H_1, of 8 ft (2.4 m) for a truck driver, and an object height, H_2, of 2 ft (600 mm) for the taillights of a vehicle. Substituting these values and S = SSD into Eqs. 3.27 and 3.28 gives

	U.S. Customary		Metric	

For SSD < L

$$L_m = \frac{A \times \text{SSD}^2}{800(H_c - 5)} \qquad L_m = \frac{A \times \text{SSD}^2}{800(H_c - 1.5)} \qquad (3.29)$$

For SSD > L

$$L_m = 2 \times \text{SSD} - \frac{800(H_c - 5)}{A} \qquad L_m = 2 \times \text{SSD} - \frac{800(H_c - 1.5)}{A} \qquad (3.30)$$

where

SSD = stopping sight distance in ft (m), and
Other terms are as defined previously.

In the case where there is an existing sag curve alignment and a new overpass structure is going to be built over it, the above equations can be rearranged to solve for the necessary clearance height, H_c, of the overpass structure to provide for the required stopping sight distance. When the clearance height is determined in this manner, it is necessary to check this value against the minimum clearance heights based on maximum vehicle height regulations and AASHTO recommendations. Maximum vehicle heights as regulated by state laws range from 13.5 to 14.5 ft (4.1 to 4.4 m). AASHTO [2001] recommends a minimum structure clearance height of 14.5 ft (4.4 m) and a desirable clearance height of 16.5 ft (5.0 m). AASHTO [2001] also recommends that clearance heights be no less than 1 ft (300 mm) greater than the maximum allowable vehicle height. This provides a margin for snow or ice accumulation, some over-height vehicles, and future roadway resurfacings. Thus, in building a new overpass structure over an existing sag curve alignment, the clearance height must be determined for both required stopping sight distance and maximum allowable vehicle height for that roadway, and the greater of the two values should be used.

EXAMPLE 3.12

An equal-tangent sag curve has an initial grade of −4.0%, a final grade of +3.0%, and a length of 1270 ft (387.096 m). An overhead guide sign is being placed directly over the *PVI* of this curve. At what height above the roadway should the bottom of this sign be placed?

SOLUTION

For this situation, Eq. 3.29 or 3.30 must be used to solve for the necessary clearance height based on stopping sight distance. Thus, the required SSD must be determined for the given sag curve specifications, based on the design speed. The design speed for the curve can be determined from the *K*-value by applying Eq. 3.10, as follows:

$$K = \frac{L}{A} = \frac{1270}{|-4-3|} = 181.4$$

From Table 3.3, this K-value corresponds approximately to a design speed of 70 mi/h ($K = 181$). For a 70-mi/h design speed, the required stopping sight distance is 730 ft. Since the curve length is greater than the required SSD (1270 > 730), Eq. 3.29 applies:

$$L = \frac{A \times \text{SSD}^2}{800(H_c - 5)}$$

Rearranging this equation to solve for the clearance height, H_c, and substituting $A = 7\%$, SSD = 730 ft, and $L = 1270$ ft gives

$$\begin{aligned} H_c &= \frac{A \times \text{SSD}^2}{800L} + 5 \\ &= \frac{7 \times 730^2}{800(1270)} + 5 \\ &= 8.67 \text{ ft} \end{aligned}$$

Although only 8.67 ft is needed for SSD requirements, AASHTO [2001] recommends a minimum clearance height of 14.5 ft to account for maximum vehicle height. Thus, the bottom of the sign should be placed at least 14.5 ft above the roadway surface (at the *PVI*), but desirably at a height of 16.5 ft according to AASHTO [2001].

3.4 HORIZONTAL ALIGNMENT

The critical aspect of horizontal alignment is the horizontal curve with the focus on design of the directional transition of the roadway in a horizontal plane. Stated differently, a horizontal curve provides a transition between two straight (or tangent) sections of roadway. A key concern in this directional transition is the ability of the vehicle to negotiate a horizontal curve. (Provision of adequate drainage is also important, but is not discussed in this book; see [AASHTO 2001].) As was the case with the straight-line vehicle performance characteristics discussed at length in Chapter 2, the highway engineer must design a horizontal alignment to accommodate the cornering capabilities of a variety of vehicles, ranging from nimble sports cars to ponderous trucks. A theoretical assessment of vehicle cornering at the level of detail given to straight-line performance in Chapter 2 is beyond the scope of this book (see [Campbell 1978] and [Wong 1978]). Instead, vehicle cornering performance is viewed only at the practical design-oriented level, with equations simplified in a manner similar to that used for the stopping-distance equation, discussed in Section 2.9.5.

3.4.1 Vehicle Cornering

Figure 3.12 illustrates the forces acting on a vehicle during cornering. Some basic horizontal curve relationships can be derived by noting that

$$W_p + F_f = F_{cp} \tag{3.31}$$

From basic physics this equation can be written as [with $F_f = f_s(W_n + F_{cn})$]

$$W \sin \alpha + f_s \left(W \cos \alpha + \frac{WV^2}{gR_v} \sin \alpha \right) = \frac{WV^2}{gR_v} \cos \alpha \tag{3.32}$$

where

f_s = coefficient of side friction (unitless),
V = vehicle speed in ft/s (m/s),
g = gravitational constant, 32.2 ft/s² (9.807 m/s²), and
Other terms are as defined in Fig 3.12.

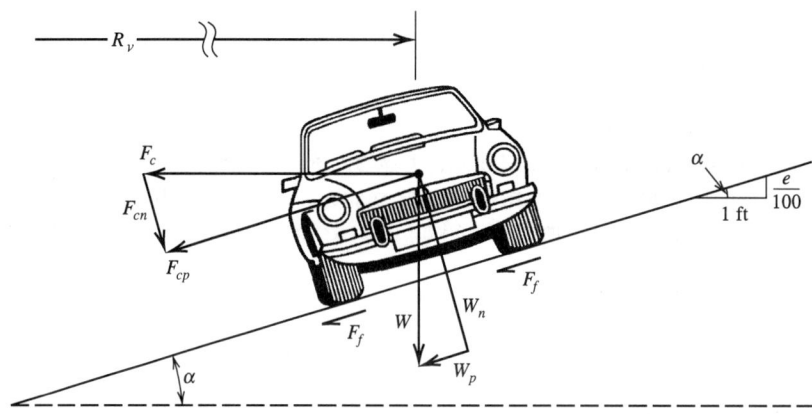

Figure 3.12 Vehicle cornering forces.

In this figure,

R_v = radius defined to the vehicle's traveled path in ft (m),
α = angle of incline in degrees,
e = number of vertical ft (m) of rise per 100 ft (m) of horizontal distance,
W = weight of the vehicle in lb (N),
W_n = vehicle weight normal to the roadway surface in lb (N),
W_p = vehicle weight parallel to the roadway surface in lb (N),

F_f = side frictional force [centripetal, in lb (N)],
F_c = centripetal force [lateral acceleration × mass, in lb (N)],
F_{cp} = centripetal force acting parallel to the roadway surface in lb (N), and,
F_{cn} = centripetal force acting normal to the roadway surface in lb (N).

Dividing both sides of Eq. 3.32 by $W \cos \alpha$ gives

$$\tan \alpha + f_s = \frac{V^2}{gR_v}(1 - f_s \tan \alpha) \tag{3.33}$$

The term $\tan \alpha$ indicates the superelevation of the curve (banking) and can be expressed in percent, and is denoted e ($e = 100 \tan \alpha$). In words, the superelevation is the number of vertical feet (meters) of rise per 100 feet (meters) of horizontal distance (see Fig. 3.12). The term $f_s \tan \alpha$ in Eq. 3.33 is conservatively set equal to zero for practical applications due to the small values that f_s and α typically assume (this is equivalent to ignoring the normal component of centripetal force). With $e = 100 \tan \alpha$, Eq. 3.33 can be arranged as

$$R_v = \frac{V^2}{g\left(f_s + \dfrac{e}{100}\right)} \tag{3.34}$$

EXAMPLE 3.13

A roadway is being designed for a speed of 70 mi/h (113 km/h). At one horizontal curve, it is known that the superelevation is 8.0% and the coefficient of side friction is 0.10. Determine the minimum radius of curve (measured to the traveled path) that will provide for safe vehicle operation.

SOLUTION

The application of Eq. 3.34 gives [with 1.467 (5280/3600) converting mi/h to ft/s]

$$R_v = \frac{V^2}{g\left(f_s + \dfrac{e}{100}\right)} = \frac{(70 \times 1.467)^2}{32.2(0.10 + 0.08)} = \underline{\underline{1819.40 \text{ ft}}}$$

This value is the minimum radius, because radii smaller than 1819.40 ft will generate centripetal forces higher than those that can be safely supported by the superelevation and the side frictional force.

In the actual design of a horizontal curve, the engineer must select appropriate values of e and f_s. The value selected for superelevation, e, is critical because high rates of superelevation can cause vehicle steering problems on the horizontal curve, and in cold climates, ice on the roadway can reduce f_s such that vehicles traveling at less than the design speed on an excessively superelevated curve could slide inward off the curve due to gravitational forces. AASHTO provides general guidelines for the selection of e and f_s for horizontal curve design, as shown in Table 3.5. The values presented in this table are grouped by five values of maximum e. The selection of any one of these five maximum e values is dependent on the type of road (for example, higher maximum e's are permitted on freeways compared with arterials and local roads) and local design practice. Limiting values of f_s are simply a function of design speed. Table 3.5 also presents calculated radii (given V, e, and f_s) by applying Eq. 3.34.

Table 3.5 Minimum Radius Using Limiting Values of e and f_s

U.S. Customary						Metric					
Design speed (mi/h)	Maximum e (%)	Limiting values of f_s	Total ($e/100 + f_s$)	Calculated radius, R_v (ft)	Rounded radius, R_v (ft)	Design speed (km/h)	Maximum e (%)	Limiting values of f_s	Total ($e/100 + f_s$)	Calculated radius, R_v (m)	Rounded radius, R_v (m)
15	4.0	0.175	0.215	70.0	70	20	4.0	0.18	0.22	14.3	15
20	4.0	0.170	0.210	127.4	125	30	4.0	0.17	0.21	33.7	35
25	4.0	0.165	0.205	203.9	205	40	4.0	0.17	0.21	60.0	60
30	4.0	0.160	0.200	301.0	300	50	4.0	0.16	0.20	98.4	100
35	4.0	0.155	0.195	420.2	420	60	4.0	0.15	0.19	149.1	150
40	4.0	0.150	0.190	563.3	565	70	4.0	0.14	0.18	214.2	215
45	4.0	0.145	0.185	732.2	730	80	4.0	0.14	0.18	279.8	280
50	4.0	0.140	0.180	929.0	930	90	4.0	0.13	0.17	375.0	375
55	4.0	0.130	0.170	1190.2	1190	100	4.0	0.12	0.16	491.9	490
60	4.0	0.120	0.160	1505.0	1505						
15	6.0	0.175	0.235	64.0	65	20	6.0	0.18	0.24	13.1	15
20	6.0	0.170	0.230	116.3	115	30	6.0	0.17	0.23	30.8	30
25	6.0	0.165	0.225	185.8	185	40	6.0	0.17	0.23	54.7	55
30	6.0	0.160	0.220	273.6	275	50	6.0	0.16	0.22	89.4	90
35	6.0	0.155	0.215	381.1	380	60	6.0	0.15	0.21	134.9	135
40	6.0	0.150	0.210	509.6	510	70	6.0	0.14	0.20	192.8	195
45	6.0	0.145	0.205	660.7	660	80	6.0	0.14	0.20	251.8	250
50	6.0	0.140	0.200	836.1	835	90	6.0	0.13	0.19	335.5	335
55	6.0	0.130	0.190	1065.0	1065	100	6.0	0.12	0.18	437.2	435
60	6.0	0.120	0.180	1337.8	1340	110	6.0	0.11	0.17	560.2	560
65	6.0	0.110	0.170	1662.4	1660	120	6.0	0.09	0.15	755.5	755
70	6.0	0.100	0.160	2048.5	2050	130	6.0	0.08	0.14	950.0	950
75	6.0	0.090	0.150	2508.4	2510						
80	6.0	0.080	0.140	3057.8	3060						
15	8.0	0.175	0.255	59.0	60	20	8.0	0.18	0.28	12.1	10
20	8.0	0.170	0.250	107.0	105	30	8.0	0.17	0.25	28.3	30
25	8.0	0.185	0.245	170.8	170	40	8.0	0.17	0.25	50.4	50
30	8.0	0.160	0.240	250.8	250	50	8.0	0.16	0.24	82.0	80
35	8.0	0.155	0.235	348.7	350	60	8.0	0.15	0.23	123.2	125
40	8.0	0.150	0.230	465.3	465	70	8.0	0.14	0.22	175.3	175
45	8.0	0.145	0.225	502.0	500	80	8.0	0.14	0.22	228.9	230
50	8.0	0.140	0.220	760.1	760	90	8.0	0.13	0.21	303.6	305
55	8.0	0.130	0.210	963.5	965	100	8.0	0.12	0.20	393.5	395
60	8.0	0.120	0.200	1204.0	1205	110	8.0	0.11	0.19	501.2	500
65	8.0	0.110	0.190	1487.4	1485	120	8.0	0.09	0.17	666.6	665
70	8.0	0.100	0.180	1820.9	1820	130	8.0	0.08	0.18	831.3	830
75	8.0	0.090	0.170	2213.3	2215						
80	8.0	0.080	0.160	2675.6	2675						

Table 3.5 Minimum Radius Using Limiting Values of e and f_s (Continued)

U.S. Customary							Metric						
Design speed (mi/h)	Maximum e (%)	Limiting values of f_s	Total $(e/100 + f_s)$	Calculated radius, R_v (ft)	Rounded radius, R_v (ft)		Design speed (km/h)	Maximum e (%)	Limiting values of f_s	Total $(e/100 + f_s)$	Calculated radius, R_v (m)	Rounded radius, R_v (m)
15	10.0	0.175	0.275	54.7	55		20	10.0	0.18	0.28	11.2	10
20	10.0	0.170	0.270	99.1	100		30	10.0	0.17	0.27	26.2	25
25	10.0	0.165	0.265	157.8	160		40	10.0	0.17	0.27	46.6	45
30	10.0	0.160	0.260	231.5	230		50	10.0	0.16	0.26	75.7	75
35	10.0	0.155	0.255	321.3	320		60	10.0	0.15	0.25	113.3	115
40	10.0	0.150	0.250	428.1	430		70	10.0	0.14	0.24	160.7	160
45	10.0	0.145	0.245	552.9	555		80	10.0	0.14	0.24	209.9	210
50	10.0	0.140	0.240	696.8	695		90	10.0	0.13	0.23	277.2	275
55	10.0	0.130	0.230	879.7	880		100	10.0	0.12	0.22	357.7	360
60	10.0	0.120	0.220	1094.6	1095		110	10.0	0.11	0.21	453.5	455
65	10.0	0.110	0.210	1345.8	1345		120	10.0	0.09	0.19	596.5	595
70	10.0	0.100	0.200	1838.8	1840		130	10.0	0.08	0.18	738.9	740
75	10.0	0.090	0.190	1980.3	1980							
80	10.0	0.080	0.180	2378.3	2380							
15	12.0	0.175	0.295	51.0	50		20	12.0	0.18	0.30	10.5	10
20	12.0	0.170	0.290	92.3	90		30	12.0	0.17	0.29	24.4	25
25	12.0	0.165	0.285	146.7	145		40	12.0	0.17	0.29	43.4	45
30	12.0	0.160	0.280	215.0	215		50	12.0	0.16	0.28	70.3	70
35	12.0	0.155	0.275	298.0	300		60	12.0	0.15	0.27	104.9	105
40	12.0	0.150	0.270	396.4	395		70	12.0	0.14	0.26	148.3	150
45	12.0	0.145	0.265	511.1	510		80	12.0	0.14	0.26	193.7	195
50	12.0	0.140	0.260	643.2	645		90	12.0	0.13	0.25	255.0	255
55	12.0	0.130	0.250	809.4	810		100	12.0	0.12	0.24	327.9	330
60	12.0	0.120	0.240	1003.4	1005		110	12.0	0.11	0.23	414.0	415
65	12.0	0.110	0.230	1228.7	1230		120	12.0	0.09	0.21	539.7	540
70	12.0	0.100	0.220	1489.8	1490		130	12.0	0.08	0.20	665.0	665
75	12.0	0.090	0.210	1791.7	1790							
80	12.0	0.080	0.200	2140.5	2140							

Note: In recognition of safety considerations, use of $e_{max} = 4.0\%$ should be limited to urban conditions.
Source: American Association of State Highway and Transportation Officials, *A Policy on Geometric Design of Highways and Streets*, Washington, DC, 2001.

3.4.2 Horizontal Curve Fundamentals

In connecting straight (tangent) sections of roadway with a horizontal curve, several options are available. The most obvious of these is the simple circular curve, which is just a curve with a single, constant radius. Other options include reverse curves, compound curves, and spiral curves. Reverse curves generally consist of two consecutive curves that turn in opposite directions. They are used to laterally shift the alignment of a highway. The curves used are usually circular and have equal radii. Reverse curves, however, are not recommended because drivers may find it difficult to stay within their lane as a result of sudden changes to the alignment. Compound curves consist of two or more curves, usually circular, in succession. Compound curves are used to fit horizontal curves to very specific alignment needs, such as interchange ramps, intersection curves, or difficult topography. In designing compound curves, care must be taken to not have successive curves with widely different radii, as this will make it difficult for drivers to maintain their lane position as they transition from one curve to the next. Spiral curves are curves with a continuously changing radius. Spiral curves are sometimes used to transition a tangent section of roadway to a circular curve. In such a case, the radius of the spiral curve is equal to infinity where it connects to the tangent section and ends with the radius value of the connecting circular curve at the other end. Because motorists usually create their own transition paths between tangent sections and circular curves by utilizing the full lane width available, spiral curves are not often used. However, there are exceptions. Spiral curves are sometimes used on high-speed roadways with sharp horizontal curves and are sometimes used to gradually introduce the superelevation of an upcoming horizontal curve. To illustrate the basic principles involved in horizontal curve design, this book will focus only on the single simple circular curve. For detailed information regarding these additional horizontal curve types, refer to standard route-surveying texts, such as Hickerson [1964].

Figure 3.13 shows the basic elements of a simple horizontal curve.

Another important term is the degree of curve, which is defined as the angle subtended by a 100-ft (30.5-m) arc along the horizontal curve. It is a measure of the sharpness of the curve and is frequently used instead of the radius in the construction of the curve. The degree of curve is directly related to the radius of the horizontal curve by

U.S. Customary   **Metric**

$$D = \frac{100\left(\frac{180}{\pi}\right)}{R} = \frac{18{,}000}{\pi R} \qquad D = \frac{30.5\left(\frac{180}{\pi}\right)}{R} = \frac{5490}{\pi R} \qquad (3.35)$$

where

D = degree of curve [angle subtended by a 100-ft (30.5-m) arc along the horizontal curve], and

Other terms are as defined in Fig. 3.13.

Note that the quantity $180/\pi$ converts from radians to degrees.

Geometric and trigonometric analyses of Fig. 3.13 reveal the following relationships:

$$T = R \tan \frac{\Delta}{2} \qquad (3.36)$$

$$E = R\left[\frac{1}{\cos(\Delta/2)} - 1\right] \qquad (3.37)$$

$$M = R\left(1 - \cos \frac{\Delta}{2}\right) \qquad (3.38)$$

$$L = \frac{\pi}{180} R \Delta \qquad (3.39)$$

where all terms are as defined in Fig. 3.13.

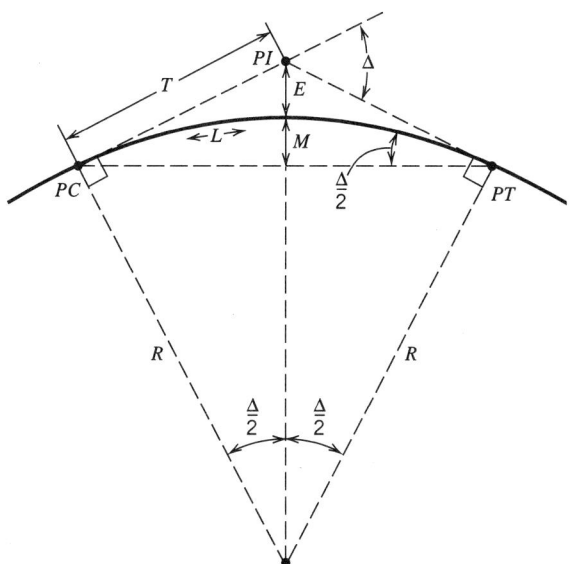

Figure 3.13 Elements of a simple circular horizontal curve.

In this figure,

R = radius, usually measured to the centerline of the road, in ft (m),
Δ = central angle of the curve in degrees,
PC = point of curve (the beginning point of the horizontal curve),
PI = point of tangent intersection,
PT = point of tangent (the ending point of the horizontal curve),
T = tangent length in ft (m),
M = middle ordinate in ft (m),
E = external distance in ft (m), and
L = length of curve in ft (m).

It is important to note that horizontal curve stationing, curve length, and curve radius (R) are usually measured to the centerline of the road. In contrast, the radius determined on the basis of vehicle forces (R_v in Eq. 3.34) is measured from the innermost vehicle path, which is assumed to be the midpoint of the innermost vehicle lane. Thus, a slight correction for lane width is required in equating the R_v of Eq. 3.34 with the R in Eqs. 3.35 to 3.39.

EXAMPLE 3.14

A horizontal curve is designed with a 2000-ft (609.600-m) radius. The curve has a tangent length of 400 ft (121.920 m) and the PI is at station 103 + 00 (3 + 139.440). Determine the stationing of the PT.

SOLUTION

Equation 3.36 is applied to determine the central angle, Δ.

$$T = R \tan \frac{\Delta}{2}$$

$$400 = 2000 \tan \frac{\Delta}{2}$$

$$\Delta = 22.62°$$

So, from Eq. 3.39, the length of the curve is

$$L = \frac{\pi}{180} R \Delta$$

$$L = \frac{3.1416}{180} 2000(22.62) = 789.58 \text{ ft}$$

Given that the tangent length is 400 ft,

stationing PC = 103 + 00 minus 4 + 00 = 99 + 00

Since horizontal curve stationing is measured along the alignment of the road,

stationing PT = stationing PC + L

= 99 + 00 plus 7 + 89.58 = $\underline{106 + 89.58}$

3.4.3 Stopping Sight Distance and Horizontal Curve Design

As is the case for vertical curve design, adequate stopping sight distance must be provided in the design of horizontal curves. Sight distance restrictions on horizontal curves occur when obstructions are present, as shown in Fig. 3.14. Such obstructions are frequently encountered in highway design due to the cost of right-of-way acquisition or the cost of moving earthen materials, such as rock outcroppings. When such an obstruction exists, the stopping sight distance is measured along the horizontal curve from the center of the traveled lane (the assumed location of the driver's eyes). As shown in Fig. 3.14, for a specified stopping distance, some distance M_s (the middle ordinate of a curve that has an arc length equal to the stopping sight distance) must be visually cleared so that the line of sight is such that sufficient stopping sight distance is available.

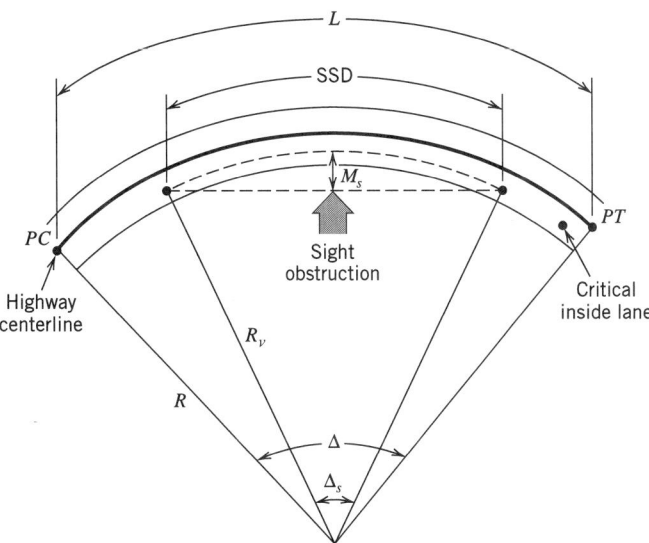

Figure 3.14 Stopping sight distance considerations for horizontal curves.

In this figure,

- L = length of curve in ft (m),
- SSD = stopping sight distance in ft (m),
- R = radius measured to the centerline of the road in ft (m),
- R_v = radius to the vehicle's traveled path (usually measured to the center of the innermost lane of the road) in ft (m),
- PC = point of curve (the beginning point of the horizontal curve),
- PT = point of tangent (the ending point of the horizontal curve),
- Δ = central angle of the curve in degrees,
- Δ_s = angle (in degrees) subtended by an arc equal in length to the required stopping sight distance (SSD), and
- M_s = middle ordinate necessary to provide adequate stopping sight distance (SSD) in ft (m).

Equations for computing stopping sight distance (SSD) relationships for horizontal curves can be derived by first determining the angle, Δ_s, for an arc length equal to the required stopping sight distance (see Fig. 3.14 and note that this is not the central angle, Δ, of the horizontal curve whose arc length is equal to L). Assuming that the length of the horizontal curve exceeds the required SSD (as shown in Fig. 3.14), we have (as with Eq. 3.39)

$$\text{SSD} = \frac{\pi}{180} R_v \Delta_s \tag{3.40}$$

Rearranging terms,

$$\Delta_s = \frac{180(\text{SSD})}{\pi R_v} \tag{3.41}$$

Substituting this into the general equation for the middle ordinate of a simple horizontal curve (Eq. 3.38) to get an expression for M_s gives

$$M_s = R_v \left(1 - \cos \frac{90(\text{SSD})}{\pi R_v}\right) \tag{3.42}$$

Solving Eq. 3.42 for SSD gives

$$\text{SSD} = \frac{\pi R_v}{90} \left[\cos^{-1}\left(\frac{R_v - M_s}{R_v}\right)\right] \tag{3.43}$$

Note that Eqs. 3.40 to 3.43 can also be applied directly to determine sight distance requirements for passing. If these equations are to be used for passing, distance values given in Table 3.4 would apply and SSD in the equations would be replaced by PSD.

EXAMPLE 3.15

A horizontal curve on a two-lane highway is designed with a 2000-ft (609.600-m) radius, 12-ft (3.6-m) lanes, and a 60-mi/h (96-km/h) design speed. Determine the distance that must be cleared from the inside edge of the inside lane to provide a sufficient stopping sight distance.

SOLUTION

Because the curve radius is usually taken to the centerline of the roadway, $R_v = R - 12/2 = 2000 - 6 = 1994$ ft, which gives the radius to the middle of the inside lane (the critical driver location). From Table 3.1, the SSD for a 60-mi/h design speed is 570 ft, so applying Eq. 3.42 gives

$$M_s = R_v \left(1 - \cos \frac{90(\text{SSD})}{\pi R_v}\right)$$

$$= 1994 \left(1 - \cos \frac{90(570)}{\pi(1994)}\right) = \underline{\underline{20.33 \text{ ft}}}$$

Therefore, 20.33 ft must be cleared, as measured from the center of the inside lane, or 14.33 ft as measured from the inside edge of the inside lane.

EXAMPLE 3.16

A two-lane highway [two 12-ft (3.6-m) lanes] has a posted speed limit of 50 mi/h (80 km/h) and, on one section, has both horizontal and vertical curves, as shown in Fig. 3.15. A recent daytime crash (driver traveling eastbound and striking a stationary roadway object) resulted in a fatality and a lawsuit alleging that the 50-mi/h (80-km/h) posted speed limit is an unsafe speed for the curves in question and was a major cause of the crash. Evaluate and comment on the roadway design.

SOLUTION

Begin with an assessment of the horizontal alignment. Two concerns must be considered: the adequacy of the curve radius and superelevation, and the adequacy of the sight distance on the eastbound (inside) lane. For the curve radius, note from Fig. 3.15 that

$$L = \text{station of } PT - \text{station of } PC$$
$$L = 32 + 75 \text{ minus } 16 + 00 = 1675 \text{ ft}$$

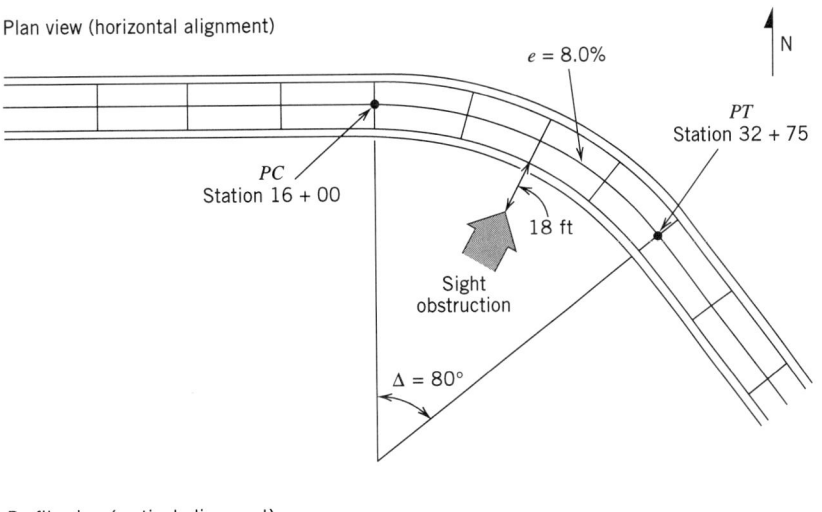

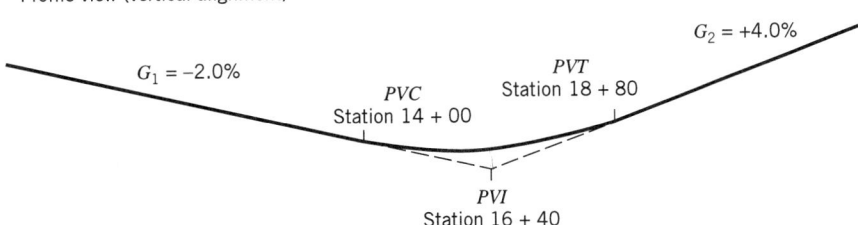

Figure 3.15 Horizontal and vertical alignment for Example 3.16.

Rearranging Eq. 3.39, we get

$$R = \frac{180}{\pi \Delta} L = \frac{180}{\pi(80)}(1675) = 1199.63 \text{ ft}$$

Using the posted speed limit of 50 mi/h with $e = 8.0\%$, we find that Eq. 3.34 can be rearranged to give (with the vehicle traveling in the middle of the inside lane, $R_v = R -$ half the lane width, or $R_v = 1199.63 - 6 = 1193.63$ ft)

$$f_s = \frac{V^2}{gR_v} - e = \frac{(50 \times 1.467)^2}{32.2(1193.63)} - 0.08 = 0.060$$

From Table 3.5, the maximum f_s for 50 mi/h is 0.14. Since 0.060 does not exceed 0.14, the radius and superelevation are sufficient for the 50-mi/h design speed.

For sight distance, the available M_s is 18 ft plus the 6-ft distance to the center of the eastbound (inside) lane, or 24 ft. Application of Eq. 3.43 gives

$$\text{SSD} = \frac{\pi R_v}{90}\left[\cos^{-1}\left(\frac{R_v - M_s}{R_v}\right)\right]$$

$$= \frac{\pi(1193.63)}{90}\left[\cos^{-1}\left(\frac{1193.63 - 24}{1193.63}\right)\right]$$

$$= 479.5 \text{ ft}$$

From Table 3.1, the required SSD at 50 mi/h is 425 ft, so the 479.5 ft of SSD provided is sufficient. Turning to the sag vertical curve, the length of curve is

$$L = \text{station of } PVT - \text{station of } PVC$$
$$L = 18 + 80 \text{ minus } 14 + 00 = 480 \text{ ft}$$

Using $A = 6\%$ (from Fig. 3.15) and applying Eq. 3.10, we obtain

$$K = \frac{L}{A} = \frac{480}{6} = 80$$

For the 50-mi/h design speed, Table 3.3 indicates a necessary K-value of 96. Thus the K-value of 80 reveals that the curve is inadequate for the 50-mi/h speed. However, because the crash occurred in daylight and sight distances on sag vertical curves are governed by nighttime conditions, this design did not contribute to the crash.

NOMENCLATURE FOR CHAPTER 3

A	absolute value of the algebraic difference in grades (in percent)
a	coefficient in the parabolic curve equation or the deceleration in the stopping distance equation
b	coefficient in the parabolic curve equation
c	elevation of the PVC
D	degree of curvature
e	rate of superelevation
F_f	frictional side force
F_c	centripetal force
F_{cn}	centripetal force normal to the roadway surface
F_{cp}	centripetal force parallel to the roadway surface
f_s	coefficient of side friction
G	grade
G_1	initial roadway grade
G_2	final roadway grade
g	gravitational constant
H	height of vehicle headlights
H_c	clearance height of structure above sag curve
H_1	height of driver's eye
H_2	height of roadway object for stopping, height of oncoming car for passing
K	horizontal distance required to effect a 1% change in slope
L	length of curve
L_m	minimum length of curve
M	middle ordinate
M_s	middle ordinate for stopping sight distance
PC	initial point of horizontal curve
PI	point of tangent intersection (horizontal curve)
PSD	passing sight distance
PT	final point of horizontal curve
PVC	initial point of vertical curve
PVI	point of tangent intersection (vertical curve)
PVT	final point of vertical curve
R	radius of curve measured to the roadway centerline
R_v	radius of curve to the vehicle's traveled path
S	sight distance
SSD	stopping sight distance
T	tangent length
V	vehicle speed
V_1	initial vehicle speed
W	vehicle weight
W_n	vehicle weight normal to the roadway surface
W_p	vehicle weight parallel to the roadway surface
x	distance from the beginning of the vertical curve to specified point
x_{hl}	distance from the beginning of the vertical curve to high or low point
Y	vertical curve offset
Y_f	end-of-curve offset (vertical curve)
Y_m	midcurve offset (vertical curve)
α	angle of superelevation
β	upward angle of headlight beam
Δ	central angle
Δ_s	angle subtended by the stopping sight distance (SSD) arc

REFERENCES

AASHTO (American Association of State Highway and Transportation Officials). *A Policy on Geometric Design of Highways and Streets*, 4th ed. Washington, DC: AASHTO, 2001.

Campbell, C. *The Sports Car: Its Design and Performance.* Cambridge, MA: Robert Bently, 1978.

Hickerson, T. F. *Route Location and Design*, 5th ed. New York: McGraw-Hill, 1964.

Wong, J. H. *Theory of Ground Vehicles*. New York: John Wiley & Sons, 1978.

PROBLEMS

3.1 A 1600-ft–long sag vertical curve (equal tangent) has a *PVC* at station 120 + 00 and elevation 1500 ft. The initial grade is −3.5% and the final grade is +6.5%. Determine the elevation and stationing of the low point, *PVI*, and *PVT*.

3.2 A 500-ft–long equal-tangent crest vertical curve connects tangents that intersect at station 340 + 00 and elevation 1322 ft. The initial grade is +4.0% and the final grade is −2.5%. Determine the elevation and stationing of the high point, *PVC*, and *PVT*.

3.3 Consider Example 3.4. Solve this problem with the parabolic equation (Eq. 3.1) rather than by using offsets.

3.4 Again consider Example 3.4. Does this curve provide sufficient stopping sight distance for a speed of 60 mi/h?

3.5 An equal-tangent sag vertical curve is designed with the *PVC* at station 109 + 00 and elevation 950 ft, the *PVI* at station 110 + 77 and elevation 947.34 ft, and the low point at station 110 + 50. Determine the design speed of the curve.

3.6 An equal-tangent vertical curve was designed in 2002 (to 2001 AASHTO guidelines) for a design speed of 70 mi/h to connect grades $G_1 = +1.0\%$ and $G_2 = -2.0\%$. The curve is to be redesigned for a 70-mi/h design speed in the year 2025. Vehicle braking technology has advanced such that the recommended design deceleration rate is 25% greater than its 2001 value used to develop Table 3.1, but due to the higher percentage of older persons in the driving population, design reaction times have increased by 20%. Also, vehicles have become smaller such that the driver's eye height is assumed to be 3.0 ft above the pavement and roadway objects are assumed to be 1.0 ft above the pavement surface. Compute the difference in design curve lengths for the 2002 and 2025 designs.

3.7 A 1200-ft equal-tangent crest vertical curve is currently designed for 50 mi/h. A civil engineering student contends that 60 mi/h is safe in a van because of the higher driver's eye height. If all other design inputs are standard, what must the driver's eye height (in the van) be for the student's claim to be valid?

3.8 A highway reconstruction project is being undertaken to reduce crash rates. The reconstruction involves a major realignment of the highway such that a 60-mi/h design speed is attained. At one point on the highway, an 800-ft equal-tangent crest vertical curve exists. Measurements show that at 3 + 52 stations from the *PVC*, the vertical curve offset is 3 ft. Assess the adequacy of this existing curve in light of the reconstruction design speed of 60 mi/h and, if the existing curve is inadequate, compute a satisfactory curve length.

3.9 Two level sections of an east-west highway ($G = 0$) are to be connected. Currently, the two sections of highway are separated by a 4000-ft (horizontal distance), 2% grade. The westernmost section of highway is the higher of the two and is at elevation 100 ft. If the highway has a 60-mi/h design speed, determine, for the crest and sag vertical curves required, the stationing and elevation of the *PVCs* and *PVTs* given that the *PVC* of the crest curve (on the westernmost level highway section) is at station 0 + 00 and elevation 100 ft. In solving this problem, assume that the curve *PVIs* are at the intersection of $G = 0$ and the 2% grade, that is, $A = 2$.

3.10 Consider Problem 3.9. Suppose it is necessary to keep the entire alignment within the 4000 ft that currently separate the two level sections. It is determined that the crest and sag curves should be connected (the *PVT* of the crest and *PVC* of the sag) with a constant-grade section that has the lowest grade possible. Again using a 60-mi/h design speed, determine, for the crest and sag vertical curves, the stationing and elevation of the *PVCs* and *PVTs* given that the westernmost level section ends at station 0 + 00 and elevation 100 ft. (Note that A must now be determined and will not be equal to 2.)

3.11 An equal-tangent crest vertical curve is designed for 60 mi/h. The initial grade is +4.0% and the final grade is negative. What is the elevation difference between the *PVC* and the high point of the curve?

3.12 An equal-tangent crest vertical curve has a 50-mi/h design speed. The initial grade is +3%. The high point is at station 33 + 37.43 and the *PVT* is at station 37 + 18.26. What is the elevation difference between the high point and the *PVT*?

3.13 A vertical curve is designed for 55 mi/h, and it has an initial grade of +2.5% and a final grade of −1.0%. The *PVT* is at station 114 + 25. It is known that a point on the curve at station 112 + 75 is at elevation 240 ft. What is the stationing and elevation of the *PVC*? What is the stationing and elevation of the high point on the curve?

3.14 An equal-tangent crest curve connects a +1.0% and a −0.5% grade. The *PVC* is at station 54 + 84 and the *PVI* is at station 57 + 44. Is this curve long enough to provide passing sight distance for a 55-mi/h design speed?

3.15 Due to crashes at a railroad crossing, an overpass (with a roadway surface 24 ft above the existing road) is to be constructed on an existing level highway. The existing highway has a design speed of 50 mi/h. The overpass structure is to be level, centered above the railroad, and 200 ft long. What length of the existing level highway must be reconstructed to provide an appropriate vertical alignment?

3.16 A section of a freeway ramp has a +4.0% grade and ends at station 127 + 00 and elevation 138 ft. It must be connected to another section of the ramp (which has a 0.0% grade) that is at station 162 + 00 and elevation 97 ft. It is determined that the crest and sag curves required to connect the ramp should be connected (the PVT of the crest and PVC of the sag) with a constant-grade section that has the lowest grade possible. Design a vertical alignment to connect between these two stations using a 50-mi/h design speed. Provide the lengths of the curves and constant-grade section.

3.17 A tangent section of highway has a -1.0% grade and ends at station 4 + 75 and elevation 82 ft. It must be connected to another section of highway that has a -1.0% grade and that begins at station 44 + 12 and elevation 131.2 ft. The connecting alignment should consist of a sag curve, constant-grade section, and crest curve, and be designed for a speed of 50 mi/h. What is the lowest grade possible for the constant-grade section that will complete this alignment?

3.18 A roadway has a design speed of 50 mi/h, and at station 100 + 00 a +3.0% grade roadway section ends and at station 130 + 00 a +2.0% grade roadway section begins. The +3.0% grade section of highway (at station 100 + 00) is at a higher elevation than the +2.0% grade section of highway (at station 130 + 00). If a -5% constant-grade section is used to connect the crest and sag vertical curves that are needed to link the +3.0 and +2.0% grade sections, what is the elevation difference between stations 100 + 00 and 130 + 00? (The entire alignment, crest and sag curves, and constant-grade section must fit between stations 100 + 00 and 130 + 00.)

3.19 A sag curve and crest curve connect a -3.0% tangent section of highway (to the west) with a +2.0% tangent section of highway (to the east). The +2.0% tangent section is at a higher elevation than the -3.0% tangent section. The two tangent sections are separated by 1275 ft of horizontal distance. If the design speed of the curves is 50 mi/h, what is the common grade between the sag and crest curves (G_2 of sag and G_1 of crest, from west to east), and what is the elevation difference between the PVC_s and PVT_c?

3.20 An overpass is being built over the PVI of an existing equal-tangent sag curve. The sag curve has a 70-mi/h design speed, and $G_1 = -6\%$, $G_2 = +3\%$. Determine the minimum necessary clearance height of the overpass and the resultant elevation of the bottom of the overpass over the PVI. (Ignore the cross-sectional width of the overpass.)

3.21 An equal-tangent sag curve has its PVI at station 10 + 00 and elevation at 138 ft. Directly above the PVI, the bottom of an overpass structure is at elevation 162 ft. The PVC is at station 4 + 00. If the initial grade is -4%, what is the highest possible value of the final grade given that a 70-mi/h design speed is to be provided in daytime conditions? What is the highest possible final grade in nighttime conditions? (*Note:* Be careful of units of A, and ignore the cross-sectional width of the overpass.)

3.22 An existing highway-railway at-grade crossing is being redesigned as grade separated to improve traffic operations. The railway must remain at the same elevation. The highway is being reconstructed to travel under the railway. The underpass will be a sag curve that connects to 2% tangent sections on both ends, and the PVI will be centered under the railway (a symmetrical alignment). The sag curve design speed is 45 mi/h. How many feet below the railway should the curve PVI be located?

3.23 You are asked to design a horizontal curve for a two-lane road. The road has 12-ft lanes. Due to expensive excavation, it is determined that a maximum of 34 ft can be cleared from the road's centerline toward the inside lane to provide for stopping sight distance. Also, local guidelines dictate a maximum superelevation of 0.08 ft/ft. What is the highest possible design speed for this curve?

3.24 A horizontal curve on a single-lane highway has its PC at station 124 + 10 and its PI at station 131 + 40. The curve has a superelevation of 0.06 ft/ft and is designed for 70 mi/h. What is the station of the PT?

3.25 A horizontal curve is being designed through mountainous terrain for a four-lane road with lanes that are 10 ft wide. The central angle (Δ) is known to be 40 degrees, the tangent distance is 510 ft, and the stationing of the tangent intersection (PI) is 2700 + 00. Under specified conditions and vehicle speed, the roadway surface is determined to have a coefficient of side friction of 0.08, and the curve's superelevation is 0.09 ft/ft. What is the stationing of the *PC* and *PT* and what is the safe vehicle speed?

3.26 A new interstate highway is being built with a design speed of 70 mi/h. For one of the horizontal curves, the radius (measured to the innermost vehicle path) is tentatively planned as 900 ft. What rate of superelevation is required for this curve?

3.27 A developer is having a single-lane raceway constructed with a 100-mi/h design speed. A curve on the raceway has a radius of 1000 ft, a central angle of 30 degrees, and *PI* stationing at 1125 + 10. If the design coefficient of side friction is 0.20, determine the superelevation required at the design speed (do not ignore the normal component of the centripetal force). Also, compute the degree of curve, length of curve, and stationing of the *PC* and *PT*.

3.28 A horizontal curve is being designed for a new two-lane highway (12-ft lanes). The *PI* is at station 250 + 50, the design speed is 65 mi/h, and a maximum superelevation of 0.08 ft/ft is to be used. If the central angle of the curve is 35 degrees, design a curve for the highway by computing the radius and stationing of the *PC* and *PT*.

3.29 You are asked to design a horizontal curve with a 40-degree central angle ($\Delta = 40$) for a two-lane road with 10-ft lanes. The design speed is 70 mi/h and superelevation is limited to 0.06 ft/ft. Give the radius, degree of curvature, and length of curve that you would recommend.

3.30 A horizontal curve on a single-lane freeway ramp is 400 ft long, and the design speed of the ramp is 45 mi/h. If the superelevation is 10% and the station of the *PC* is 17 + 35, what is the station of the *PI* and how much distance must be cleared from the center of the lane to provide adequate stopping sight distance?

3.31 A freeway exit ramp has a single lane and consists entirely of a horizontal curve with a central angle of 90 degrees and a length of 628 ft. If the distance cleared from the centerline for sight distance is 19.4 ft, what design speed was used?

3.32 A horizontal curve on a two-lane highway (12-ft lanes) has *PC* at station 123 + 50 and *PT* at station 129 + 34. The central angle is 34 degrees, the superelevation is 0.08, and 20.3 ft is cleared (for sight distance) from the inside edge of the innermost lane. Determine a maximum safe speed (assuming current design standards) to the nearest 5 mi/h.

3.33 For the horizontal curve in Problem 3.29, what distance must be cleared from the inside edge of the inside lane to provide adequate stopping sight distance?

3.34 A horizontal curve was designed for a four-lane highway for adequate SSD. Lane widths are 12 ft, and the superelevation is 0.06 and was set assuming maximum f_s. If the necessary sight distance required 52 ft of lateral clearance from the roadway centerline, what design speed was used for the curve?

3.35 A section of highway has vertical and horizontal curves with the same design speed. A vertical curve on this highway connects a +1% and a +3% grade and is 420 ft long. If a horizontal curve on this highway is on a two-lane section with 12-ft lanes, has a central angle of 37 degrees, and has a superelevation of 6%, what is the length of the horizontal curve?

3.36 A section of a two-lane highway (12-ft lanes) is designed for 75 mi/h. At one point a vertical curve connects a −2.5% and +1.5% grade. The *PVT* of this curve is at station 25 + 10. It is known that a horizontal curve starts (has *PC*) 292 ft before the vertical curve's *PVC*. If the superelevation of the horizontal curve is 0.08 and the central angle is 38 degrees, what is the station of the *PT*?

3.37 Two straight sections of freeway cross at a right angle. At the point of crossing, the east-west highway is at elevation 150 ft and has a constant +5.0% grade (upgrade in the east direction), and the north-south highway is at elevation 125 ft and has a constant −3.0% grade (downgrade in the north direction). Design a 90-degree ramp that connects the northbound direction of travel to the eastbound direction of travel. Design the ramp for the highest design speed (to nearest 5 mi/h) with the constraint that the minimum allowable value of *D* is 8.0. (Assume that the *PC* of the horizontal curve is at station 15 + 00, and the vertical curve *PVI*s are at the *PC* and *PT*.) Give the stationing and elevations of the *PC*, *PT*, *PVC*s, and *PVT*s.

Chapter 4

Pavement Design

4.1 INTRODUCTION

Pavements are among the costliest items associated with highway construction and maintenance, and are largely responsible for making the U.S. highway system the most expensive public works project undertaken by any society. Because the pavement and associated shoulder structures are the most expensive items to construct and maintain, it is important for highway engineers to have a basic understanding of pavement design principles.

In the United States, there are over 3 million miles (4.8 million kilometers) of highways. Surprisingly, about 45% of these roads are not paved but are composed of either gravel or a stabilized material (consisting of an aggregate material bound together with a cementing agent such as portland cement, lime fly ash, or asphaltic cement). Highways that carry higher volumes of traffic with heavy axle loads require surfaces with asphalt concrete or portland cement concrete to provide for all-weather operations and prevent permanent deformation of the highway surface. These types of pavements can cost upward of several million dollars per mile to construct. Some states, such as Pennsylvania, Texas, Illinois, and California, have pavement construction and rehabilitation budgets that approach a billion dollars per year, and when these are coupled with their maintenance budgets, it is easy to see why construction and maintenance of pavement infrastructure must be done in a cost-effective manner.

Pavement provides two basic functions. First, it helps guide the driver and delineate the roadway by giving a visual perspective of the horizontal and vertical alignment of the traveled path. Consequently, pavement gives the driver information about the driving task and the steering control of the vehicle. The second function of pavement is to support vehicle loads. This second function is the focus of this chapter.

4.2 PAVEMENT TYPES

In general, there are two types of pavement structures: flexible pavements and rigid pavements. There are, however, many variations of these pavement types, including some with soil cement and stabilized bases that have cemented aggregate. Composite pavements (which are made of both rigid and flexible layers), continuously reinforced pavements, and post-tensioned pavements are other types, which usually require specialized designs and are not covered in this chapter.

As with any structure, the underlying soil must ultimately carry the load that is placed on it. A pavement's function is to distribute the traffic load stresses to the soil (subgrade) at a magnitude that will not shear or distort the soil. Typical soil-bearing capacities can be less than 50 lb/in^2 (345 kPa) and in some cases as low as 2 to 3 lb/in^2 (14 to 21 kPa). When soil is saturated with water, the bearing capacity can be very low, and in these cases it is very important for pavement to distribute tire loads to the soil in such a way as to prevent failure of the pavement structure.

A typical automobile weighs approximately 3500 lb (15.5 kN), with tire pressures around 35 lb/in^2 (241 kPa). These loads are small compared with a typical tractor–semi-trailer truck, which can weigh up to 80,000 lb (355.8 kN)—the legal limit, in many states, on five axles with tire pressures of 100 lb/in^2 (690 kPa) or higher. Truck loads such as these represent the standard type of loading used in pavement design. In this chapter, attention is directed toward an accepted procedure that can be used to design pavement structures for high–traffic-volume highway facilities subjected to heavy truck traffic. The design of lower-volume facilities, which may have stabilized-soil and gravel-surfaced pavements, can be found in other references [Yoder and Witczak, 1975].

4.2.1 Flexible Pavements

A flexible pavement is constructed with asphaltic cement and aggregates and usually consists of several layers, as shown in Fig. 4.1. The lower layer is called the subgrade (the soil itself). The upper 6 to 8 inches (152 to 203 mm) of the subgrade is usually scarified and blended to provide a uniform material before it is compacted to maximum density. The next layer is the subbase, which usually consists of crushed aggregate (rock). This material has better engineering properties (higher modulus values) than the subgrade material, in terms of its bearing capacity. The next layer is the base layer and is also often made of crushed aggregates (of a higher strength than those used in the subbase), which are either unstabilized or stabilized with a cementing material. The cementing material can be portland cement, lime fly ash, or asphaltic cement.

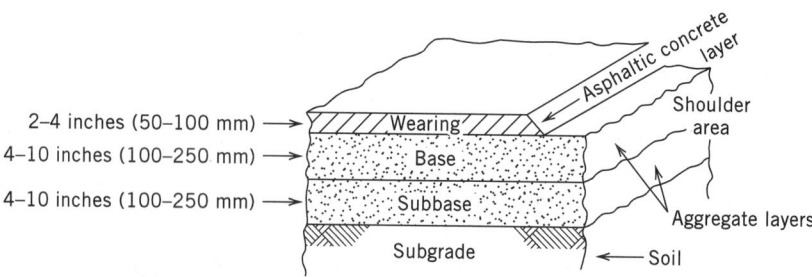

Figure 4.1 Typical flexible-pavement cross section.

The top layer of a flexible pavement is referred to as the wearing surface. It is usually made of asphaltic concrete, which is a mixture of asphalt cement and aggregates. The purpose of the wearing layer is to protect the base layer from wheel abrasion and to waterproof the entire pavement structure. It also provides a skid-resistant surface that is important for safe vehicle stops. Typical thicknesses of the individual layers are shown in Fig. 4.1. These thicknesses vary with the type of axle loading, available materials, and expected pavement design life, which is the number of years the pavement is expected to provide adequate service before it must undergo major rehabilitation.

4.2.2 Rigid Pavements

A rigid pavement is constructed with portland cement concrete (PCC) and aggregates, as shown in Fig. 4.2. As is the case with flexible pavements, the subgrade (the lower layer) is often scarified, blended, and compacted to maximum density. In rigid pavements, the base layer (see Fig. 4.2) is optional, depending on the engineering properties of the subgrade. If the subgrade soil is poor and erodable, then it is advisable to use a base layer. However, if the soil has good engineering properties and drains well, a base layer need not be used. The top layer (wearing surface) is the portland cement concrete slab. Slab length varies from a spacing of 10 to 13 ft (3.05 to 3.96 m) to a spacing of 40 ft (12.19 m) or more.

Transverse contraction joints are built into the pavement to control cracking due to shrinkage of the concrete during the curing process. Load transfer devices, such as dowel bars, are placed in the joints to minimize deflections and reduce stresses near the edges of the slabs. Slab thicknesses for PCC highway pavements usually vary from 8 to 12 inches (roughly 200 to 300 mm), as shown in Fig. 4.2.

4.3 PAVEMENT SYSTEM DESIGN: PRINCIPLES FOR FLEXIBLE PAVEMENTS

The primary function of the pavement structure is to reduce and distribute the surface stresses (contact tire pressure) to an acceptable level at the subgrade (to a level that prevents permanent deformation). A flexible pavement reduces the stresses by distributing the traffic wheel loads over greater and greater areas, through the individual

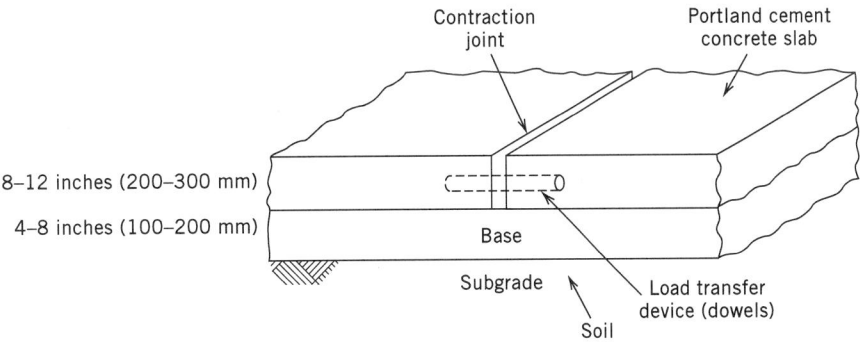

Figure 4.2 Typical rigid-pavement cross section.

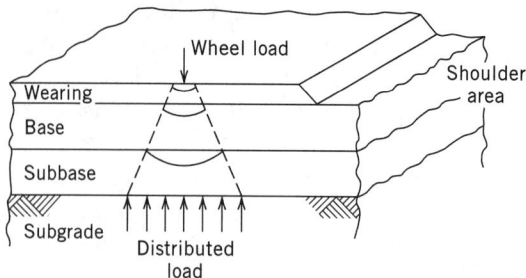

Figure 4.3 Distribution of load on a flexible pavement.

layers, until the stress at the subgrade is at an acceptably low level. The traffic loads are transmitted to the subgrade by aggregate-to-aggregate particle contact. Confining pressures (lateral forces due to material weight) in the subbase and base layers increase the bearing strength of these materials. A cone of distributed loads reduces and spreads the stresses to the subgrade, as shown in Fig. 4.3.

4.3.1 Calculation of Flexible Pavement Stresses and Deflections

To design a pavement structure, one must be able to calculate the stresses and deflections in the pavement system. In the simplest case, the wheel load can be assumed to consist of a point load on a single-layer system, as shown in Fig. 4.4. This type of load and configuration can be analyzed with the Boussinesq solutions that were derived for soils analysis. The Boussinesq theory assumes that the pavement is one layer thick and the material is elastic, homogeneous, and isotropic. The basic equation for the stress at a point in the system is

U.S. Customary	Metric	
$\sigma_z = K \dfrac{P}{z^2}$	$\sigma_z = 1000 K \dfrac{P}{z^2}$	(4.1)

where

σ_z = stress at a point in lb/in² (kPa),
P = wheel load in lb (N),
z = depth of the point in question in inches (mm), and
K = variable defined as

$$K = \frac{3}{2\pi} \frac{1}{[1 + (r/z)^2]^{5/2}} \qquad (4.2)$$

4.3 Pavement System Design: Principles for Flexible Pavements

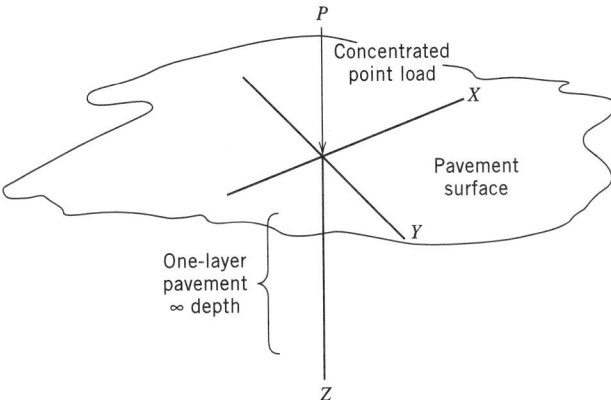

Figure 4.4 Point load on a one-layer pavement.

where r = radial distance in inches (mm) from the centerline of the point load to the point in question.

Although the Boussinesq theory is useful for beginning the study of pavement stress calculations, it is not very representative of pavement system loading and configuration because it applies to a point load on one layer. A more realistic approach is to expand the point load to an elliptical area that represents a tire footprint. The tire footprint can be defined by an equivalent circular area with a radius calculated by

$$\textbf{U.S. Customary} \qquad \textbf{Metric}$$
$$a = \sqrt{\frac{P}{p\pi}} \qquad a = \sqrt{\frac{P}{p\pi/1000}} \qquad (4.3)$$

where

a = equivalent load radius of the tire footprint in inches (mm),
P = tire load in lb (N), and
p = tire pressure in lb/in² (kPa).

The integration of the load from a point to a circular area can be used to determine the stresses and deflections in a one-layer pavement system.

Several researchers have developed influence charts, graphical solutions, and equations for the calculation of stresses and deflections, which take into account a circular load. Ahlvin and Ulery [1962] provided solutions for the evaluation of stresses, strains, and deflections at any point in a homogeneous half-space. Their work makes it easier to analyze a more complex pavement system than that considered in the Boussinesq example. The one-layer equations by Ahlvin and Ulery can be used for a material with any Poisson ratio, μ, which describes the change in width relative to length when a load is applied along the vertical axis. Based on Ahlvin and Ulery's work, the equation for the calculation of vertical stress is

$$\sigma_z = p(A + B) \tag{4.4}$$

The equation for radial-horizontal stress (which is a cause of pavement cracking) is

$$\sigma_r = p[2\mu A + C + (1 - 2\mu)F] \tag{4.5}$$

The equation for deflection is

$$\Delta_z = \frac{p(1+\mu)a}{E}\left[\frac{z}{a}A + (1-\mu)H\right] \tag{4.6}$$

where

σ_z = vertical stress in lb/in^2 (kPa),
σ_r = radial-horizontal stress in lb/in^2 (kPa),
Δ_z = deflection, at depth z, in inches (mm)
p = pressure due to the tire load in lb/in^2 (kPa),
μ = Poisson ratio,
E = modulus of elasticity (known as Young's modulus, the ratio of stress to strain as a load is applied to a material) in lb/in^2 (kPa), and
$A, B, C, F,$ and H = function values, as presented in Table 4.1, that depend on z/a and r/a, the depth in radii and offset distance in radii, respectively,

where

z = depth of the point in question in inches (mm),
r = radial distance in inches (mm) from the centerline of the point load to the point in question, and
a = equivalent load radius of the tire footprint in inches (mm).

Table 4.1 One–Layer Elastic Function Values

	Function A																
	Offset (r/a)																
Depth (z/a)	0	0.2	0.4	0.6	0.8	1	1.2	1.5	2	3	4	5	6	8	10	12	14
0	1.0	1.0	1.0	1.0	1.0	0.5	0	0	0	0	0	0	0	0	0	0	0
0.1	0.90050	0.89748	0.88679	0.86126	0.78797	0.43015	0.09645	0.02787	0.00856	0.00211	0.00084	0.00042	0				
0.2	0.80388	0.79824	0.77884	0.73483	0.63014	0.38269	0.15433	0.05251	0.01680	0.00419	0.00167	0.00083	0.00048	0.0020			
0.3	0.71265	0.70518	0.68316	0.62690	0.52081	0.34375	0.17964	0.07199	0.02440	0.00622	0.00250						
0.4	0.62861	0.62015	0.59241	0.53767	0.44329	0.31048	0.18709	0.08593	0.03118								
0.5	0.55279	0.54403	0.51622	0.46448	0.38390	0.28156	0.18556	0.09499	0.03701	0.01013	0.00407	0.00209	0.00118	0.0053	0.00025	0.00014	0.00009
0.6	0.48550	0.47691	0.45078	0.40427	0.33676	0.25588	0.17952	0.10010									
0.7	0.42654	0.41874	0.39491	0.35428	0.29833	0.21727	0.17124	0.10228	0.04558								
0.8	0.37531	0.36832	0.34729	0.31243	0.26581	0.21297	0.16206	0.10236									
0.9	0.33104	0.32492	0.30669	0.27707	0.23832	0.19488	0.15253	0.10094									
1	0.29289	0.28763	0.27005	0.24697	0.21468	0.17868	0.14329	0.09849	0.05185	0.01742	0.00761	0.00393	0.00226	0.00097	0.00050	0.00029	0.00018
1.2	0.23178	0.22795	0.21662	0.19890	0.17626	0.15101	0.12570	0.09192	0.05260	0.01935	0.00871	0.00459	0.00269	0.00115			
1.5	0.16795	0.16552	0.15877	0.14804	0.13436	0.11892	0.10296	0.08048	0.05116	0.02142	0.01013	0.00548	0.00325	0.00141	0.00073	0.00043	0.00027
2	0.10557	0.10453	0.10140	0.09647	0.09011	0.08269	0.07471	0.06275	0.04496	0.02221	0.01160	0.00659	0.00399	0.00180	0.00094	0.00056	0.00036
2.5	0.07152	0.07098	0.06947	0.06698	0.06373	0.05974	0.05555	0.04880	0.03787	0.02143	0.01221	0.00732	0.00463	0.00214	0.00115	0.00068	0.00043
3	0.05132	0.05101	0.05022	0.04886	0.04707	0.04487	0.04241	0.03839	0.03150	0.01980	0.01220	0.00770	0.00505	0.00242	0.00132	0.00079	0.00051
4	0.02986	0.02976	0.02907	0.02832	0.02802	0.02749	0.02651	0.02490	0.02193	0.01592	0.01109	0.00768	0.00536	0.00282	0.00160	0.00099	0.00065
5	0.01942	0.01938				0.01835			0.01573	0.01249	0.00949	0.00708	0.00527	0.00298	0.00179	0.00113	0.00075
6	0.01361					0.01307			0.01168	0.00983	0.00795	0.00628	0.00492	0.00299	0.00188	0.00124	0.00084
7	0.01005					0.00976			0.00894	0.00784	0.00661	0.00548	0.00445	0.00291	0.00193	0.00130	0.00091
8	0.00772					0.00755			0.00703	0.00635	0.00554	0.00472	0.00398	0.00276	0.00189	0.00134	0.00094
9	0.00612					0.00600			0.00566	0.00520	0.00466	0.00409	0.00353	0.00256	0.00184	0.00133	0.00094
10								0.00477	0.00465	0.00438	0.00397	0.00352	0.00326	0.00241	0.00184	0.00133	0.00096

Table 4.1 One-Layer Elastic Function Values (*continued*)

Function B

Depth (z/a)	Offset (r/a)														
	0	0.2	0.4	0.6	0.8	1	1.2	1.5	2	3	4	5	6	8	10	12	14
0	0	0	0	0	0	0	0	0	0	0	0	0	0	0	0	0	0
0.1	0.09852	0.10140	0.11138	0.13424	0.18796	0.05388	−0.07899	−0.02672	−0.00845	−0.00210	−0.00084	−0.00042					
0.2	0.18857	0.19306	0.20772	0.23524	0.25983	0.08513	−0.07759	−0.04448	−0.01593	−0.00412	−0.00166	−0.00083	−0.00024	−0.00010			
0.3	0.28362	0.26787	0.28018	0.29483	0.27257	0.10757	−0.04316	−0.04999	−0.02166	−0.00599	−0.00245						
0.4	0.32016	0.32259	0.32748	0.32273	0.26925	0.12404	−0.00766	−0.04535	−0.02522								
0.5	0.35777	0.35752	0.35323	0.33106	0.26236	0.13591	0.02165	−0.03455	−0.02651	−0.00991	−0.00388	−0.00199	−0.00116	−0.00049			
0.6	0.37831	0.37531	0.36308	0.32822	0.25411	0.14440	0.04457	−0.02101									
0.7	0.38487	0.37962	0.36072	0.31929	0.24638	0.14986	0.06209	−0.00702	−0.02329								
0.8	0.38091	0.37408	0.35133	0.30699	0.23779	0.15292	0.07530	0.00614									
0.9	0.36962	0.36275	0.33734	0.29299	0.22891	0.15404	0.08507	0.01795									
1	0.35355	0.34553	0.32075	0.27819	0.21978	0.15355	0.09210	0.02814	−0.01005	−0.01115	−0.00608	−0.00344	−0.00210	−0.00092	−0.00048	−0.00028	−0.00018
1.2	0.31485	0.30730	0.28481	0.24836	0.20113	0.14915	0.10002	0.04378	0.00023	−0.00995	−0.00632	−0.00378	−0.00236	−0.00107			
1.5	0.25602	0.25025	0.23338	0.20694	0.17368	0.13732	0.10193	0.05745	0.01385	−0.00669	−0.00600	−0.00401	−0.00265	−0.00126	−0.00068	−0.00040	−0.00026
2	0.17889	0.18144	0.16644	0.15198	0.13375	0.11331	0.09254	0.06371	0.02836	0.00028	−0.00410	−0.00371	−0.00278	−0.00148	−0.00084	−0.00050	−0.00033
2.5	0.12807	0.12633	0.12126	0.11327	0.10298	0.09130	0.07869	0.06022	0.03429	0.00661	−0.00130	−0.00271	−0.00250	−0.00156	−0.00094	−0.00059	−0.00039
3	0.09487	0.09394	0.09099	0.08635	0.08033	0.07325	0.06551	0.05354	0.03511	0.01112	0.00157	−0.00134	−0.00192	−0.00151	−0.00099	−0.00065	−0.00046
4	0.05707	0.05666	0.05562	0.05383	0.05145	0.04773	0.04532	0.03995	0.03066	0.01515	0.00595	0.00155	−0.00029	−0.00109	−0.00094	−0.00068	−0.00050
5	0.03772	0.03760				0.03384			0.02474	0.01522	0.00810	0.00371	0.00132	−0.00043	−0.00070	−0.00068	−0.00049
6	0.02666					0.02468			0.01968	0.01380	0.00867	0.00496	0.00254	0.00028	−0.00037	−0.00047	−0.00045
7	0.01980					0.01868			0.01577	0.01204	0.00842	0.00547	0.00332	0.00093	−0.00002	−0.00029	−0.00037
8	0.01526					0.01459			0.01279	0.01034	0.00779	0.00554	0.00372	0.00141	0.00035	−0.00008	−0.00025
9	0.01212					0.01170			0.01054	0.00888	0.00705	0.00533	0.00386	0.00178	0.00066	0.00012	−0.00012
10								0.00924	0.00879	0.00764	0.00631	0.00501	0.00382	0.00199			

Table 4.1 One-Layer Elastic Function Values (*continued*)

	Function C																
Depth (z/a)	Offset (r/a)																
	0	0.2	0.4	0.6	0.8	1	1.2	1.5	2	3	4	5	6	8	10	12	14
0	0	0	0	0	0	0	0	0	0	0	0	0	0				
0.1	−0.04926	−0.05142	−0.05903	−0.07708	−0.12108	0.02247	0.12007	0.04475	0.01536	0.00403	0.00164	0.00082	0				
0.2	−0.09429	−0.09755	−0.10872	−0.12977	−0.14552	0.02419	0.14896	0.07892	0.02951	0.00796	0.00325	0.00164	0.00094				
0.3	−0.13181	−0.13484	−0.14415	−0.15023	−0.12990	0.01988	0.13394	0.09816	0.04148	0.01169	0.00483						
0.4	−0.16008	−0.16188	−0.16519	−0.15985	−0.11168	0.01292	0.11014	0.10422	0.05067					0.00039			
0.5	−0.17889	−0.17835	−0.17497	−0.15625	−0.09833	0.00483	0.08730	0.10125	0.05690	0.01824	0.00778	0.00399	0.00231	0.00098	0.00050	0.00029	0.00018
0.6	−0.18915	−0.18633	−0.17336	−0.14934	−0.08967	−0.00304	0.06731	0.09313	0.06129								
0.7	−0.19244	−0.18831	−0.17393	−0.14147	−0.08409	−0.01061	0.05028	0.08253									
0.8	−0.19046	−0.18481	−0.16784	−0.13393	−0.08066	−0.01744	0.03582	0.07114									
0.9	−0.18481	−0.17841	−0.16024	−0.12664	−0.07828	−0.02337	0.02359	0.05993									
1	−0.17678	−0.17050	−0.15188	−0.11995	−0.07634	−0.02843	0.01331	0.04939	0.05429	0.02726	0.01333	0.00726	0.00433	0.00188	0.00098	0.00057	0.00036
1.2	−0.15742	−0.15117	−0.13467	−0.10763	−0.07289	−0.03575	−0.00245	0.03107	0.04522	0.02791	0.01467	0.00824	0.00501	0.00221	0.00141	0.00083	0.00039
1.5	−0.12801	−0.12277	−0.11101	−0.09145	−0.06711	−0.04124	−0.01702	0.01088	0.03154	0.02652	0.01570	0.00933	0.00585	0.00266	0.00179	0.00107	0.00069
2	−0.08944	−0.08491	−0.07976	−0.06925	−0.05560	−0.04144	−0.02687	−0.00782	0.01267	0.02070	0.01527	0.01013	0.00585	0.00327	0.00179	0.00107	
2.5	−0.06403	−0.06068	−0.05839	−0.05259	−0.04522	−0.03605	−0.02800	−0.01536	0.00103	0.01340	0.00987	0.00707	0.00569	0.00209	0.00128	0.00083	
3	−0.04744	−0.04560	−0.04339	−0.04089	−0.03642	−0.03130	−0.02587	−0.01748	−0.00528	0.00792	0.01030	0.00888	0.00689	0.00392	0.00232	0.00145	0.00096
4	−0.02854	−0.02737	−0.02562	−0.02585	−0.02421	−0.02112	−0.01964	−0.01586	−0.00956	0.00038	0.00492	0.00602	0.00561	0.00389	0.00254	0.00168	0.00155
5	−0.01886	−0.01810				−0.01568			−0.00939	−0.00293	−0.00128	0.00329	0.00391	0.00341	0.00250	0.00177	0.00127
6	−0.01333					−0.01118			−0.00819	−0.00405	−0.00079	0.00129	0.00234	0.00272	0.00227	0.00173	0.00130
7	−0.00990					−0.00902			−0.00678	−0.00417	−0.00180	−0.00004	0.00113	0.00200	0.00193	0.00161	0.00128
8	−0.00763					−0.00699			−0.00552	−0.00393	−0.00225	−0.00077	0.00029	0.00134	0.00157	0.00143	0.00120
9	−0.00607					−0.00423			−0.00452	−0.00353	−0.00235	−0.00118	−0.00027	0.00082	0.00124	0.00122	0.00110
10								−0.00381	−0.00373	−0.00314	−0.00233	−0.00137	−0.00630	0.00040			

Table 4.1 One-Layer Elastic Function Values (*continued*)

Depth (z/a)	Function F Offset (r/a)														
	0	0.2	0.4	0.6	0.8	1	1.2	1.5	2	3	4	5	6	8	10	12	14
0	0.5	0.5	0.5	0.5	0.5	0	-0.34722	-0.22222	-0.12500	-0.05556	-0.03125	-0.02000	-0.01389	-0.00781	-0.00500	-0.00347	-0.00255
0.1	0.45025	0.44794	0.43981	0.41954	0.35789	0.03817	-0.20800	-0.17612	-0.10950	-0.05151	-0.02961	-0.01917					
0.2	0.40194	0.39781	0.38294	0.34823	0.26215	0.05466	-0.11165	-0.13381	-0.09441	-0.04750	-0.02798	-0.01835	-0.01295	-0.00742			
0.3	0.35633	0.35094	0.34508	0.29016	0.20503	0.06372	-0.05346	-0.09768	-0.08010	-0.04356	-0.02636						
0.4	0.31431	0.30801	0.28681	0.24469	0.17086	0.06848	-0.01818	-0.06835	-0.06684								
0.5	0.27639	0.26997	0.24890	0.20937	0.14752	0.07037	0.00388	-0.04529	-0.05479	-0.03595	-0.02320	-0.01590	-0.00154	-0.00681	-0.00450	-0.00318	-0.00237
0.6	0.24275	0.23444	0.21667	0.18138	0.13042	0.07068	0.01797	-0.02479									
0.7	0.21327	0.20762	0.18956	0.15903	0.11740	0.06963	0.02704	-0.01392	-0.03469								
0.8	0.18765	0.18287	0.16679	0.14053	0.10604	0.06774	0.03277	-0.00365									
0.9	0.16552	0.16158	0.14747	0.12528	0.09664	0.06533	0.03619	0.00408									
1	0.14645	0.14280	0.12395	0.11225	0.08850	0.06256	0.03819	0.00984	-0.01367	-0.01994	-0.01591	-0.01209	-0.00931	-0.00587	-0.00400	-0.00289	-0.00219
1.2	0.11589	0.11360	0.10460	0.09449	0.07486	0.05670	0.03913	0.01716	-0.00452	-0.01491	-0.01337	-0.01068	-0.00844	-0.00550			
1.5	0.08398	0.08196	0.07719	0.06918	0.05919	0.04804	0.03686	0.02177	0.00413	-0.00879	-0.00995	-0.00870	-0.00723	-0.00495	-0.00353	-0.00261	-0.00201
2	0.05279	0.05348	0.04994	0.04614	0.04162	0.03593	0.03029	0.02197	0.01043	-0.00189	-0.00546	-0.00589	-0.00544	-0.00410	-0.00307	-0.00233	-0.00183
2.5	0.03576	0.03673	0.03459	0.03263	0.03014	0.02762	0.02406	0.01927	0.01188	0.00198	-0.00226	-0.00364	-0.00386	-0.00332	-0.00263	-0.00208	-0.00166
3	0.02566	0.02586	0.02255	0.02595	0.02263	0.02097	0.01911	0.01623	0.01144	0.00396	-0.00010	-0.00192	-0.00258	-0.00263	-0.00223	-0.00183	-0.00150
4	0.01493	0.01536	0.01412	0.01259	0.01386	0.01331	0.01256	0.01134	0.00912	0.00508	0.00209	0.00026	-0.00076	-0.00148	-0.00153	-0.00137	-0.00120
5	0.00971	0.01011				0.00905			0.00700	0.00475	0.00277	0.00129	0.00031	-0.00066	-0.00096	-0.00099	-0.00093
6	0.00680					0.00675			0.00538	0.00409	0.00278	0.00170	0.00088	-0.00010	-0.00053	-0.00066	-0.00070
7	0.00503					0.00483			0.00428	0.00346	0.00258	0.00178	0.00114	0.00027	-0.00020	-0.00041	-0.00049
8	0.00386					0.00380			0.00350	0.00291	0.00229	0.00174	0.00125	0.00048	0.00003	-0.00020	-0.00033
9	0.00306					0.00374			0.00291	0.00247	0.00203	0.00163	0.00124	0.00062	0.00020	-0.00005	-0.00019
10								0.00267	0.00246	0.00213	0.00176	0.00149	0.00126	0.00070			

Table 4.1 One-Layer Elastic Function Values (*continued*)

Depth (z/a)	Function H Offset (r/a)														
	0	0.2	0.4	0.6	0.8	1	1.2	1.5	2	3	4	5	6	8	10	12	14
0	2.0	1.97987	1.91751	1.80575	1.62553	1.27319	0.93676	0.71185	0.51671	0.33815	0.25200	0.20045	0.16626	0.12576	0.09918	0.08346	0.07023
0.1	1.80998	1.79018	1.72886	1.61961	1.44711	1.18107	0.92670	0.70888	0.51627	0.33794	0.25184	0.20081					
0.2	1.63961	1.62068	1.56242	1.46001	1.30614	1.09996	0.90098	0.70074	0.51382	0.33726	0.25162	0.20072	0.16688	0.12512			
0.3	1.48806	1.47044	1.40979	1.32442	1.19210	1.02740	0.86726	0.68823	0.50966	0.33638	0.25124						
0.4	1.35407	1.33802	1.28963	1.20822	1.09555	0.96202	0.83042	0.67238	0.50412								
0.5	1.23607	1.22176	1.17894	1.10830	1.01312	0.90298	0.79308	0.65429	0.49278	0.33293	0.24996	0.19982	0.16668	0.12493	0.09996	0.08295	0.07123
0.6	1.13238	1.11998	1.08350	1.02154	0.94120	0.84917	0.75653	0.63469									
0.7	1.04131	1.03037	0.99794	0.91049	0.87742	0.80030	0.72143	0.61442	0.48061								
0.8	0.96125	0.95175	0.92386	0.87928	0.82136	0.75571	0.68809	0.59398									
0.9	0.89072	0.88251	0.85856	0.82616	0.77950	0.71495	0.65677	0.57361									
1	0.82843	0.85005	0.80465	0.76809	0.72587	0.67769	0.62701	0.55364	0.45122	0.31877	0.24386	0.19673	0.16516	0.12394	0.09952	0.08292	0.07104
1.2	0.72410	0.71882	0.70370	0.67937	0.64814	0.61187	0.57329	0.51552	0.43013	0.31162	0.24070	0.19520	0.16369	0.12350	0.09876	0.08270	
1.5	0.60555	0.60233	0.57246	0.57633	0.55559	0.53138	0.50496	0.46379	0.39872	0.29945	0.23495	0.19053	0.16199	0.12281	0.09792	0.08196	0.07064
2	0.47214	0.47022	0.44512	0.45656	0.44502	0.43202	0.41702	0.39242	0.35054	0.27740	0.22418	0.18618	0.15846	0.12124	0.09700	0.08115	0.07026
2.5	0.38518	0.38403	38098	0.37608	0.36940	0.36155	0.35243	0.33698	0.30913	0.25550	0.21208	0.17898	0.15395	0.11928	0.09558	0.08061	0.06980
3	0.32457	0.32403	0.32184	0.31887	0.31464	0.30969	0.30381	0.29364	0.27453	0.23487	0.19977	0.17154	0.14919	0.11694	0.09300	0.07864	0.06897
4	0.24620	0.24588	0.24820	0.25128	0.24168	0.23932	0.23668	0.23164	0.22188	0.19908	0.17640	0.15596	0.13864	0.11172	0.09558	0.07675	0.06848
5	0.19805	0.19785				0.19455			0.18450	0.17080	0.15575	0.14130	0.12785	0.10585	0.08915	0.07452	0.06695
6	0.16554					0.16326			0.15750	0.14868	0.13842	0.12792	0.11778	0.09990	0.08562	0.07452	0.06522
7	0.14217					0.14077			0.13699	0.13097	0.12404	0.11620	0.10843	0.09387	0.08197	0.07210	0.06377
8	0.12448					0.12352			0.12112	0.11680	0.11176	0.10600	0.09976	0.08848	0.07800	0.06928	0.06200
9	0.11079					0.10989			0.10854	0.10548	0.10161	0.09702	0.09234	0.08298	0.07407	0.6678	0.05976
10								0.09900	0.09820	0.09510	0.09290	0.08980	0.08300	0.07710			

101

EXAMPLE 4.1

A tire with 100-lb/in² (689-kPa) air pressure distributes a load over an area with a circular contact radius, a, of 5 inches (127 mm). The pavement was constructed with a material that has a modulus of elasticity of 50,000 lb/in² (345,000 kPa) and a Poisson ratio of 0.45. Calculate the radial-horizontal stress and deflection at a point on the pavement surface under the center of the tire load. Also, calculate the radial-horizontal stress and deflection at a point at a depth of 20 inches (508 mm) and a radial distance of 10 inches (254 mm) from the center of the tire load. (Use Ahlvin and Ulery equations.)

SOLUTION

With $z = 0$ inches and $r = 0$ inches,

$$\frac{z}{a} = \frac{0}{5} = 0 \quad \text{and} \quad \frac{r}{a} = \frac{0}{5} = 0$$

From Table 4.1, for the above values,

$$A = 1.0, \quad B = 0, \quad C = 0, \quad F = 0.5, \quad \text{and} \quad H = 2.0$$

The radial-horizontal stress as calculated from Eq. 4.5 is

$$\sigma_r = p[2\mu A + C + (1 - 2\mu)F]$$
$$= 100\{2(0.45)(1.0) + 0 + [1 - 2(0.45)](0.5)\}$$
$$= \underline{95.0 \text{ lb/in}^2}$$

The deflection as calculated from Eq. 4.6 is

$$\Delta_z = \frac{p(1+\mu)a}{E}\left[\frac{z}{a}A + (1-\mu)H\right]$$
$$= \frac{100(1+0.45)5}{50,000}[0(1) + (1 - 0.45)2.0]$$
$$= \underline{0.0159 \text{ inches}}$$

With $z = 20$ inches and $r = 10$ inches,

$$\frac{z}{a} = \frac{20}{5} = 4 \quad \text{and} \quad \frac{r}{a} = \frac{10}{5} = 2$$

From Table 4.1, for the above values,

$$A = 0.02193, \quad B = 0.03066, \quad C = -0.00956, \quad F = 0.00912, \quad \text{and} \quad H = 0.22188$$

The radial-horizontal stress is

$$\sigma_r = p[2\mu A + C + (1 - 2\mu)F]$$
$$= 100\{2(0.45)(0.02193) + (-0.00956) + [1 - 2(0.45)](0.00912)\}$$
$$= \underline{1.109 \text{ lb/in}^2}$$

The deflection is

$$\Delta_z = \frac{p(1+\mu)a}{E}\left[\frac{z}{a}A + (1-\mu)H\right]$$

$$= \frac{100(1+0.45)5}{50{,}000}[4(0.02193) + (1-0.45)(0.22188)]$$

$$= \underline{\underline{0.003 \text{ inches}}}$$

The Boussinesq theory and the Ahlvin and Ulery equations can be used to calculate the stresses and deflections in a simple pavement system. The utility of these is that they serve as a basis for more complex pavement analysis. With the availability of computers, there have been significant advances in mechanistic pavement analysis methodologies. Pavement systems are represented as multilayer systems that are homogeneous and isotropic with linear responses (the material returns to its original shape after the load is removed). Computer programs allow engineers to input various axle loads, tire pressures, and material properties, and program outputs consist of stresses, strains, and deflections.

4.4 THE AASHTO FLEXIBLE-PAVEMENT DESIGN PROCEDURE

There are several accepted flexible-pavement design procedures available, including the Asphalt Institute method, the National Stone Association procedure, and the Shell procedure. Most of the procedures have been field verified and used by highway agencies for several years. The selection of one procedure over another is usually based on a highway agency's experience and satisfaction with design results.

A widely accepted flexible-pavement design procedure is presented in the *AASHTO Guide for Design of Pavement Structures,* which is published by the American Association of State Highway and Transportation Officials (AASHTO). The procedure was first published in 1972, with the latest revisions in 1993. The 1993 AASHTO design procedure is the same as the 1986 AASHTO procedure, except the new procedure has a revised section for overlay designs. Test data, used for the development of the design procedure, were collected at the AASHO Road Test in Illinois from 1958 to 1960 (AASHO, which stands for American Association of State Highway Officials, was the prior name of AASHTO).

A pavement can be subjected to a number of detrimental effects, including fatigue failures (cracking), which are the result of repeated loading caused by traffic passing over the pavement. The pavement is also placed in an uncontrolled environment that produces temperature extremes and moisture variations. The combination of the environment, traffic loads, material variations, and construction variations requires a comparatively complex set of design procedures to incorporate all of the variables. The AASHTO pavement design procedure meets most of the demands placed on a flexible-pavement design procedure. It considers environment, load, and materials in a methodology that is relatively easy to use. The AASHTO pavement design procedure has been widely accepted throughout the United States and around the world. Details of this procedure are presented in the following sections.

4.4.1 Serviceability Concept

Prior to the AASHO Road Test, there was no real consensus as to the definition of pavement failure. In the eyes of an engineer, pavement failure occurred whenever cracking, rutting, or other surface distresses became visible. In contrast, the motoring public usually associated pavement failure with poor ride quality. Pavement engineers conducting the AASHO Road Test were faced with the task of combining the two failure definitions so that a single design procedure could be used to satisfy both critics. The Pavement Serviceability-Performance Concept was developed by Carey and Irick [1962] to handle the question concerning pavement failure. Carey and Irick considered pavement performance histories and noted that pavements usually begin their service life in excellent condition and deteriorate as traffic loading is applied in conjunction with prevailing environmental conditions. The performance curve is the historical record of the performance of the pavement. Pavement performance, at any point in time, is known as the present serviceability index, or PSI. Examples of pavement performance (or PSI trends) are shown in Fig. 4.5.

At any time, the present serviceability index of a pavement can be measured. It is usually measured by a panel of raters who drive over the pavement section and rate the pavement performance on a scale of 1 to 5, with 5 being the smoothest ride. The accumulation of traffic loads causes the pavement to deteriorate, and, as expected, the serviceability rating drops. At some point, a terminal serviceability index (TSI) is reached. At this point, most raters feel that the pavement is in need of rehabilitation.

Having raters evaluate all of the nation's highways on a continuous basis would be an overwhelming task. Instead, correlations have been made between panel opinions and measured variables such as pavement roughness, rutting, and cracking. Consequently, mechanical devices are now commonly used to determine PSI. It has been found that new pavements usually have an initial PSI rating of approximately 4.2 to 4.5. The point at which pavements are considered to have failed (the TSI) varies by type of highway. Highway facilities such as interstate highways or principal arterials usually have TSIs of 2.5 or 3.0, whereas local roads can have TSIs of 2.0.

4.4.2 Flexible-Pavement Design Equation

At the conclusion of the AASHO Road Test, a regression analysis (see Chapter 8 for a discussion of regression analysis) was performed to determine the interactions of traffic loadings, material properties, layer thickness, and climate. The relationship between axle loads and the thickness index of the pavement system is shown in Fig. 4.6. The thickness index represents a combination of layer thickness and strength coefficients. The term "thickness index" is the same as the term "structural number," which will be discussed later.

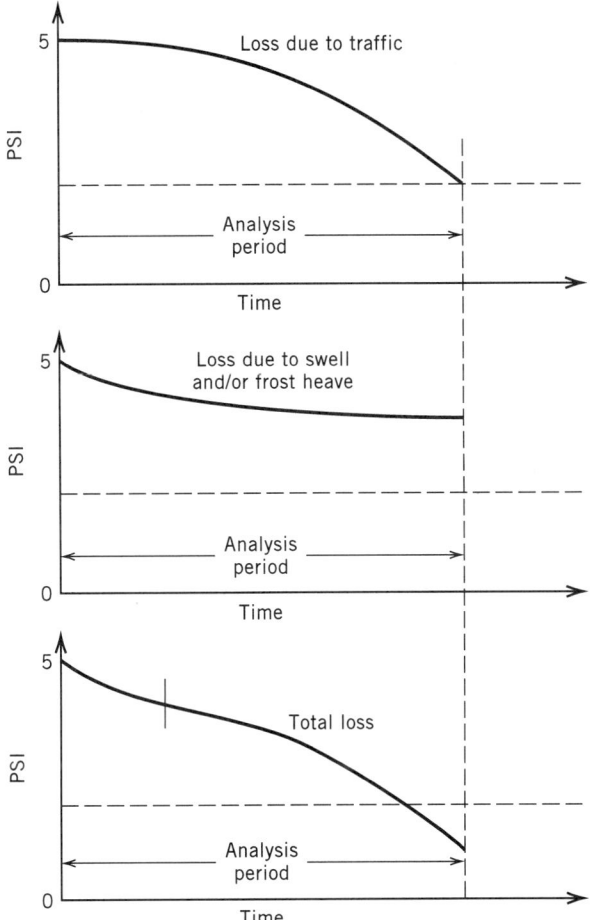

Figure 4.5 Pavement performance trends.
Redrawn from *AASHTO Guide for Design of Pavement Structures*, Washington, DC, The American Association of State Highway and Transportation Officials, copyright 1993. Used by permission.

The relationship shown in Fig. 4.6 can be used to determine the required thickness index of a flexible pavement. For example, assume that a new pavement must withstand one million applications of a 12,000-lb (12-kip, 53.4-kN) single-axle load. Based on the curves, it can be seen that a thickness index of approximately 2.5 is needed to sustain that type of loading. There are many combinations of pavement materials and thicknesses that can provide a thickness index value of 2.5; however, it is the responsibility of the pavement engineer to select a practical and economic combination of the materials to satisfy design inputs. Because of the large number of variables involved in real-world pavement design, the graph shown in Fig. 4.6 quickly loses its utility. Consequently, an equation was developed for flexible-pavement design, which replaces the graph in Fig. 4.6.

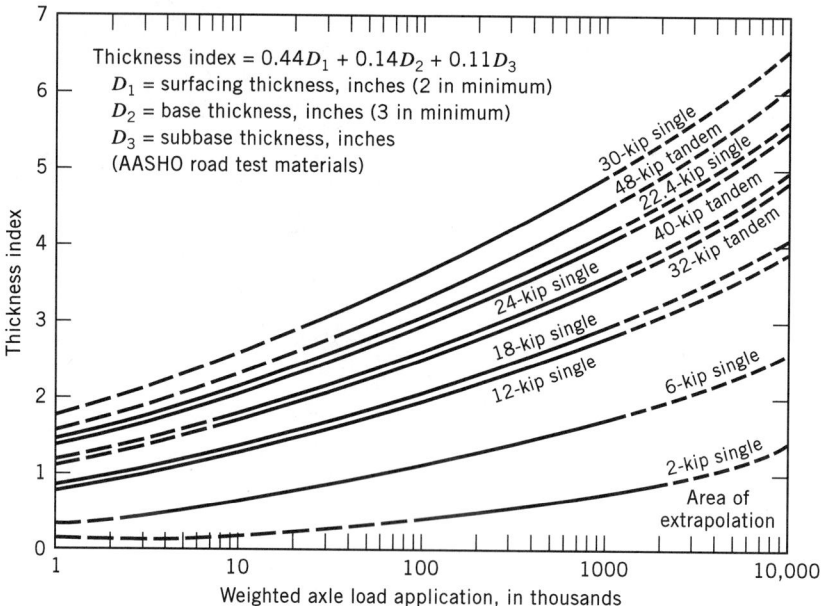

Figure 4.6 AASHO road test thickness index versus axle loads.
Redrawn from *AASHO Road Test Report 5*, Highway Research Board, Special Report 61E, Washington, DC, 1962. Used by permission.

The basic equation for flexible-pavement design given in the 1993 AASHTO design guide permits engineers to determine a structural number (thickness index) necessary to carry a designated traffic loading. The AASHTO equation is

$$\log_{10} W_{18} = Z_R S_o + 9.36[\log_{10}(SN + 1)] - 0.20 \\ + \frac{\log_{10}[\Delta PSI/2.7]}{0.40 + [1094/(SN + 1)^{5.19}]} \\ + 2.32 \log_{10} M_R - 8.07 \quad (4.7)$$

where

W_{18} = 18-kip–equivalent single-axle load,
Z_R = reliability (z-statistic from the standard normal curve),
S_o = overall standard deviation of traffic,
SN = structural number,
ΔPSI = loss in serviceability from the time the pavement is new until it reaches its TSI, and
M_R = soil resilient modulus of the subgrade in lb/in².

4.4 The AASHTO Flexible-Pavement Design Procedure   107

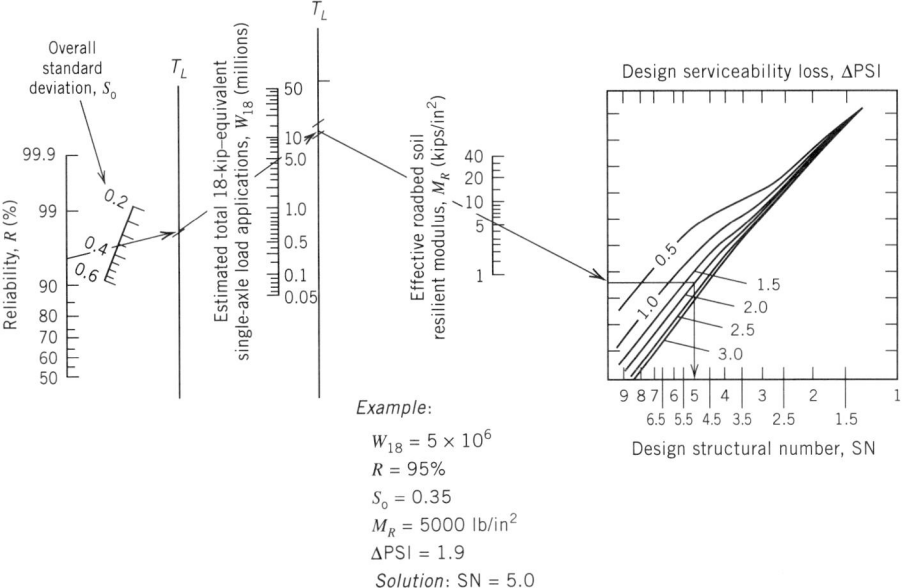

Figure 4.7 Design chart for flexible pavements based on the use of mean values for each input.
Redrawn from *AASHTO Guide for Design of Pavement Structures*, Washington, DC, The American Association of State Highway and Transportation Officials, copyright 1993. Used by permission.

A graphical solution to Eq. 4.7 is shown in Fig. 4.7. Details on the variables that serve as inputs to Eq. 4.7 and Fig. 4.7 are as follows:

W_{18} Automobiles and truck traffic provide a wide range of vehicle axle types and axle loads. If one were to attempt to account for the variety of traffic loadings encountered on a pavement, this input variable would require a significant amount of data collection and design evaluation. Instead, the problem of handling mixed traffic loading is solved with the adoption of a standard 18-kip (80.1-kN)–equivalent single-axle load (ESAL). The idea is to determine the impact of any axle load on the pavement in terms of the equivalent amount of pavement impact that an 18-kip single-axle load would have. For example, if a 44-kip (195.7-kN) tandem-axle (double-axle) load has 2.88 times the impact on pavement structure as an 18-kip single-axle load, 2.88 would be the W_{18} value assigned to this tandem-axle load. The AASHO Road Test also found that the 18-kip (80.1-kN)–equivalent axle load is a function of the terminal serviceability index of the pavement structure. The axle-load equivalency factors for flexible pavement design, with a TSI of 2.5, are presented in Tables 4.2 (for single axles), 4.3 (for tandem axles), and 4.4 (for triple axles).

Table 4.2 Axle-Load Equivalency Factors for Flexible Pavements, Single Axles, and TSI = 2.5

Axle load (kips)	Pavement structural number (SN)					
	1	2	3	4	5	6
2	0.0004	0.0004	0.0003	0.0002	0.0002	0.0002
4	0.003	0.004	0.004	0.003	0.002	0.002
6	0.011	0.017	0.017	0.013	0.010	0.009
8	0.032	0.047	0.051	0.041	0.034	0.031
10	0.078	0.102	0.118	0.102	0.088	0.080
12	0.168	0.198	0.229	0.213	0.189	0.176
14	0.328	0.358	0.399	0.388	0.360	0.342
16	0.591	0.613	0.646	0.645	0.623	0.606
18	1.00	1.00	1.00	1.00	1.00	1.00
20	1.61	1.57	1.49	1.47	1.51	1.55
22	2.48	2.38	2.17	2.09	2.18	2.30
24	3.69	3.49	3.09	2.89	3.03	3.27
26	5.33	4.99	4.31	3.91	4.09	4.48
28	7.49	6.98	5.90	5.21	5.39	5.98
30	10.3	9.5	7.9	6.8	7.0	7.8
32	13.9	12.8	10.5	8.8	8.9	10.0
34	18.4	16.9	13.7	11.3	11.2	12.5
36	24.0	22.0	17.7	14.4	13.9	15.5
38	30.9	28.3	22.6	18.1	17.2	19.0
40	39.3	35.9	28.5	22.5	21.1	23.0
42	49.3	45.0	35.6	27.8	25.6	27.7
44	61.3	55.9	44.0	34.0	31.0	33.1
46	75.5	68.8	54.0	41.4	37.2	39.3
48	92.2	83.9	65.7	50.1	44.5	46.5
50	112.0	102.0	79.0	60.0	53.0	55.0

Source: AASHTO Guide for Design of Pavement Structures, The American Association of State Highway and Transportation Officials, Washington, DC, copyright 1993. Used by permission.

Z_R Represents the probability that serviceability will be maintained at adequate levels from a user's point of view throughout the design life of the facility. This factor estimates the likelihood that the pavement will perform at or above the TSI level during the design period, and accounts for the inherent uncertainty in design. Equation 4.7 uses the z-statistic, which is obtained from the cumulative probabilities of the standard normal distribution (a normal distribution with mean equal to 0 and variance equal to 1). The z-statistics corresponding to various probability levels are given in Table 4.5. In the flexible-pavement-design nomograph (Fig. 4.7), the probabilities (in percent) are used directly (instead of Z_R as in the case of Eq. 4.7), and these percent probabilities are denoted R, the reliability (see Table 4.5).

Highways such as interstates and major arterials, which are costly to reconstruct (have their pavements rehabilitated) because of resulting traffic delay and disruption, require a high reliability level, whereas local roads, which will have lower impacts on users in the event of pavement rehabilitation, do not. Typical reliability values for interstate highways are 90% or higher, whereas local roads can have a reliability as low as 50%.

4.4 The AASHTO Flexible-Pavement Design Procedure

Table 4.3 Axle-Load Equivalency Factors for Flexible Pavements, Tandem Axles, and TSI = 2.5

Axle load (kips)	Pavement structural number (SN)					
	1	2	3	4	5	6
2	0.0001	0.0001	0.0001	0.0000	0.0000	0.0000
4	0.0005	0.0005	0.0004	0.0003	0.0003	0.0002
6	0.002	0.002	0.002	0.001	0.001	0.001
8	0.004	0.006	0.005	0.004	0.003	0.003
10	0.008	0.013	0.011	0.009	0.007	0.006
12	0.015	0.024	0.023	0.018	0.014	0.013
14	0.026	0.041	0.042	0.033	0.027	0.024
16	0.044	0.065	0.070	0.057	0.047	0.043
18	0.070	0.097	0.109	0.092	0.077	0.070
20	0.107	0.141	0.162	0.141	0.121	0.110
22	0.160	0.198	0.229	0.207	0.180	0.166
24	0.231	0.273	0.315	0.292	0.260	0.242
26	0.327	0.370	0.420	0.401	0.364	0.342
28	0.451	0.493	0.548	0.534	0.495	0.470
30	0.611	0.648	0.703	0.695	0.658	0.633
32	0.813	0.843	0.889	0.887	0.857	0.834
34	1.06	1.08	1.11	1.11	1.09	1.08
36	1.38	1.38	1.38	1.38	1.38	1.38
38	1.75	1.73	1.69	1.68	1.70	1.73
40	2.21	2.16	2.06	2.03	2.08	2.14
42	2.76	2.67	2.49	2.43	2.51	2.61
44	3.41	3.27	2.99	2.88	3.00	3.16
46	4.18	3.98	3.58	3.40	3.55	3.79
48	5.08	4.80	4.25	3.98	4.17	4.49
50	6.12	5.76	5.03	4.64	4.86	5.28
52	7.33	6.87	5.93	5.38	5.63	6.17
54	8.72	8.14	6.95	6.22	6.47	7.15
56	10.3	9.6	8.1	7.2	7.4	8.2
58	12.1	11.3	9.4	8.2	8.4	9.4
60	14.2	13.1	10.9	9.4	9.6	10.7
62	16.5	15.3	12.6	10.7	10.8	12.1
64	19.1	17.6	14.5	12.2	12.2	13.7
66	22.1	20.3	16.6	13.8	13.7	15.4
68	25.3	23.3	18.9	15.6	15.4	17.2
70	29.0	26.6	21.5	17.6	17.2	19.2
72	33.0	30.3	24.4	19.8	19.2	21.3
74	37.5	34.4	27.6	22.2	21.3	23.6
76	42.5	38.9	31.1	24.8	23.7	26.1
78	48.0	43.9	35.0	27.8	26.2	28.8
80	54.0	49.4	39.2	30.9	29.0	31.7
82	60.6	55.4	43.9	34.4	32.0	34.8
84	67.8	61.9	49.0	38.2	35.3	38.1
86	75.7	69.1	54.5	42.3	38.8	41.7
88	84.3	76.9	60.6	46.8	42.6	45.6
90	93.7	85.4	67.1	51.7	46.8	49.7

Source: AASHTO Guide for Design of Pavement Structures, The American Association of State Highway and Transportation Officials, Washington, DC, copyright 1993. Used by permission.

Table 4.4 Axle-Load Equivalency Factors for Flexible Pavements, Triple Axles, and TSI = 2.5

Axle load (kips)	Pavement structural number (SN)					
	1	2	3	4	5	6
2	0.0000	0.0000	0.0000	0.0000	0.0000	0.0000
4	0.0002	0.0002	0.0002	0.0001	0.0001	0.0001
6	0.0006	0.0007	0.0005	0.0004	0.0003	0.0003
8	0.001	0.002	0.001	0.001	0.001	0.001
10	0.003	0.004	0.003	0.002	0.002	0.002
12	0.005	0.007	0.006	0.004	0.003	0.003
14	0.008	0.012	0.010	0.008	0.006	0.006
16	0.012	0.019	0.018	0.013	0.011	0.010
18	0.018	0.029	0.028	0.021	0.017	0.016
20	0.027	0.042	0.042	0.032	0.027	0.024
22	0.038	0.058	0.060	0.048	0.040	0.036
24	0.053	0.078	0.084	0.068	0.057	0.051
26	0.072	0.103	0.114	0.095	0.080	0.072
28	0.098	0.133	0.151	0.128	0.109	0.099
30	0.129	0.169	0.195	0.170	0.145	0.133
32	0.169	0.213	0.247	0.220	0.191	0.175
34	0.219	0.266	0.308	0.281	0.246	0.228
36	0.279	0.329	0.379	0.352	0.313	0.292
38	0.352	0.403	0.461	0.436	0.393	0.368
40	0.439	0.491	0.554	0.533	0.487	0.459
42	0.543	0.594	0.661	0.644	0.597	0.567
44	0.666	0.714	0.781	0.769	0.723	0.692
46	0.811	0.854	0.918	0.911	0.868	0.838
48	0.979	1.015	1.072	1.069	1.033	1.005
50	1.17	1.20	1.24	1.25	1.22	1.20
52	1.40	1.41	1.44	1.44	1.43	1.41
54	1.66	1.66	1.66	1.66	1.66	1.66
56	1.95	1.93	1.90	1.90	1.91	1.93
58	2.29	2.25	2.17	2.16	2.20	2.24
60	2.67	2.60	2.48	2.44	2.51	2.58
62	3.09	3.00	2.82	2.76	2.85	2.95
64	3.57	3.44	3.19	3.10	3.22	3.36
66	4.11	3.94	3.61	3.47	3.62	3.81
68	4.71	4.49	4.06	3.88	4.05	4.30
70	5.38	5.11	4.57	4.32	4.52	4.84
72	6.12	5.79	5.13	4.80	5.03	5.41
74	6.93	6.54	5.74	5.32	5.57	6.04
76	7.84	7.37	6.41	5.88	6.15	6.71
78	8.83	8.28	7.14	6.49	6.78	7.43
80	9.92	9.28	7.95	7.15	7.45	8.21
82	11.1	10.4	8.8	7.9	8.2	9.0
84	12.4	11.6	9.8	8.6	8.9	9.9
86	13.8	12.9	10.8	9.5	9.8	10.9
88	15.4	14.3	11.9	10.4	10.6	11.9
90	17.1	15.8	13.2	11.3	11.6	12.9

Source: AASHTO Guide for Design of Pavement Structures, The American Association of State Highway and Transportation Officials, Washington, DC, copyright 1993. Used by permission.

Table 4.5 Cumulative Percent Probabilities of Reliability, R, of the Standard Normal Distribution, and Corresponding Z_R

R	0	1	2	3	4	5	6	7	8	9	9.5	9.9
90	−1.282	−1.341	−1.405	−1.476	−1.555	−1.645	−1.751	−1.881	−2.054	−2.326	−2.576	−3.080
80	−0.842	−0.878	−0.915	−0.954	−0.994	−1.036	−1.080	−1.126	−1.175	−1.227	−1.253	−1.272
70	−0.524	−0.553	−0.583	−0.613	−0.643	−0.675	−0.706	−0.739	−0.772	−0.806	−0.824	−0.838
60	−0.253	−0.279	−0.305	−0.332	−0.358	−0.385	−0.412	−0.440	−0.468	−0.496	−0.510	−0.522
50	0	−0.025	−0.050	−0.075	−0.100	−0.125	−0.151	−0.176	−0.202	−0.228	−0.241	−0.251

Example: To be 95% confident that the pavement will remain at or above its TSI ($R = 95$ for use in Fig. 4.7), a Z_R value of −1.645 would be used in Eq. 4.7 (and in Eq. 4.19).

S_o The overall standard deviation, S_o, takes into account the designers' inability to accurately estimate the variation in future 18-kip (80.1-kN)–equivalent axle loads, and the statistical error in the equations resulting from variability in materials and construction practices. Typical values of S_o are on the order of 0.30 to 0.50.

SN The structural number, SN, represents the overall structural requirement needed to sustain the design's traffic loadings. The structural number is discussed further in Section 4.4.3.

ΔPSI The amount of serviceability loss over the life of the pavement, ΔPSI, is determined during the pavement design process. The engineer must decide on the final PSI level for a particular pavement. Loss of serviceability is caused by pavement roughness, cracking, patching, and rutting. As pavement distress increases, serviceability decreases. If the design is for a pavement with heavy traffic loads, such as an interstate highway, then the serviceability loss may only be 1.2 (an initial PSI of 4.2 and a TSI of 3.0), whereas a low-volume road can be allowed to deteriorate further, with a possible total serviceability loss of 2.7 or more.

M_R The soil resilient modulus, M_R, is used to reflect the engineering properties of the subgrade (the soil). Each time a vehicle passes over pavement, stresses are developed in the subgrade. After the load passes, the subgrade soil relaxes and the stress is relieved. The resilient modulus test is used to determine the properties of the soil under this repeated load. The resilient modulus can be determined by AASHTO test method T274. Measurement of the resilient modulus is not performed by all transportation agencies; therefore, a relationship between M_R and the California bearing ratio (CBR) has been determined. The CBR has been widely used to determine the supporting characteristics of soils since the mid-1930s, and a significant amount of historical information is available. The CBR is the ratio of the load-bearing capacity of the soil to the load-bearing capacity of a high-quality aggregate, multiplied by 100. The relationship, used to provide a very basic approximation of M_R (in lb/in²) from a known CBR, is

$$M_R = 1500 \times \text{CBR} \qquad (4.8)$$

The coefficient of 1500 in Eq. 4.8 is used for CBR values less than 10. Caution must be exercised in applying this equation to higher CBRs because the coefficient (the value 1500 shown in Eq. 4.8) has a range of 750 to 3000.

4.4.3 Structural Number

The objective of Eq. 4.7 and the nomograph in Fig. 4.7 is to determine a required structural number for given axle loadings, reliability, overall standard deviation, change in PSI, and soil resilient modulus. As previously mentioned, there are many pavement material combinations and thicknesses that will provide satisfactory pavement service life. The following equation can be used to relate individual material types and thicknesses to the structural number:

$$SN = a_1 D_1 + a_2 D_2 M_2 + a_3 D_3 M_3 \qquad (4.9)$$

where

a_1, a_2, a_3 = structural-layer coefficients of the wearing surface, base, and subbase layers, respectively,

D_1, D_2, D_3 = thickness of the wearing surface, base, and subbase layers in inches, respectively, and

M_2, M_3 = drainage coefficients for the base and subbase, respectively.

Values for the structural-layer coefficients for various types of material are presented in Table 4.6. Drainage coefficients are used to modify the thickness of the lower pavement layers (base and subbase) to take into account a material's drainage characteristics. A value of 1.0 for a drainage coefficient represents a material with good drainage characteristics (a sandy material). A soil such as clay does not drain very well and, consequently, will have a lower drainage coefficient (less than 1.0) than a sandy material. The reader is referred to [AASHTO 1993] for further information on drainage coefficients.

Because there are many combinations of structural-layer coefficients and thicknesses that solve Eq. 4.9, some guidelines are used to narrow the number of solutions. Experience has shown that wearing layers are typically 2 to 4 inches (50.8 to 101.6 mm) thick, whereas subbases and bases range from 4 to 10 inches (101.6 to 254.0 mm) thick.

Table 4.6 Structural-Layer Coefficients

Pavement component	Coefficient
Wearing surface	
Sand-mix asphaltic concrete	0.35
Hot-mix asphaltic concrete	0.44
Base	
Crushed stone	0.14
Dense-graded crushed stone	0.18
Soil cement	0.20
Emulsion/aggregate-bituminous	0.30
Portland cement/aggregate	0.40
Lime-pozzolan/aggregate	0.40
Hot-mix asphaltic concrete	0.40
Subbase	
Crushed stone	0.11

4.4 The AASHTO Flexible-Pavement Design Procedure

Knowing which of the materials is the most costly per inch (mm) of depth will assist with the determination of an initial layer thickness.

EXAMPLE 4.2

A pavement is to be designed to last 10 years. The initial PSI is 4.2 and the TSI (the final PSI) is determined to be 2.5. The subgrade has a soil resilient modulus of 15,000 lb/in^2 (103,430 kPa). Reliability is 95% with an overall standard deviation of 0.4. For design, the daily car, pickup truck, and light van traffic is 30,000, and the daily truck traffic consists of 1000 passes of single-unit trucks with two single axles and 350 passes of tractor semi-trailer trucks with single, tandem, and triple axles. The axle weights are

cars, pickups, light vans = two 2000-lb (8.9-kN) single axles
single-unit truck = 8000-lb (35.6-kN) steering, single axle
= 22,000-lb (97.9-kN) drive, single axle
tractor semi-trailer truck = 10,000-lb (44.5-kN) steering, single axle
= 16,000-lb (71.2-kN) drive, tandem axle
= 44,000-lb (195.7-kN) trailer, triple axle

M_2 and M_3 are equal to 1.0 for the materials in the pavement structure. Four inches (101.6 mm) of hot-mix asphalt is to be used as the wearing surface and 10 inches (254 mm) of crushed stone as the subbase. Determine the thickness required for the base if soil cement is the material to be used.

SOLUTION

Because the axle-load equivalency factors presented in Tables 4.2, 4.3, and 4.4 are a function of the structural number (SN), we have to assume an SN to start the problem (later we will arrive at a structural number and check to make sure that it is consistent with our assumed value). A typical assumption is to let SN = 4. Given this, the 18-kip–equivalent single-axle load (18-kip ESAL) for cars, pickups, and light vans is

2-kip single-axle equivalent = 0.0002 (Table 4.2)

This gives an 18-kip ESAL total of 0.0004 for each vehicle. For single-unit trucks,

8-kip single-axle equivalent = 0.041 (Table 4.2)
22-kip single-axle equivalent = 2.090 (Table 4.2)

This gives an 18-kip ESAL total of 2.131 for single-unit trucks. For tractor semi-trailer trucks,

10-kip single-axle equivalent = 0.102 (Table 4.2)
16-kip tandem-axle equivalent = 0.057 (Table 4.3)
44-kip triple-axle equivalent = 0.769 (Table 4.4)

This gives an 18-kip ESAL total of 0.928 for tractor semi-trailer trucks. Note the comparatively small effect of cars and other light vehicles in terms of the 18-kip ESAL. This small effect underscores the nonlinear relationship between axle loads and pavement damage. For example, looking at Table 4.2 with SN = 4, a 36-kip single-axle load has 14.4 times the impact on pavement as an 18-kip single-axle load (twice the weight has 14.4 times the impact).

Given the computed 18-kip ESAL, the daily traffic on this highway produces an 18-kip ESAL total of 2467.8 ($0.0004 \times 30{,}000 + 2.131 \times 1000 + 0.928 \times 350$). Traffic (total axle accumulations) over the 10-year design period will be

$$2467.8 \times 365 \times 10 = 9{,}007{,}470 \text{ 18-kip ESAL}$$

With an initial PSI of 4.2 and a TSI of 2.5, ΔPSI = 1.7. Solving Eq. 4.7 for SN (using an equation solver on a calculator or computer) with $Z_R = -1.645$ (which corresponds to R = 95%, as shown in Table 4.5) gives SN = 3.94 (Fig. 4.7 can also be used to arrive at an approximate solution for SN). Note that this is very close to the value that was assumed (SN = 4.0) to get the load equivalency factors from Tables 4.2, 4.3, and 4.4. If Eq. 4.7 gave SN = 5, we would go back and recompute total axle accumulations using the SN of 5 to read the axle-load equivalency factors in Tables 4.2, 4.3, and 4.4. Usually one iteration of this type is all that is needed. Later, Example 4.5 will provide a demonstration of this type of iteration.

Given that SN = 3.94, Eq. 4.9 can be applied with $a_1 = 0.44$ (surface course, hot-mix asphalt, Table 4.6), $a_2 = 0.20$ (base course, soil cement, Table 4.6), and $a_3 = 0.11$ (subbase, crushed stone, Table 4.6), $M_2 = 1.0$ (given), $M_3 = 1.0$ (given), $D_1 = 4.0$ inches (given), and $D_3 = 10.0$ inches (given).

$$SN = a_1 D_1 + a_2 D_2 M_2 + a_3 D_3 M_3$$

$$3.94 = 0.44(4) + 0.20 D_2(1.0) + 0.11(10.0)(1.0)$$

Solving for D_2 gives $D_2 = 5.4$ inches. Using $D_2 = \underline{5.5 \text{ inches}}$ would be a conservative estimate and allow for variations in construction. Rounding up to the nearest 0.5 inch is a safe practice.

EXAMPLE 4.3

A flexible pavement is constructed with 4 inches (101.6 mm) of hot-mix asphalt wearing surface, 8 inches (203.2 mm) of emulsion/aggregate-bituminous base, and 8 inches (203.2 mm) of crushed stone subbase. The subgrade has a soil resilient modulus of 10,000 lb/in^2 (68,950 kPa), and M_2 and M_3 are equal to 1.0 for the materials in the pavement structure. The overall standard deviation is 0.5, the initial PSI is 4.5, and the TSI is 2.5. The daily traffic has 1080 20-kip (89.0-kN) single axles, 400 24-kip (106.8-kN) single axles, and 680 40-kip (177.9-kN) tandem axles. How many years would you estimate this pavement would last (how long before its PSI drops below a TSI of 2.5) if you wanted to be 90% confident that your estimate was not too high, and if you wanted to be 99% confident that your estimate was not too high?

SOLUTION

The pavement's structural number is determined from Eq. 4.9 (using Table 4.6) as

$$SN = a_1 D_1 + a_2 D_2 M_2 + a_3 D_3 M_3$$
$$5.04 = 0.44(4) + 0.30(8)(1.0) + 0.11(8.0)(1.0)$$

For the daily axle loads, the equivalency factors (reading axle equivalents from Tables 4.2 and 4.3 while using SN = 5, which is very close to the 5.04 computed above) are

20-kip single-axle equivalent = 1.51 (Table 4.2)
24-kip single-axle equivalent = 3.03 (Table 4.2)
40-kip tandem-axle equivalent = 2.08 (Table 4.3)

Thus the total daily 18-kip ESAL is

$$\text{daily } W_{18} = 1.51(1080) + 3.03(400) + 2.08(680) = 4257.2 \text{ 18-kip ESAL}$$

Applying Eq. 4.7, with $S_o = 0.5$, SN = 5.04, ΔPSI = 2.0 (4.5 − 2.5), and $M_R = 10{,}000$ lb/in², we find that at $R = 90\%$ ($Z_R = -1.282$ for purposes of Eq. 4.7, as shown in Table 4.5), W_{18} is 26,128,077. Therefore, the number of years is

$$\text{years} = \frac{26{,}128{,}077}{365 \times 4257.2}$$
$$= \underline{\underline{16.82}}$$

Similarly, with $R = 99\%$ ($Z_R = -2.326$ for purposes of Eq. 4.7, as shown in Table 4.5), W_{18} is 7,854,299, so the number of years is

$$\text{years} = \frac{7{,}854{,}299}{365 \times 4257.2}$$
$$= \underline{\underline{5.05}}$$

These results show that you can be 99% confident that the pavement will last (have a PSI above 2.5) at least 5.05 years, and you can be 90% confident that it would have a PSI above 2.5 for 16.82 years. This example demonstrates the large impact that the chosen reliability value can have on pavement design.

4.5 PAVEMENT SYSTEM DESIGN: PRINCIPLES FOR RIGID PAVEMENTS

Rigid pavements distribute wheel loads by the beam action of the portland cement concrete (PCC) slab, which is made of a material that has a high modulus of elasticity, on the order of 4 to 5 million lb/in² (27.6 to 34.5 GPa). This beam action (see

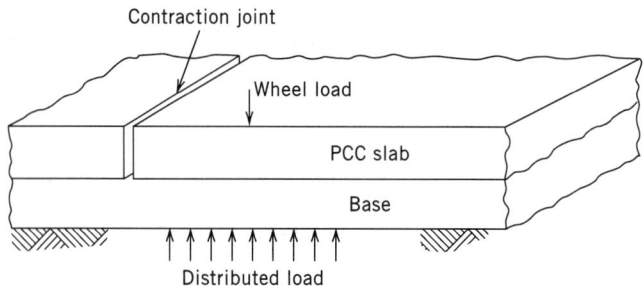

Figure 4.8 Beam action of a rigid pavement.

Fig. 4.8) distributes the wheel loads over a large area of the pavement, thus reducing the high stresses experienced at the surface of the pavement to a level that is acceptable to the subgrade soil.

4.5.1 Calculation of Rigid-Pavement Stresses and Deflections

H. M. Westergaard [1926] presented an important theoretical analysis for rigid pavements. He assumed that the PCC slabs act as homogeneous, isotropic, and elastic solids, and that subgrades react much like a liquid. As a result, the deflections of the slab at any point are a function of the load and the modulus of the subgrade reaction (which is related to the subgrade CBR and soil resilient modulus). Using U.S. Customary units (the system in which the equations were originally derived),

$$\Delta = \frac{k}{p} \qquad (4.10)$$

where

Δ = slab deflection in inches,
k = modulus of subgrade reaction in lb/in^3, and
p = reactive pressure in lb/in^2.

The modulus k is assumed to be constant at each point under the slab and independent of deflection. The Westergaard equations were developed for three loading cases: an interior load, an edge load, and a corner load, as shown in Fig. 4.9. The equations have been referenced throughout the pavement design literature to the point that some equations have been miscopied and misused. Even Dr. Westergaard had to correct his equations in later publications. In more recent work, Ioannides, Thompson, and Barenberg [1985] reconsidered the Westergaard solutions and compared them with finite element analysis. Their work has refined and corrected the original solutions. These closed-form solutions provide the foundation for rigid-pavement analysis.

4.5 Pavement System Design: Principles for Rigid Pavements

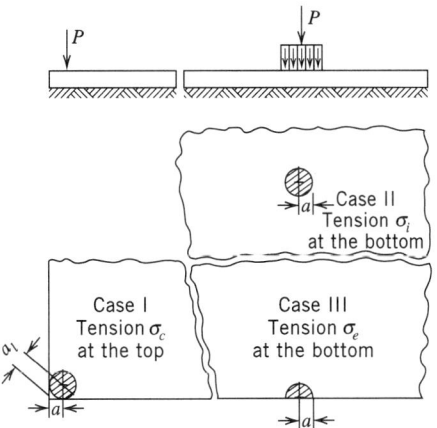

Figure 4.9 Westergaard loading cases. Redrawn from "Computation of Stresses in Concrete Roads," by H. M. Westergaard, Highway Research Board Proceedings of the Fifth Annual Meeting, Washington, DC, 1926. Used by permission.

The Westergaard solutions for stresses and deflections as revised by Ioannides, Thompson, and Barenberg are presented here. For the interior loading,

$$\sigma_i = \frac{3P(1+\mu)}{2\pi h^2}\left[\ln\left(\frac{2l}{a}\right) + 0.5 - \gamma\right] + \frac{3P(1+\mu)}{64h^2}\left(\frac{a}{l}\right)^2 \quad (4.11)$$

$$\Delta_i = \frac{P}{8kl^2}\left\{1 + \left(\frac{1}{2\pi}\right)\left[\ln\left(\frac{a}{2l}\right) + \gamma - \frac{5}{4}\right]\left(\frac{a}{l}\right)^2\right\} \quad (4.12)$$

where

σ_i = interior bending stress in lb/in^2,
Δ_i = interior deflection in inches,
P = total load in lb,
μ = Poisson ratio,
h = slab thickness in inches,
k = modulus of subgrade reaction in lb/in^3,
a = radius of circular load in inches (the tire footprint radius),
γ = Euler's constant, equal to 0.577215, and
l = radius of relative stiffness (a measure of the slab thickness in inches), defined as

$$l = \left[\frac{Eh^3}{12(1-\mu^2)k}\right]^{0.25} \quad (4.13)$$

where

E = modulus of elasticity in lb/in^2, and
Other terms are as defined previously.

Note that the radius of the tire footprint, in U.S. Customary units, is

$$a = \sqrt{\frac{P}{p\pi}} \tag{4.14}$$

where

a = equivalent load radius of the tire footprint in inches,
P = tire load in lb, and
p = tire pressure in lb/in².

For the edge loading,

$$\sigma_e = 0.529(1 + 0.54\mu)\left(\frac{P}{h^2}\right)\left[\log_{10}\left(\frac{Eh^3}{ka^4}\right) - 0.71\right] \tag{4.15}$$

$$\Delta_e = 0.408(1 + 0.4\mu)\left(\frac{P}{kl^2}\right) \tag{4.16}$$

For the corner loading,

$$\sigma_c = \frac{3P}{h^2}\left[1 - \left(\frac{a_l}{l}\right)^{0.72}\right] \tag{4.17}$$

$$\Delta_c = \frac{P}{kl^2}\left[1.205 - 0.69\left(\frac{a_l}{l}\right)\right] \tag{4.18}$$

where

a_l = distance to the point of action of the resulting load on a common angle bisection at the slab corner as shown in Fig. 4.9, equal to $\sqrt{2} \times a$, and
Other terms are as defined previously.

EXAMPLE 4.4

A 15,000-lb (66.72-kN) wheel load is placed on a portland cement concrete (PCC) slab that is 10.0 inches (254.0 mm) thick. The concrete has a modulus of elasticity of 4,500,000 lb/in² (31.0 GPa) with a Poisson ratio of 0.18. The modulus of subgrade reaction is 200 lb/in³ (0.0543 N/mm³). Tire pressure is 100 lb/in² (689.5 kPa). Using the revised Westergaard equations, calculate the stress and deflection if the load is placed on the corner of the slab.

SOLUTION

Begin by computing the radius of relative stiffness, l, using Eq. 4.13:

$$l = \left[\frac{Eh^3}{12(1-\mu^2)k}\right]^{0.25}$$

$$= \left[\frac{4,500,000(10)^3}{12(1-0.18^2)200}\right]^{0.25}$$

$$= 37.31$$

To determine a_l, Eq. 4.14 is used:

$$a = \sqrt{\frac{P}{p\pi}} = \sqrt{\frac{15,000}{100\pi}} = 6.91 \text{ inches}$$

which gives $a_l = 9.77$ inches ($\sqrt{2} \times 6.91$). Substituting values into Eq. 4.17 to get the corner stress,

$$\sigma_c = \frac{3P}{h^2}\left[1-\left(\frac{a_l}{l}\right)^{0.72}\right]$$

$$= \frac{3(15,000)}{10^2}\left[1-\left(\frac{9.77}{37.31}\right)^{0.72}\right]$$

$$= \underline{278.51 \text{ lb/in}^2}$$

Substituting values into Eq. 4.18 to get the corner deflection,

$$\Delta_c = \frac{P}{kl^2}\left[1.205 - 0.69\left(\frac{a_l}{l}\right)\right]$$

$$= \frac{15,000}{200(37.31)^2}\left[1.205 - 0.69\left(\frac{9.77}{37.31}\right)\right]$$

$$= \underline{0.055 \text{ inches}}$$

4.6 THE AASHTO RIGID-PAVEMENT DESIGN PROCEDURE

The design procedure for rigid pavements presented in the AASHTO design guide is also based on the field results of the AASHO Road Test. The AASHTO design procedure is applicable to jointed plain (no steel reinforcement in the slab), reinforced (with welded wire fabric reinforcement), and continuously reinforced (steel bars for longitudinal and transverse reinforcement) pavements. Because faulting, which is a distress due to different slab elevations, was not a failure consideration in the AASHO Road Test, the design of nondoweled joints must be checked with a procedure other than the one presented here.

The design procedure for rigid pavements is based on a selected reduction in serviceability and is similar to the procedure for flexible pavements. However, instead of measuring pavement strength by using a structural number, the thickness of the PCC slab is the measure of strength. The regression equation that is used (in U.S. Customary units) to determine the thickness of a rigid-pavement PCC slab is

$$\log_{10} W_{18} = Z_R S_o + 7.35[\log_{10}(D+1)] - 0.06$$
$$+ \frac{\log_{10}[\Delta \text{PSI}/3.0]}{1 + [1.624 \times 10^7/(D+1)^{8.46}]} \quad (4.19)$$
$$+ (4.22 - 0.32\,\text{TSI})\log_{10}\left(\frac{S'_c C_d [D^{0.75} - 1.132]}{215.63 J\{D^{0.75} - [18.42/(E_c/k)^{0.25}]\}}\right)$$

where

W_{18} = 18-kip–equivalent single-axle load,

Z_R = reliability (z-statistic from the standard normal curve),

S_o = overall standard deviation of traffic,

D = PCC slab thickness in inches,

TSI = pavement's terminal serviceability index,

ΔPSI = loss in serviceability from the time when the pavement is new until it reaches its TSI,

S'_c = concrete modulus of rupture in lb/in^2,

C_d = drainage coefficient,

J = load transfer coefficient,

E_c = concrete elastic modulus in lb/in^2, and

k = modulus of subgrade reaction in lb/in^3.

A graphic solution to Eq. 4.19 is shown in Figs. 4.10 and 4.11. The terms used in Eq. 4.19 and Figs. 4.10 and 4.11 are defined as follows:

W_{18} The 18-kip (80.1-kN)–equivalent single-axle load is the same concept as discussed for the flexible-pavement design procedure. However, instead of being a function of the structural number, this value is a function of slab thickness. The axle-load equivalency factors used in rigid-pavement design are presented in Tables 4.7 (for single axles), 4.8 (for tandem axles), and 4.9 (for triple axles).

Z_R As with flexible-pavement design, the reliability, Z_R, is defined as the probability that serviceability will be maintained at adequate levels from a user's point of view throughout the design life of the facility (the PSI will stay above the TSI). In the rigid-pavement design nomograph (Figs. 4.10 and 4.11), the probabilities (in percent) are used directly (instead of Z_R as in Eq. 4.19), and these percent probabilities are denoted R (see Table 4.5, which still applies).

S_o As was the case in flexible-pavement design, the overall standard deviation, S_o, takes into account designers' inability to accurately estimate future 18-kip (80.1-kN)–equivalent axle loads and the statistical error in the equations resulting from variability in materials and construction practices.

4.6 The AASHTO Rigid-Pavement Design Procedure

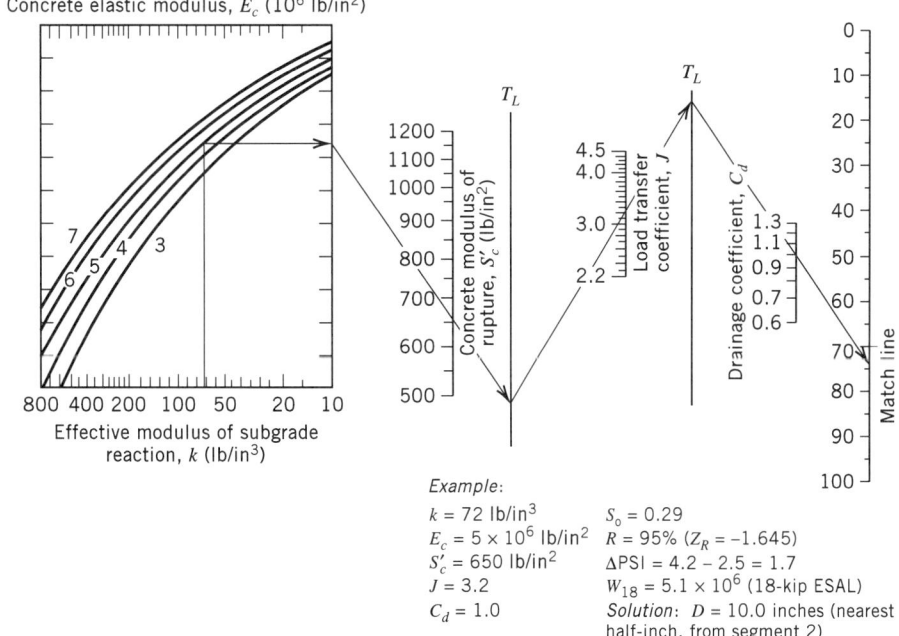

Figure 4.10 Segment 1 of the design chart for rigid pavement based on the use of mean values for each input variable.
Redrawn from *AASHTO Guide for Design of Pavement Structures*, The American Association of State Highway and Transportation Officials, Washington, DC, copyright 1993. Used by permission.

TSI The pavement's terminal serviceability index, TSI, is the point at which the pavement can no longer perform in a serviceable manner, as discussed previously for the flexible-pavement design procedure.

ΔPSI The amount of serviceability loss, ΔPSI, over the life of the pavement is the difference between the initial PSI and the TSI, as discussed for the flexible-pavement design procedure.

S'_c The concrete modulus of rupture, S'_c, is a measure of the tensile strength of the concrete and is determined by loading a beam specimen, at the third points, to failure. The test method is ASTM C78, Flexural Strength of Concrete. Because concrete gains strength with age, the average 28-day strength is used for design purposes. Typical values are 500 to 1200 lb/in^2 (3450 to 8270 kPa).

C_d The drainage coefficient, C_d, is slightly different from the value used in flexible-pavement design. In rigid-pavement design, it accounts for the drainage characteristics of the subgrade. A value of 1.0 for the drainage coefficient represents a material with good drainage characteristics (such as a sandy material). Soils with less-than-ideal drainage characteristics will have drainage coefficients less than 1.0.

122 Chapter 4 Pavement Design

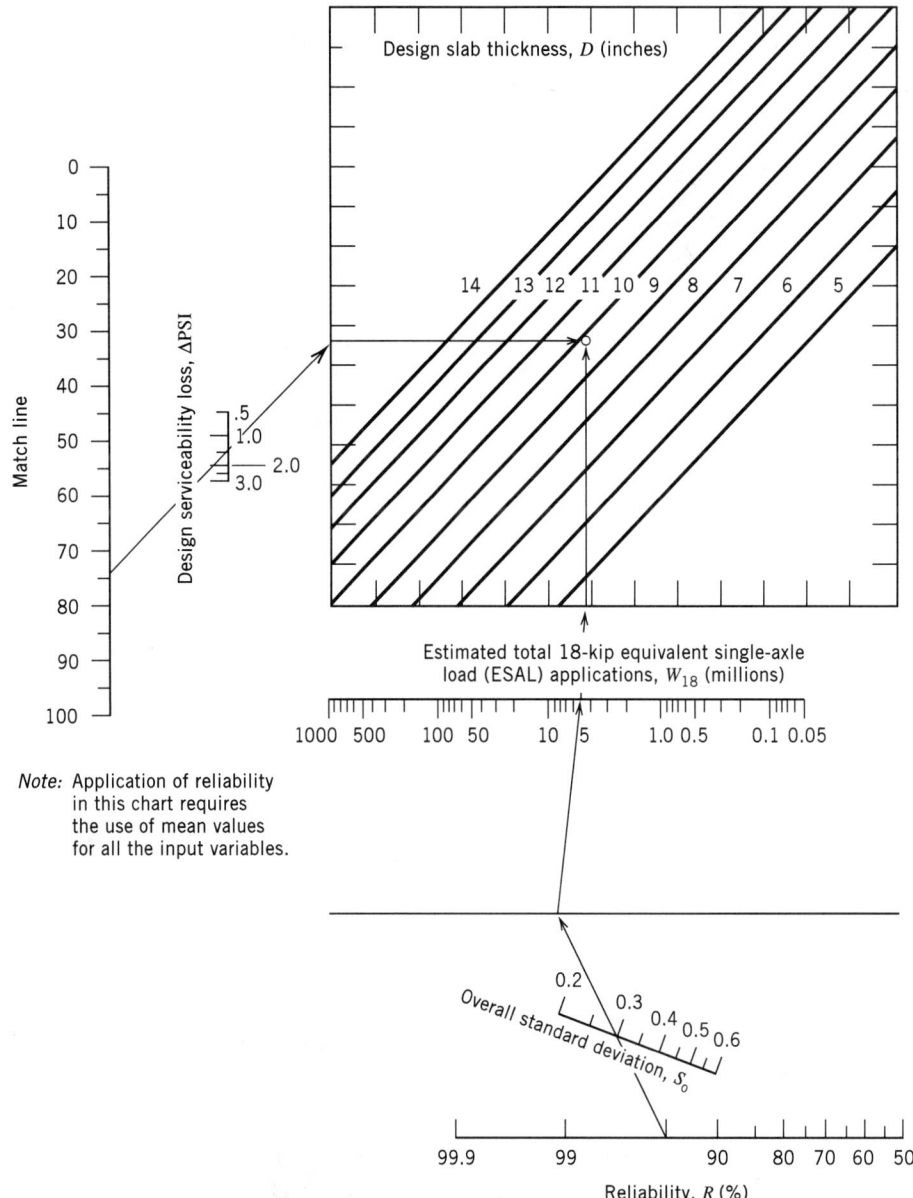

Figure 4.11 Segment 2 of the design chart for rigid pavements based on the use of mean values for each input variable.
Redrawn from *AASHTO Guide for Design of Pavement Structures*, The American Association of State Highway and Transportation Officials, Washington, DC, copyright 1993. Used by permission.

J The load transfer coefficient, *J*, is a factor that is used to account for the ability of pavement to transfer a load from one PCC slab to another across the slab joints. Many rigid pavements have dowel bars across the joints to transfer loads between slabs. Pavements with dowel bars at the joints are typically designed with a *J* value of 3.2.

E_c The concrete modulus of elasticity, E_c, is derived from the stress-strain curve as taken in the elastic region. As discussed previously, the modulus of elasticity is also known as Young's modulus. Typical values of E_c for portland cement concrete are between 3 and 7 million lb/in^2 (20.7 and 48.3 GPa).

k The modulus of subgrade reaction, *k*, depends upon several different factors, including the moisture content and density of the soil. It should be noted that most highway agencies do not perform testing to measure the *k* value of the soil. At best, the agency will have a CBR value for the subgrade. Typical values for *k* range from 100 to 800 lb/in^3 (0.0271 to 0.2165 N/mm^3). Table 4.10 indicates the relationship between CBR and *k* values.

Table 4.7 Axle-Load Equivalency Factors for Rigid Pavements, Single Axles, and TSI = 2.5

Axle load (kips)	Slab thickness, *D* (inches)								
	6	7	8	9	10	11	12	13	14
2	0.0002	0.0002	0.0002	0.0002	0.0002	0.0002	0.0002	0.0002	0.0002
4	0.003	0.002	0.002	0.002	0.002	0.002	0.002	0.002	0.002
6	0.012	0.011	0.010	0.010	0.010	0.010	0.010	0.010	0.010
8	0.039	0.035	0.033	0.032	0.032	0.032	0.032	0.032	0.032
10	0.097	0.089	0.084	0.082	0.081	0.080	0.080	0.080	0.080
12	0.203	0.189	0.181	0.176	0.175	0.174	0.174	0.174	0.173
14	0.376	0.360	0.347	0.341	0.338	0.337	0.336	0.336	0.336
16	0.634	0.623	0.610	0.604	0.601	0.599	0.599	0.599	0.598
18	1.00	1.00	1.00	1.00	1.00	1.00	1.00	1.00	1.00
20	1.51	1.52	1.55	1.57	1.58	1.58	1.59	1.59	1.59
22	2.21	2.20	2.28	2.34	2.38	2.40	2.41	2.41	2.41
24	3.16	3.10	3.22	3.36	3.45	3.50	3.53	3.54	3.55
26	4.41	4.26	4.42	4.67	4.85	4.95	5.01	5.04	5.05
28	6.05	5.76	5.92	6.29	6.61	6.81	6.92	6.98	7.01
30	8.16	7.67	7.79	8.28	8.79	9.14	9.35	9.46	9.52
32	10.8	10.1	10.1	10.7	11.4	12.0	12.3	12.6	12.7
34	14.1	13.0	12.9	13.6	14.6	15.4	16.0	16.4	16.5
36	18.2	16.7	16.4	17.1	18.3	19.5	20.4	21.0	21.3
38	23.1	21.1	20.6	21.3	22.7	24.3	25.6	26.4	27.0
40	29.1	26.5	25.7	26.3	27.9	29.9	31.6	32.9	33.7
42	36.2	32.9	31.7	32.2	34.0	36.3	38.7	40.4	41.6
44	44.6	40.4	38.8	39.2	41.0	43.8	46.7	49.1	50.8
46	54.5	49.3	47.1	47.3	49.2	52.3	55.9	59.0	61.4
48	66.1	59.7	56.9	56.8	58.7	62.1	66.3	70.3	73.4
50	79.4	71.7	68.2	67.8	69.6	73.3	78.1	83.0	87.1

Source: AASHTO Guide for Design of Pavement Structures, The American Association of State Highway and Transportation Officials, Washington, DC, copyright 1993. Used by permission.

Table 4.8 Axle-Load Equivalency Factors for Rigid Pavements, Tandem Axles, and TSI = 2.5

Axle load (kips)	Slab thickness, D (inches)								
	6	7	8	9	10	11	12	13	14
2	0.0001	0.0001	0.0001	0.0001	0.0001	0.0001	0.0001	0.0001	0.0001
4	0.0006	0.0006	0.0005	0.0005	0.0005	0.0005	0.0005	0.0005	0.0005
6	0.002	0.002	0.002	0.002	0.002	0.002	0.002	0.002	0.002
8	0.007	0.006	0.006	0.005	0.005	0.005	0.005	0.005	0.005
10	0.015	0.014	0.013	0.013	0.012	0.012	0.012	0.012	0.012
12	0.031	0.028	0.026	0.026	0.025	0.025	0.025	0.025	0.025
14	0.057	0.052	0.049	0.048	0.047	0.047	0.047	0.047	0.047
16	0.097	0.089	0.084	0.082	0.081	0.081	0.080	0.080	0.080
18	0.155	0.143	0.136	0.133	0.132	0.131	0.131	0.131	0.131
20	0.234	0.220	0.211	0.206	0.204	0.203	0.203	0.203	0.203
22	0.340	0.325	0.313	0.308	0.305	0.304	0.303	0.303	0.303
24	0.475	0.462	0.450	0.444	0.441	0.440	0.439	0.439	0.439
26	0.644	0.637	0.627	0.622	0.620	0.619	0.618	0.618	0.618
28	0.855	0.854	0.852	0.850	0.850	0.850	0.849	0.849	0.849
30	1.11	1.12	1.13	1.14	1.14	1.14	1.14	1.14	1.14
32	1.43	1.44	1.47	1.49	1.50	1.51	1.51	1.51	1.51
34	1.82	1.82	1.87	1.92	1.95	1.96	1.97	1.97	1.97
36	2.29	2.27	2.35	2.43	2.48	2.51	2.52	2.52	2.53
38	2.85	2.80	2.91	3.03	3.12	3.16	3.18	3.20	3.20
40	3.52	3.42	3.55	3.74	3.87	3.94	3.98	4.00	4.01
42	4.32	4.16	4.30	4.55	4.74	4.86	4.91	4.95	4.96
44	5.26	5.01	5.16	5.48	5.75	5.92	6.01	6.06	6.09
46	6.36	6.01	6.14	6.53	6.90	7.14	7.28	7.36	7.40
48	7.64	7.16	7.27	7.73	8.21	8.55	8.75	8.86	8.92
50	9.11	8.50	8.55	9.07	9.68	10.14	10.42	10.58	10.66
52	10.8	10.0	10.0	10.6	11.3	11.9	12.3	12.5	12.7
54	12.8	11.8	11.7	12.3	13.2	13.9	14.5	14.8	14.9
56	15.0	13.8	13.6	14.2	15.2	16.2	16.8	17.3	17.5
58	17.5	16.0	15.7	16.3	17.5	18.6	19.5	20.1	20.4
60	20.3	18.5	18.1	18.7	20.0	21.4	22.5	23.2	23.6
63	23.5	21.4	20.8	21.4	22.8	24.4	25.7	26.7	27.3
64	27.0	24.6	23.8	24.4	25.8	27.7	29.3	30.5	31.3
66	31.0	28.1	27.1	27.6	29.2	31.3	33.2	34.7	35.7
68	35.4	32.1	30.9	31.3	32.9	35.2	37.5	39.3	40.5
70	40.3	36.5	35.0	35.3	37.0	39.5	42.1	44.3	45.9
72	45.7	41.4	39.6	39.8	41.5	44.2	47.2	49.8	51.7
74	51.7	46.7	44.6	44.7	46.4	49.3	52.7	55.7	58.0
76	58.3	52.6	50.2	50.1	51.8	54.9	58.6	62.1	64.8
78	65.5	59.1	56.3	56.1	57.7	60.9	65.0	69.0	72.3
80	73.4	66.2	62.9	62.5	64.2	67.5	71.9	76.4	80.2
82	82.0	73.9	70.2	69.6	71.2	74.7	79.4	84.4	88.8
84	91.4	82.4	78.1	77.3	78.9	82.4	87.4	93.0	98.1
86	102.0	92.0	87.0	86.0	87.0	91.0	96.0	102.0	108.0
88	113.0	102.0	96.0	95.0	96.0	100.0	105.0	112.0	119.0
90	125.0	112.0	106.0	105.0	106.0	110.0	115.0	123.0	130.0

Source: AASHTO Guide for Design of Pavement Structures, The American Association of State Highway and Transportation Officials, Washington, DC, copyright 1993. Used by permission.

Table 4.9 Axle-Load Equivalency Factors for Rigid Pavements, Triple Axles, and TSI = 2.5

Axle load (kips)	Slab thickness, D (inches)								
	6	7	8	9	10	11	12	13	14
2	0.0001	0.0001	0.0001	0.0001	0.0001	0.0001	0.0001	0.0001	0.0001
4	0.0003	0.0003	0.0003	0.0003	0.0003	0.0003	0.0003	0.0003	0.0003
6	0.001	0.001	0.001	0.001	0.001	0.001	0.001	0.001	0.001
8	0.003	0.002	0.002	0.002	0.002	0.002	0.002	0.002	0.002
10	0.006	0.005	0.005	0.005	0.005	0.005	0.005	0.005	0.005
12	0.011	0.010	0.010	0.009	0.009	0.009	0.009	0.009	0.009
14	0.020	0.018	0.017	0.017	0.016	0.016	0.016	0.016	0.016
16	0.033	0.030	0.029	0.028	0.027	0.027	0.027	0.027	0.027
18	0.053	0.048	0.045	0.044	0.044	0.043	0.043	0.043	0.043
20	0.080	0.073	0.069	0.067	0.066	0.066	0.066	0.066	0.066
22	0.116	0.107	0.101	0.099	0.098	0.097	0.097	0.097	0.097
24	0.163	0.151	0.144	0.141	0.139	0.139	0.138	0.138	0.138
26	0.222	0.209	0.200	0.195	0.194	0.193	0.192	0.192	0.192
28	0.295	0.281	0.271	0.265	0.263	0.262	0.262	0.262	0.262
30	0.384	0.371	0.359	0.354	0.351	0.350	0.349	0.349	0.349
32	0.490	0.480	0.468	0.463	0.460	0.459	0.458	0.458	0.458
34	0.616	0.609	0.601	0.596	0.594	0.593	0.592	0.592	0.592
36	0.765	0.762	0.759	0.757	0.756	0.755	0.755	0.755	0.755
38	0.939	0.941	0.946	0.948	0.950	0.951	0.951	0.951	0.951
40	1.14	1.15	1.16	1.17	1.18	1.18	1.18	1.18	1.18
42	1.38	1.38	1.41	1.44	1.45	1.46	1.46	1.46	1.46
44	1.65	1.65	1.70	1.74	1.77	1.78	1.78	1.78	1.78
46	1.97	1.96	2.03	2.09	2.13	2.15	2.16	2.16	2.16
48	2.34	2.31	2.40	2.49	2.55	2.58	2.59	2.60	2.60
50	2.76	2.71	2.81	2.94	3.02	3.07	3.09	3.10	3.11
52	3.24	3.15	3.27	3.44	3.56	3.62	3.66	3.68	3.68
54	3.79	3.66	3.79	4.00	4.16	4.26	4.30	4.33	4.34
56	4.41	4.23	4.37	4.63	4.84	4.97	5.03	5.07	5.09
58	5.12	4.87	5.00	5.32	5.59	5.76	5.85	5.90	5.93
60	5.91	5.59	5.71	6.08	6.42	6.64	6.77	6.84	6.87
62	6.80	6.39	6.50	6.91	7.33	7.62	7.79	7.88	7.93
64	7.79	7.29	7.37	7.82	8.33	8.70	8.92	9.04	9.11
66	8.90	8.28	8.33	8.83	9.42	9.88	10.17	10.33	10.42
68	10.1	9.4	9.4	9.9	10.6	11.2	11.5	11.7	11.9
70	11.5	10.6	10.6	11.1	11.9	12.6	13.0	13.3	13.5
72	13.0	12.0	11.8	12.4	13.3	14.1	14.7	15.0	15.2
74	14.6	13.5	13.2	13.8	14.8	15.8	16.5	16.9	17.1
76	16.5	15.1	14.8	15.4	16.5	17.6	18.4	18.9	19.2
78	18.5	16.9	16.5	17.1	18.2	19.5	20.5	21.1	21.5
80	20.6	18.8	18.3	18.9	20.2	21.6	22.7	23.5	24.0
82	23.0	21.0	20.3	20.9	22.2	23.8	25.2	26.1	26.7
84	25.6	23.3	22.5	23.1	24.5	26.2	27.8	28.9	29.6
86	28.4	25.8	24.9	25.4	26.9	28.8	30.5	31.9	32.8
88	31.5	28.6	27.5	27.9	29.4	31.5	33.5	35.1	36.1
90	34.8	31.5	30.3	30.7	32.2	34.4	36.7	38.5	39.8

Source: AASHTO Guide for Design of Pavement Structures, The American Association of State Highway and Transportation Officials, Washington, DC, copyright 1993. Used by permission.

Table 4.10 Relationship between California Bearing Ratio (CBR) and Modulus of Subgrade Reaction, k

CBR	k, lb/in^3
2	100
10	200
20	250
25	290
40	420
50	500
75	680
100	800

EXAMPLE 4.5

A rigid pavement is to be designed to provide a service life of 20 years and has an initial PSI of 4.4 and a TSI of 2.5. The modulus of subgrade reaction is determined to be 300 lb/in^3 (0.0813 N/mm^3). For design, the daily car, pickup truck, and light van traffic is 20,000; and the daily truck traffic consists of 200 passes of single-unit trucks with single and tandem axles, and 410 passes of tractor semi-trailer trucks with single, tandem, and triple axles. The axle weights are

$$\begin{aligned}
\text{cars, pickups, light vans} &= \text{two 2000-lb (8.9-kN) single axles}\\
\text{single-unit trucks} &= \text{10,000-lb (44.5-kN) steering, single axle}\\
&= \text{22,000-lb (97.9-kN) drive, tandem axle}\\
\text{tractor semi-trailer trucks} &= \text{12,000-lb (53.4-kN) steering, single axle}\\
&= \text{18,000-lb (80.1-kN) drive, tandem axle}\\
&= \text{50,000-lb (222.4-kN) trailer, triple axle}
\end{aligned}$$

Reliability is 95%, the overall standard deviation is 0.45, the concrete's modulus of elasticity is 4.5 million lb/in^2 (31.03 GPa), the concrete's modulus of rupture is 900 lb/in^2 (6210 kPa), the load transfer coefficient is 3.2, and the drainage coefficient is 1.0. Determine the required slab thickness.

SOLUTION

Because the axle-load equivalency factors presented in Tables 4.7, 4.8, and 4.9 are a function of the slab thickness (D), we have to assume a D value to start the problem (later we will arrive at a slab thickness and check to make sure that it is consistent with our assumed value). A typical assumption is to let $D = 10$ inches. Given this, the 18-kip–equivalent single-axle load (18-kip ESAL) for cars, pickups, and light vans is

$$\text{2-kip single-axle equivalent} = 0.0002 \text{ (Table 4.7)}$$

This gives an 18-kip ESAL total of 0.0004 for each vehicle. For single-unit trucks,

$$\text{10-kip single-axle equivalent} = 0.081 \text{ (Table 4.7)}$$
$$\text{22-kip tandem-axle equivalent} = 0.305 \text{ (Table 4.8)}$$

This gives an 18-kip ESAL total of 0.386 for single-unit trucks. For tractor semi-trailer trucks,

12-kip single-axle equivalent = 0.175 (Table 4.7)
18-kip tandem-axle equivalent = 0.132 (Table 4.8)
50-kip triple-axle equivalent = 3.020 (Table 4.9)

This gives an 18-kip ESAL total of 3.327 for tractor semi-trailer trucks.

Given the computed 18-kip ESAL, the daily traffic on this highway produces an 18-kip ESAL total of 1449.27 (0.0004 × 20,000 + 0.386 × 200 + 3.327 × 410). Traffic (total axle accumulations) over the 20-year design period will be

$$1449.27 \times 365 \times 20 = 10{,}579{,}671 \text{ 18-kip ESAL}$$

With an initial PSI of 4.4 and a TSI of 2.5, ΔPSI = 1.9. Solving Eq. 4.19 for D (using an equation solver on a calculator or computer) with $Z_R = -1.645$ (which corresponds to $R = 95\%$ as shown in Table 4.5) gives $D = 9.21$ inches (233.9 mm). (Figures 4.10 and 4.11 can also be used to arrive at an approximate solution for D.) Note that this value differs from the slab thickness assumed to derive the axle-load equivalency factors. Recomputing the axle-load equivalency factors with $D = 9$ inches (for Tables 4.7, 4.8, and 4.9) for cars, pickups, and light vans gives

2-kip single-axle equivalent = 0.0002 (Table 4.7)

This gives an 18-kip ESAL total of 0.0004 (same as before) for each vehicle. For single-unit trucks,

10-kip single-axle equivalent = 0.082 (Table 4.7)
22-kip tandem-axle equivalent = 0.308 (Table 4.8)

This gives an 18-kip ESAL total of 0.390 (up from 0.386) for single-unit trucks. For tractor semi-trailer trucks,

12-kip single-axle equivalent = 0.176 (Table 4.7)
18-kip tandem-axle equivalent = 0.133 (Table 4.8)
50-kip triple-axle equivalent = 2.940 (Table 4.9)

This gives an 18-kip ESAL total of 3.249 (down from 3.327) for tractor semi-trailer trucks.

Given the computed 18-kip ESAL, the daily traffic on this highway produces an 18-kip ESAL of 1418.09 (0.0004 × 20,000 + 0.390 × 200 + 3.249 × 410). Traffic (total axle accumulations) over the 20-year design period will be

$$1418.09 \times 365 \times 20 = 10{,}352{,}057 \text{ 18-kip ESAL}$$

Again solving Eq. 4.18 for D gives $D = 9.17$ inches. (Figs. 4.10 and 4.11 can also be used to arrive at an approximate solution for D.) This is very close to the assumed $D = 9.0$ inches and is only a minor change from the 9.21 inches previously obtained. To be conservative, we would round up to the nearest 0.5 inch and make the slab 9.5 inches.

Table 4.11 Proportion of Directional W_{18} Assumed to Be in the Design Lane

Number of directional lanes	Proportion of directional W_{18} in the design lane (*PDL*)
1	1.00
2	0.80–1.00
3	0.60–0.80
4	0.50–0.75

The final point to be covered with regard to pavement design relates to the case where there are multiple lanes of a highway (such as an interstate) in one direction. Because traffic tends to be distributed among the lanes, in some instances the pavement can be designed using a fraction of the total directional W_{18}. However, because traffic tends to concentrate in the right lane (particularly heavy vehicles), this fraction is not as simple as dividing W_{18} by the number of lanes. In equation form,

$$\text{design-lane } W_{18} = PDL(\text{directional } W_{18}) \tag{4.20}$$

where

W_{18} = 18-kip–equivalent single-axle load (ESAL) and
PDL = proportion of directional W_{18} assumed to be in the design lane.

AASHTO-recommended values for *PDL* are given in Table 4.11.

As an example, suppose the computed directional W_{18} is an 18-kip ESAL of 10,000,000 and there are three lanes in the direction of travel. If the highway is conservatively designed, Table 4.11 shows that 80% of the axle loads can be assumed to be in the design lane ($PDL = 0.8$). So the design W_{18} would be 8,000,000 (0.8 × 10,000,000), and this value would be used in the equations and nomographs. This design procedure applies to both flexible and rigid pavements.

EXAMPLE 4.6

In 1986, a rigid pavement on a northbound section of interstate highway was designed with a 12-inch (304.8-mm) PCC slab, an E_c of 6×10^6 lb/in² (48.27 GPa), a concrete modulus of rupture of 800 lb/in² (5,520 kPa), a load transfer coefficient of 3.0, an initial PSI of 4.5, and a terminal serviceability index of 2.5. The overall standard deviation was 0.45, the modulus of subgrade reaction was 190 lb/in³ (0.05149 N/mm³), and a reliability of 95% was used along with a drainage coefficient of 1.0. The pavement was designed for a 20-year life, and traffic was assumed to be composed entirely of tractor semi-trailer trucks with one 16-kip (71.2-kN) single axle, one 20-kip (88.9-kN) single axle, and one 35-kip (155.7-kN) tandem axle (the effect of all other vehicles was ignored). The interstate has four northbound lanes and was conservatively designed. How many tractor semi-trailer trucks, per day, were assumed to be traveling in the northbound direction?

SOLUTION

Given that the slab thickness D is 12 inches, for the tractor semi-trailer trucks we have

16-kip single-axle equivalent = 0.599 (Table 4.7)
20-kip single-axle equivalent = 1.590 (Table 4.7)
35-kip tandem-axle equivalent = 2.245 (Table 4.8)

Note that the value of 2.245 for the 35,000-lb tandem-axle linear interpolation uses 34-kip and 36-kip values [(1.97 + 2.52)/2]. The summation of these axle equivalents gives 4.434 18-kip ESAL per truck.

With an initial PSI of 4.5 and a TSI of 2.5, ΔPSI = 2.0. Solving Eq. 4.19 for W_{18} with $Z_R = -1.645$ (which corresponds to $R = 95\%$ as shown in Table 4.5) gives $W_{18} =$ 39,740,309 18-kip ESAL. Figures 4.10 and 4.11 can also be used to arrive at an approximate solution for W_{18}. Thus the total daily truck traffic in the design lane is

$$\text{traffic} = \frac{39{,}740{,}309 \text{ 18-kip ESAL}}{365 \text{ days/year} \times 20 \text{ years} \times 4.434 \text{ 18-kip ESAL/truck}}$$

$$= \underline{1227.76 \text{ trucks/day}}$$

To determine the total directional volume (total number of northbound trucks), we note from Table 4.11 that the *PDL* for a conservative design on a four-lane highway is 0.75, and the application of Eq. 4.20 gives

$$\text{directional } W_{18} = \frac{\text{design-lane } W_{18}}{PDL}$$

$$= \frac{1227.76}{0.75}$$

$$= \underline{1637.01 \text{ trucks/day}}$$

NOMENCLATURE FOR CHAPTER 4

a	equivalent tire radius	r	radial distance from the center of the load
a_l	distance to point of action of the resulting load on a common angle bisection	R	reliability
		S'_c	concrete modulus of rupture
a_1, a_2, a_3	structural-layer coefficients for wearing surface, base, and subbase	S_o	overall standard deviation for AASHTO design equations structural number
C_d	drainage coefficient for rigid-pavement design	SN	structural number
		TSI	terminal serviceability index
CBR	California bearing ratio	W_{18}	18-kip (80.1-kN)–equivalent single-axle load
D	slab thickness, AASHTO design equation	z	depth in pavement
D_1, D_2, D_3	structural-layer thicknesses for wearing surface, base, and subbase	Z_R	reliability for AASHTO design equations
		Δ	deflection
E	modulus of elasticity (Young's modulus)	Δ_c	corner deflection
E_c	concrete modulus of elasticity	Δ_e	edge deflection
h	slab thickness, Westergaard solutions	Δ_i	interior deflection
l	radius of relative stiffness	Δ_z	deflection at depth z
k	modulus of subgrade reaction	ΔPSI	change in the present serviceability index
M_2, M_3	drainage coefficients for base and subbase	σ_c	bending stress, corner
M_R	soil resilient modulus	σ_e	bending stress, edge
p	tire pressure	σ_i	bending stress, interior
P	wheel load	σ_r	radial-horizontal stress
PDL	proportion of directional W_{18} assumed in the design lane	σ_z	vertical stress at depth z
		μ	Poisson ratio
PSI	present serviceability index	γ	Euler's constant

REFERENCES

AASHTO (American Association of State Highway and Transportation Officials). *AASHTO Guide for Design of Pavement Structures*. Washington, DC: AASHTO, 1993.

Ahlvin, R. G., and H. H. Ulery. "Tabulated Values for Determining the Complete Pattern of Stresses, Strains, and Deflections Beneath a Uniform Circular Load on a Homogeneous Half Space." *Highway Research Board Bulletin* 342, 1962.

Carey, W., and P. Irick. *The Pavement Serviceability-Performance Concept*. Highway Research Board Special Report 61E, AASHO Road Test, 1962.

Highway Research Board. *AASHTO Road Test Report 5*. Special Report G. Washington, DC, 1962.

Ioannides, A. M., M. R. Thompson, and E. J. Barenberg. *Westergaard Solutions Reconsidered*. Transportation Research Record 1043. Washington, DC, 1985.

Westergaard, H. M. "Computation of Stresses in Concrete Roads." *Proceedings of the Fifth Annual Meeting of the Highway Research Board*. Washington, DC, 1926.

Yoder, E. J., and M. W. Witczak. *Principles of Pavement Design*, 2nd ed. New York: Wiley, 1975.

PROBLEMS

4.1 A tire carries a 5000-lb load and has a pressure of 100 lb/in^2. The pavement that the tire is on is constructed with a modulus of elasticity of 43,500 lb/in^2. A deflection of 0.016 inches is observed at a point at the pavement surface, 0.8 inches from the center of the tire load. Using the Ahlvin and Ulery equations, what is the radial-horizontal stress at this point?

4.2 A wheel carrying a 6700-lb load generates a deflection of 0.035 inches, 2 inches below the center of the tire load. The contact area is measured at 80 in^2 and the Poisson ratio is 0.5. Using the Ahlvin and Ulery equations, what is the pavement's modulus of elasticity?

4.3 A wheel load produces a circular contact radius of 3.5 inches and the pavement material has a modulus of elasticity of 43,500 lb/in^2. At a point on the pavement surface under the center of the tire load, the radial-horizontal stress is 87 lb/in^2 and the deflection is 0.165 inches. Using the Ahlvin and Ulery equations, what is the load applied to the wheel?

4.4 A pavement is 25 inches thick and has a modulus of elasticity of 36,250 lb/in^2 with a Poisson ratio of 0.40. A wheel load is applied 50 inches from the edge of the pavement. The wheel's tire has a pressure of 101.5 lb/in^2 and a circular contact radius of 12.7 inches. Using the Ahlvin and Ulery equations, determine the vertical stress, radial-horizontal stress, and deflection at a point at the bottom of the pavement, at the pavement's edge.

4.5 Truck A has two single axles. One axle weighs 12,000 lb and the other weighs 23,000 lb. Truck B has an 8000-lb single axle and a 43,000-lb tandem axle. On a flexible pavement with a 3-inch hot-mix asphalt wearing surface, a 6-inch soil-cement base, and an 8-inch crushed stone subbase, which truck will cause more pavement damage? (Assume drainage coefficients are 1.0.)

4.6 A flexible pavement has a 4-inch hot-mix asphalt wearing surface, a 7-inch dense-graded crushed stone base, and a 10-inch crushed stone subbase. The pavement is on a soil with a resilient modulus of 5000 lb/in^2. The pavement was designed with 90% reliability, an overall standard deviation of 0.4, and a ΔPSI of 2.0 (a TSI of 2.5). The drainage coefficients are 0.9 and 0.8 for the base and subbase, respectively. How many 25-kip single-axle loads can be carried before the pavement reaches its TSI (with given reliability)?

4.7 A highway has the following pavement design daily traffic: 300 single axles at 10,000 lb each, 120 single axles at 18,000 lb each, 100 single axles at 23,000 lb each, 100 tandem axles at 32,000 lb each, 30 single axles at 32,000 lb each, and 100 triple axles at 40,000 lb each. A flexible pavement is designed to have 4 inches of sand-mix asphalt wearing surface, 6 inches of soil-cement base, and 7 inches of crushed stone subbase. The pavement has a 10-year design life, a reliability of 85%, an overall standard deviation of 0.30, drainage coefficients of 1.0, an initial PSI of 4.7, and a TSI of 2.5. What is the minimum acceptable soil resilient modulus?

4.8 Consider the conditions in Problem 4.7. Suppose the state has relaxed truck weight limits and the impact has been to reduce the number of 18,000-lb single-axle loads from 120 to 20 and increase the number of 32,000-lb single-axle loads from 30 to 90 (all other traffic is unaffected). Under these revised daily counts, what is the minimum acceptable soil resilient modulus?

4.9 A flexible pavement was designed for the following daily traffic with a 12-year design life: 1300 single axles at 8,000 lb each, 900 tandem axles at 15,000 lb each, 20 single axles at 40,000 lb each, and 200 tandem axles at 40,000 lb each. The highway was designed with 4 inches of hot-mix asphalt wearing surface, 4 inches of hot-mix asphaltic base, and 8 inches of crushed stone subbase. The reliability was 70%, overall standard deviation was 0.5, ΔPSI was 2.0 (with a TSI of 2.5), and all drainage coefficients were 1.0. What was the soil resilient modulus of the subgrade used in design?

4.10 A flexible pavement has a structural number of 3.8 (all drainage coefficients are equal to 1.0). The initial PSI is 4.7 and the terminal serviceability is 2.5. The soil has a CBR of 9. The overall standard deviation is 0.40 and the reliability is 95%. The pavement is currently designed for 1800 equivalent 18-kip single-axle loads per day. If the number of 18-kip single-axle loads were to increase by 30%, by how many years would the pavement's design life be reduced?

4.11 An engineer plans to replace the rigid pavement in Example 4.5 with a flexible pavement. The chosen design has 6 inches of sand-mix asphalt wearing surface, 9 inches of soil-cement base, and 10 inches of crushed stone subbase. All drainage coefficients are 1.0 and the soil resilient modulus is 5000 lb/in^2. If the highway's traffic is the same (same axle loadings per vehicle as in Example 4.5), for how many years could you be 95% sure that this pavement will last? (Assume that any parameters not given in this problem are the same as those given in Example 4.5.)

4.12 You have been asked to design a flexible pavement, and the following daily traffic is expected for design: 5000 single axles at 10,000 lb each, 400 single axles at 24,000 lb each, 1000 tandem axles at 30,000 lb each, and 100 tandem axles at 50,000 lb each. There are three lanes in the design direction (conservative design is to be used). Reliability is 90%, overall standard deviation is 0.40, ΔPSI is 1.8, and the design life is 15 years. The soil has a resilient modulus of 13,750 lb/in^2. If the TSI is 2.5, what is the required structural number?

4.13 A flexible pavement is designed with 5 inches of hot-mix asphalt wearing surface, 6 inches of hot-mix asphaltic base, and 10 inches of crushed stone subbase. All drainage coefficients are 1.0. Daily traffic is 200 passes of a 20-kip single axle, 200 passes of a 40-kip tandem axle, and 80 passes of a 22-kip single axle. If the initial minus the terminal PSI is 2.0 (the TSI is 2.5), the soil resilient modulus is 3000 lb/in^2, and the overall standard deviation is 0.6, what is the probability (reliability) that this pavement will last 20 years before reaching its terminal serviceability?

4.14 A flexible pavement is designed with 4 inches of sand-mix asphalt wearing surface, 6 inches of dense-graded crushed stone base, and 8 inches of crushed stone subbase. All drainage coefficients are 1.0. The pavement is designed for 18-kip single-axle loads (1290 per day). The initial PSI is 4.5 and the TSI is 2.5. The soil has a resilient modulus of 12,000 lb/in^2. If the overall standard deviation is 0.40, what is the probability that this pavement will have a PSI greater than 2.5 after 20 years?

4.15 A flexible pavement has a 4-inch sand-mix asphalt wearing surface, 10-inch soil cement base, and a 10-inch crushed stone subbase. It is designed to withstand 400 20-kip single-axle loads and 900 35-kip tandem-axle loads per day. The subgrade CBR is 8, the overall standard deviation is 0.45, the initial PSI is 4.2, and the final PSI is 2.5. What is the probability that this pavement will have a PSI above 2.5 after 25 years? (Drainage coefficients are 1.0.)

4.16 A three-lane northbound section of interstate (with the design lane conservatively designed) has rigid pavement (PCC) and was designed with a 10-inch slab, 90% reliability, 700 lb/in^2 concrete modulus of rupture, 4.5 million lb/in^2 modulus of elasticity, 3.0 load transfer coefficient, and an overall standard deviation of 0.35. The initial PSI is 4.6 and the TSI is 2.5. The CBR is 2 with a drainage coefficient of 1.0. The road was designed exclusively for trucks that have one 24-kip tandem axle and one 12-kip single axle. It is known from weigh-in-motion scales that there have been 13 million 18-kip–equivalent single-axle loads in the entire northbound direction of this freeway so far. If a section of flexible pavement is used to replace a section of the PCC that was removed for utility work, what structural number should be used so that the PCC and flexible pavements have the same life expectancy (the new life of the flexible pavement and the remaining life of the PCC)?

4.17 A tire carries a 10,000-lb load and the tire's pressure is 90 lb/in^2. The tire is on a pavement with a modulus of elasticity of 4.2 million lb/in^2 and a Poisson ratio of 0.25. The modulus of subgrade reactions is 150 lb/in^3. If, for edge loading, the edge stress is 218.5 lb/in^2, what is the pavement's slab thickness? (Use the revised Westergaard equations.)

4.18 A pavement has an 8-inch slab with a modulus of elasticity of 3.5 million lb/in^2 and a Poisson ratio of 0.30. The radius of relative stiffness is 30.106. A 12,000-lb wheel load is applied (interior loading) and produces an interior deflection of 0.008195 inches. What is the interior stress? (Use the revised Westergaard equations.)

4.19 A pavement has a 10-inch slab with a Poisson ratio of 0.36. The pavement is on a subgrade with a soil resilient modulus of 250 lb/in^3. A 17,000-lb load is applied (corner loading), and a_l is 7 inches. If the corner deflection is 0.05 inches, what is the modulus of elasticity of the pavement? (Use the revised Westergaard equations.)

4.20 A 12-inch pavement slab has a modulus of elasticity of 4 million lb/in^2 and a Poisson ratio of 0.40. The pavement is on a soil with a modulus of subgrade reaction equal to 300 lb/in^3. A wheel load of 9000 lb is applied and the radius of circular load is 5 inches. What would the interior and edge stresses be, and what would the interior and edge slab deflections be? (Use the revised Westergaard equations.)

4.21 Consider the two trucks in Problem 4.5. Which truck will cause more pavement damage on a rigid pavement with a 10-inch slab?

4.22 A rigid pavement is designed with an 11-inch slab thickness, 90% reliability, E_c = 4 million lb/in^2, modulus of rupture of 600 lb/in^2, modulus of subgrade reaction of 150 lb/in^3, a 2.8 load transfer coefficient, initial

PSI of 4.8, final PSI of 2.5, overall standard deviation of 0.35, and a drainage coefficient of 0.8. The pavement has a 20-year design life. The pavement has three lanes and is conservatively designed for trucks that have one 20,000-lb single axle, one 26,000-lb tandem axle, and one 34,000-lb triple axle. What is the daily estimated truck traffic on the three lanes?

4.23 A rigid pavement is on a highway with two lanes in one direction, and the pavement is conservatively designed. The pavement has an 11-inch slab with a modulus of elasticity of 5,000,000 lb/in^2 and a concrete modulus of rupture of 700 lb/in^2, and it is on a soil with a CBR of 25. The design drainage coefficient is 1.0, the overall standard deviation is 0.3, and the load transfer coefficient is 3.0. The pavement was designed to last 20 years (initial PSI of 4.7 and a final PSI of 2.5) with 95% reliability carrying trucks with one 18-kip single axle and one 28-kip tandem axle. However, after the pavement was designed, one more lane was added in the design direction (conservative design still used), and the weight limits on the trucks were increased to a 20-kip single and a 34-kip tandem axle (the slab thickness was unchanged from the original two-lane design with lighter trucks). If El Niño has caused the drainage coefficient to drop to 0.8, how long will the pavement last with the new loading and the additional lane (same volume of truck traffic)?

4.24 You have been asked to design the pavement for an access highway to a major truck terminal. The design daily truck traffic consists of the following: 80 single axles at 22,500 lb each, 570 tandem axles at 25,000 lb each, 50 tandem axles at 39,000 lb each, and 80 triple axles at 48,000 lb each. The highway is to be designed with rigid pavement having a modulus of rupture of 600 lb/in^2 and a modulus of elasticity of 5 million lb/in^2. The reliability is to be 95%, the overall standard deviation is 0.4, the drainage coefficient is 0.9, ΔPSI is 1.7 (with a TSI of 2.5), and the load transfer coefficient is 3.2. The modulus of subgrade reaction is 200 lb/in^3. If a 20-year design life is to be used, determine the required slab thickness.

4.25 A rigid pavement is being designed with the same parameters as used in Problem 4.9. The modulus of subgrade reaction is 300 lb/in^3 and the slab thickness is determined to be 8.5 inches. The load transfer coefficient is 3.0, the drainage coefficient is 1.0, and the modulus of elasticity is 4 million lb/in^2. What is the design modulus of rupture? (Assume that any parameters not given in this problem are the same as those given in Problem 4.9.)

4.26 A rigid pavement is designed with a 10-inch slab, an E_c of 6 million lb/in^2, a concrete modulus of rupture of 432 lb/in^2, a load transfer coefficient of 3.0, an initial PSI of 4.7, and a terminal serviceability index of 2.5. The overall standard deviation is 0.35, the modulus of subgrade reaction is 190 lb/in^3, and a reliability of 90% is used along with a drainage coefficient of 0.8. The pavement is designed assuming traffic is composed entirely of trucks (100 per day). Each truck has one 20-kip single axle and one 42-kip tandem axle (the effect of all other vehicles is ignored). A section of this road is to be replaced (due to different subgrade characteristics) with a flexible pavement having a structural number of 4 and is expected to last the same number of years as the rigid pavement. What is the assumed soil resilient modulus? (Assume all other factors are the same as for the rigid pavement.)

4.27 A four-lane northbound section of interstate has rigid pavement and was designed with an 8-inch slab, 90% reliability, a 700 lb/in^2 concrete modulus of rupture, a 5 million lb/in^2 modulus of elasticity, a 3.0 load transfer coefficient, and an overall standard deviation of 0.3. The initial PSI is 4.6 and the TSI is 2.5. The pavement was conservatively designed (assuming the upper limit of the W_{18} design lane load) to last 20 years, and the CBR is 25 with a drainage coefficient of 1.0. A design mistake was made that ignored 1000 total northbound (daily) passes of trucks with 22-kip single and 30-kip tandem axles. What slab thickness should have been used?

4.28 Consider the loading conditions in Problem 4.7. A rigid pavement is used with a modulus of subgrade reaction of 200 lb/in^3, a slab thickness of 8 inches, a load transfer coefficient of 3.2, a modulus of elasticity of 5 million lb/in^2, a modulus of rupture of 600 lb/in^2, and a drainage coefficient of 1.0. How many years would the pavement be expected to last using the same reliability as in Problem 4.7? (Assume all other factors are as in Problem 4.7.)

4.29 Consider Problem 4.28. How long would the rigid pavement be expected to last if you wanted to be 95% sure that the pavement would stay above the 2.5 TSI?

4.30 Consider the traffic conditions in Example 4.5. Suppose a 10-inch slab was used and all other parameters are as described in Example 4.5. What would the design life be if the drainage coefficient was 0.8, and what would it be if it was 0.6?

4.31 Consider the conditions in Example 4.6. Suppose all of the parameters are the same, but further soil tests found that the modulus of subgrade reaction was only 150 lb/in^3. In light of this new soil finding, how would the design life of the pavement change?

4.32 Consider the conditions in Example 4.6. Suppose all of the parameters are the same, but a quality control problem resulted in a modulus of rupture of 600 lb/in^2 instead of 800 lb/in^2. How would the design life of the pavement change?

Chapter 5

Fundamentals of Traffic Flow and Queuing Theory

5.1 INTRODUCTION

It is important to realize that the primary function of a highway is to provide mobility. This mobility must be provided with safety in mind while achieving an acceptable level of performance (such as acceptable vehicle speeds). Many of the safety-related aspects of highway design were discussed in Chapter 3, and focus is now shifted to measures of performance.

The analysis of vehicle traffic provides the basis for measuring the operating performance of highways. In undertaking such an analysis, the various dimensions of traffic, such as number of vehicles per unit time (flow), vehicle types, vehicle speeds, and the variation in traffic flow over time, must be addressed because they all influence highway design (the selection of the number of lanes, pavement types, and geometric design) and highway operations (selection of traffic control devices, including signs, markings, and traffic signals), both of which impact the performance of the highway. In light of this, it is important for the analysis of traffic to begin with theoretically consistent quantitative techniques that can be used to model traffic flow, speed, and temporal fluctuations. The intent of this chapter is to focus on models of traffic flow and queuing, thus providing the groundwork for quantifying measures of performance (and levels of service, which will be discussed in Chapters 6 and 7).

5.2 TRAFFIC STREAM PARAMETERS

Traffic streams can be characterized by a number of different operational performance measures. Before commencing with a discussion of the specific measures, it is important to provide definitions for the contexts in which these measures apply. A traffic stream that operates free from the influence of such traffic control devices as signals and stop signs is classified as uninterrupted flow. This type of traffic flow is influenced primarily by roadway characteristics and the interactions of the vehicles in the traffic stream. Freeways, multilane highways, and two-lane highways often operate under uninterrupted flow conditions. Traffic streams that operate under the influence of signals and stop signs are classified as interrupted flow. Although all the concepts in this chapter are generally applicable to both types of flow, there are some additional complexities involved with the analysis of traffic flow at signalized and unsignalized intersections. Chapter 7 will address the additional complexities relating to the analysis of traffic flow at signalized intersections. For details on the analysis of traffic flow at unsignalized intersections, refer to other sources [Transportation Research Board 1975, 2000]. It should be noted that environmental conditions (day vs. night, sunny vs. rainy, etc.) can also affect the flow of traffic, but this issue is beyond the scope of this book.

5.2.1 Traffic Flow, Speed, and Density

Traffic flow, speed, and density are variables that form the underpinnings of traffic analysis. To begin the study of these variables, the basic definitions of traffic flow, speed, and density must be presented. Traffic flow is defined as

$$q = \frac{n}{t} \tag{5.1}$$

where

q = traffic flow in vehicles per unit time,
n = number of vehicles passing some designated roadway point during time t, and
t = duration of time interval.

Flow is often measured over the course of an hour, in which case the resulting value is typically referred to as volume. Thus, when the term "volume" is used, it is generally understood that the corresponding value is in units of vehicles per hour (veh/h). The definition of flow is more generalized to account for the measurement of vehicles over any period of time. In practice, the analysis flow rate is usually based on the peak 15-minute flow within the hour of interest. This aspect will be described in more detail in Chapter 6.

Aside from knowing the total number of vehicles passing a point in some time interval, the amount of time between the passing of successive vehicles (or time between the arrival of successive vehicles) is also of interest. The time between the passage of the front bumpers of successive vehicles, at some designated highway point, is known as the time headway. The time headway is related to t, as defined in Eq. 5.1, by

$$t = \sum_{i=1}^{n} h_i \tag{5.2}$$

where

t = duration of time interval,
h_i = time headway of the ith vehicle (the elapsed time between the arrivals of vehicles i and $i - 1$), and
n = number of measured vehicle time headways at some designated roadway point.

Substituting Eq. 5.2 into Eq. 5.1 gives

$$q = \frac{n}{\sum_{i=1}^{n} h_i} \qquad (5.3)$$

or

$$q = \frac{1}{\bar{h}} \qquad (5.4)$$

where $\bar{h}$ = average time headway ($\sum h_i / n$) in unit time per vehicle. The importance of time headways in traffic analysis will be given additional attention in forthcoming sections of this chapter.

The average traffic speed is defined in two ways. The first is the arithmetic mean of the vehicle speeds observed at some designated point along the roadway. This is referred to as the time-mean speed and is expressed as

$$\bar{u}_t = \frac{\sum_{i=1}^{n} u_i}{n} \qquad (5.5)$$

where

$\bar{u}_t$ = time-mean speed in unit distance per unit time,
u_i = spot speed (the speed of the vehicle at the designated point on the highway, as might be obtained using a radar gun) of the ith vehicle, and
n = number of measured vehicle spot speeds.

The second definition of speed is more useful in the context of traffic analysis and is determined on the basis of the time necessary for a vehicle to travel some known length of roadway. This measure of average traffic speed is referred to as the space-mean speed and is expressed as (assuming that the travel time for all vehicles is measured over the same length of roadway)

$$\bar{u}_s = \frac{l}{\bar{t}} \qquad (5.6)$$

where

$\bar{u}_s$ = space-mean speed in unit distance per unit time,
l = length of roadway used for travel time measurement of vehicles, and
$\bar{t}$ = average vehicle travel time, defined as

$$\bar{t} = \frac{1}{n} \sum_{i=1}^{n} t_i \qquad (5.7)$$

where

t_i = time necessary for vehicle i to travel a roadway section of length l, and
n = number of measured vehicle travel times.

Substituting Eq. 5.7 into Eq. 5.6 yields

$$\bar{u}_s = \frac{l}{\dfrac{1}{n}\sum_{i=1}^{n} t_i} \tag{5.8}$$

or

$$\bar{u}_s = \frac{1}{\dfrac{1}{n}\sum_{i=1}^{n}\left[\dfrac{1}{(l/t_i)}\right]} \tag{5.9}$$

which is the harmonic mean of speed (space-mean speed). Space-mean speed is the speed variable used in traffic models.

EXAMPLE 5.1

The speeds of five vehicles were measured (with radar) at the midpoint of a 0.5-mile (0.8-km) section of roadway. The speeds for vehicles 1, 2, 3, 4, and 5 were 44, 42, 51, 49, and 46 mi/h (70.8, 67.6, 82.1, 78.8, and 74 km/h), respectively. Assuming all vehicles were traveling at constant speed over this roadway section, calculate the time-mean and space-mean speeds.

SOLUTION

For the time-mean speed, Eq. 5.5 is applied, giving

$$\bar{u}_t = \frac{\sum_{i=1}^{n} u_i}{n}$$

$$= \frac{44 + 42 + 51 + 49 + 46}{5}$$

$$= 46.4 \text{ mi/h}$$

For the space-mean speed, Eq. 5.9 will be applied. This equation is based on travel time; however, because it is known that the vehicles were traveling at constant speed, we can rearrange this equation to utilize the measured speed, knowing that distance, l, divided by travel time, t, is equal to speed ($l/t_i = u$).

$$\bar{u}_s = \cfrac{1}{\cfrac{1}{n}\sum_{i=1}^{n}\left[\cfrac{1}{(l/t_i)}\right]} = \cfrac{1}{\cfrac{1}{n}\sum_{i=1}^{n}\left[\cfrac{1}{u_i}\right]}$$

$$= \cfrac{1}{\cfrac{1}{5}\left(\cfrac{1}{44}+\cfrac{1}{42}+\cfrac{1}{51}+\cfrac{1}{49}+\cfrac{1}{46}\right)}$$

$$= \cfrac{1}{0.02166}$$

$$= \underline{46.17 \text{ mi/h}}$$

Note that the space-mean speed will always be lower than the time-mean speed, unless all vehicles are traveling at the exact same speed, in which case the two measures will be equal.

Finally, traffic density is defined as

$$k = \frac{n}{l} \qquad (5.10)$$

where

k = traffic density in vehicles per unit distance,
n = number of vehicles occupying some length of roadway at some specified time, and
l = length of roadway.

The density can also be related to the individual spacing between successive vehicles (measured from front bumper to front bumper). The roadway length, l, in Eq. 5.10 can be defined as

$$l = \sum_{i=1}^{n} s_i \qquad (5.11)$$

where

s_i = spacing of the ith vehicle (the distance between vehicles i and $i-1$, measured from front bumper to front bumper), and
n = number of measured vehicle spacings.

Substituting Eq. 5.11 into Eq. 5.10 gives

$$k = \frac{n}{\sum_{i=1}^{n} s_i} \qquad (5.12)$$

or

$$k = \frac{1}{\bar{s}} \tag{5.13}$$

where $\bar{s}$ = average spacing ($\sum s_i / n$) in unit distance per vehicle.

Time headway and spacing are referred to as microscopic measures because they describe characteristics specific to individual pairs of vehicles within the traffic stream. Measures that describe the traffic stream as a whole, such as flow, average speed, and density, are referred to as macroscopic measures. As indicated by the preceding equations, the microscopic measures can be aggregated and related to the macroscopic measures.

Based on the definitions presented, a simple identity provides the basic relationship among traffic flow, speed (space-mean), and density (denoting space-mean speed, $\bar{u}_s$, as simply u for notational convenience):

$$q = uk \tag{5.14}$$

where

q = flow, typically in units of veh/h,
u = speed (space-mean speed), typically in units of mi/h (km/h), and
k = density, typically in units of veh/mi (veh/km).

EXAMPLE 5.2

Vehicle time headways and spacings were measured at a point along a highway, from a single lane, over the course of an hour. The average values were calculated as 2.5 s/veh for headway and 200 ft/veh (61 m/veh) for spacing. Calculate the average speed of the traffic.

SOLUTION

To calculate the average speed of the traffic, the fundamental relationship in Eq. 5.14 is used. To begin, the flow and density need to be calculated from the headway and spacing data. Flow is determined from Eq. 5.4 as

$$q = \frac{1}{2.5 \text{ s/veh}}$$
$$= 0.40 \text{ veh/s}$$

or, because the data were collected for an hour,

$$q = 0.40 \text{ veh/s} \times 3600 \text{ s/h}$$
$$= 1440 \text{ veh/h}$$

Density is determined from Eq. 5.13 as

$$k = \frac{1}{200 \text{ ft/veh}}$$
$$= 0.005 \text{ veh/ft}$$

or, applying this spacing over the course of one mile,

$$k = 0.005 \text{ veh/ft} \times 5280 \text{ ft/mi}$$
$$= 26.4 \text{ veh/mi}$$

Now applying Eq. 5.14, after rearranging to solve for speed, gives

$$u = \frac{q}{k}$$
$$= \frac{1440 \text{ veh/h}}{26.4 \text{ veh/mi}}$$
$$= \underline{54.5 \text{ mi/h}}$$

Note that the average speed of traffic can be determined directly from the average headway and spacing values, as follows:

$$u = \frac{\bar{s}}{\bar{h}}$$
$$= \frac{200 \text{ ft/veh}}{2.5 \text{ s/veh}}$$
$$= \underline{80 \text{ ft/s } (54.5 \text{ mi/h})}$$

5.3 BASIC TRAFFIC STREAM MODELS

While the preceding definitions and relationships provide the basis for the measurement and calculation of traffic stream parameters, it is essential to also understand the interaction of the individual macroscopic measures in order to fully analyze the operational performance of the traffic stream. The models that describe these interactions are discussed in the following sections, and it will be shown that Eq. 5.14 serves the important function of linking specific models of traffic into a consistent, generalized model.

5.3.1 Speed-Density Model

The most intuitive starting point for developing a consistent, generalized traffic model is to focus on the relationship between speed and density. To begin, consider a section of highway with only a single vehicle on it. Under these conditions, the density (veh/mi) will be very low and the driver will be able to travel freely at a speed close to the design speed of the highway. This speed is referred to as the free-flow speed because vehicle speed is not inhibited by the presence of other vehicles. As more and more vehicles begin to use a section of highway, the traffic density will increase and the average operating speed of vehicles will decline from the free-flow value as drivers slow to allow for the maneuvers of other vehicles. Eventually, the highway section will become so congested (will have such a high density) that the traffic will come to a stop ($u = 0$), and the density will be determined by the length of the vehicles and the spaces that drivers leave between them. This high-density condition is referred to as the jam density.

One possible representation of the process described above is the linear relationship shown in Fig. 5.1. Mathematically, such a relationship can be expressed as

$$u = u_f\left(1 - \frac{k}{k_j}\right) \tag{5.15}$$

where

u = space-mean speed in mi/h (km/h),
u_f = free-flow speed in mi/h (km/h),
k = density in veh/mi (veh/km), and
k_j = jam density in veh/mi (veh/km).

The advantage of using a linear representation of the speed-density relationship is that it provides a basic insight into the relationships among traffic flow, speed, and density interactions without having these insights clouded by the additional complexity that a nonlinear speed-density relationship introduces. However, it is important to note that field studies have shown that the speed-density relationship tends to be nonlinear at low densities and high densities (those that approach the jam density). In fact, the overall speed-density relationship is better represented by three relationships: (1) a nonlinear relationship at low densities that has speed slowly declining from the free-flow value, (2) a linear relationship over the large medium-

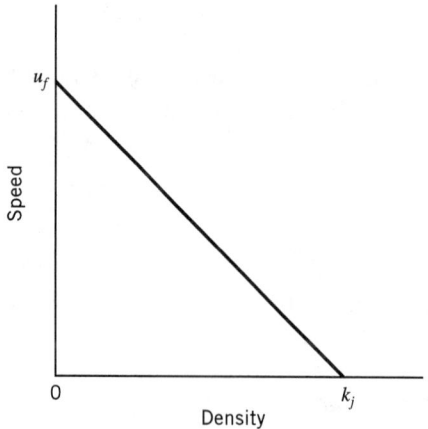

Figure 5.1 Illustration of a typical linear speed-density relationship.

density region (speed declining linearly with density as shown in Eq. 5.15), and (3) a nonlinear relationship near the jam density as the speed asymptotically approaches zero with increasing density. For the purposes of exposition, we present only traffic stream models that are based on the assumption of a linear speed-density relationship. Examples of nonlinear speed-density relationships are provided elsewhere [Pipes 1967; Drew 1965].

5.3.2 Flow-Density Model

Using the assumption of a linear speed-density relationship as shown in Eq. 5.15, a parabolic flow-density model can be obtained by substituting Eq. 5.15 into Eq. 5.14:

$$q = u_f \left(k - \frac{k^2}{k_j} \right) \qquad (5.16)$$

where all terms are as defined previously.

The general form of Eq. 5.16 is shown in Fig. 5.2. Note in this figure that the maximum flow rate, q_{cap}, represents the highest rate of traffic flow that the highway is capable of handling. This is referred to as the traffic flow at capacity, or simply the capacity of the roadway. The traffic density that corresponds to this capacity flow rate is k_{cap}, and the corresponding speed is u_{cap}. Equations for q_{cap}, k_{cap}, and u_{cap} can be derived by differentiating Eq. 5.16, because at maximum flow

$$\frac{dq}{dk} = u_f \left(1 - \frac{2k}{k_j} \right) = 0 \qquad (5.17)$$

and because the free-flow speed (u_f) is not equal to zero,

$$k_{cap} = \frac{k_j}{2} \qquad (5.18)$$

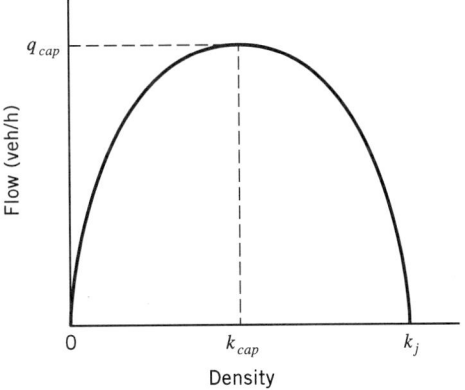

Figure 5.2 Illustration of the parabolic flow-density relationship.

Substituting Eq. 5.18 into Eq. 5.15 gives

$$u_{cap} = u_f\left(1 - \frac{k_j}{2k_j}\right) \qquad (5.19)$$
$$= \frac{u_f}{2}$$

and using Eq. 5.18 and Eq. 5.19 in Eq. 5.14 gives

$$q_{cap} = u_{cap} k_{cap}$$
$$= \frac{u_f k_j}{4} \qquad (5.20)$$

5.3.3 Speed-Flow Model

Again returning to the linear speed-density model (Eq. 5.15), a corresponding speed-flow model can be developed by rearranging Eq. 5.15 to

$$k = k_j\left(1 - \frac{u}{u_f}\right) \qquad (5.21)$$

and by substituting Eq. 5.21 into Eq. 5.14,

$$q = k_j\left(u - \frac{u^2}{u_f}\right) \qquad (5.22)$$

The speed-flow model defined by Eq. 5.22 again gives a parabolic function, as shown in Fig. 5.3. Note that Fig. 5.3 shows that two speeds are possible for flows, q, up to the highway's capacity, q_{cap} (this follows from the two densities possible for given flows as shown in Fig. 5.2). It is desirable, for any given flow, to keep the average space-mean speed on the upper portion of the speed-flow curve (above u_{cap}). When speeds drop below u_{cap}, traffic is in a highly congested and unstable condition.

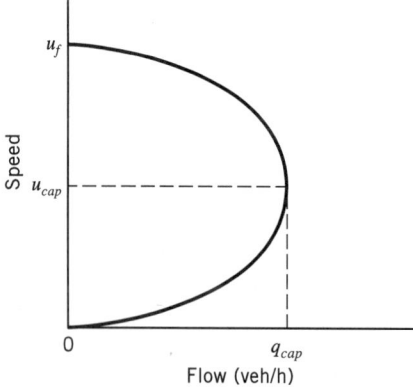

Figure 5.3 Illustration of the parabolic speed-flow relationship.

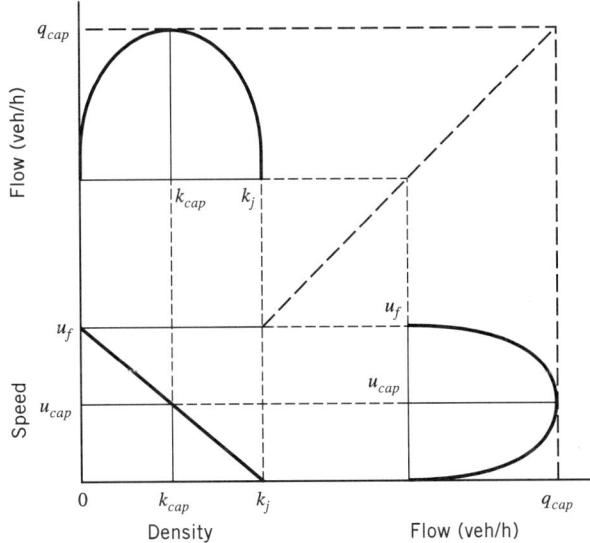

Figure 5.4 Flow-density, speed-density, and speed-flow relationships (assuming a linear speed-density model).

All of the flow, speed, and density relationships and their interactions are graphically represented in Fig. 5.4.

EXAMPLE 5.3

A section of highway is known to have a free-flow speed of 55 mi/h (88.5 km/h) and a capacity of 3300 veh/h. In a given hour, 2100 vehicles were counted at a specified point along this highway section. If the linear speed-density relationship shown in Eq. 5.15 applies, what would you estimate the space-mean speed of these 2100 vehicles to be?

SOLUTION

The jam density is first determined from Eq. 5.20 as

$$k_j = \frac{4q_{cap}}{u_f}$$

$$= \frac{4 \times 3300}{55}$$

$$= 240.0 \text{ veh/mi}$$

Rearranging Eq. 5.22 to solve for u,

$$\frac{k_j}{u_f}u^2 - k_j u + q = 0$$

Substituting,

$$\frac{240.0}{55}u^2 - 240.0u + 2100 = 0$$

which gives $u = \underline{44.08 \text{ mi/h}}$ or $\underline{10.92 \text{ mi/h}}$. Both of these speeds are feasible, as shown in Fig. 5.3.

5.4 MODELS OF TRAFFIC FLOW

With the basic relationships among traffic flow, speed, and density formalized, attention can now be directed toward a more microscopic view of traffic flow. That is, instead of simply modeling the number of vehicles passing a specified point on a highway in some time interval, there is considerable analytic value in modeling the time between the arrivals of successive vehicles (the concept of vehicle time headway presented earlier). The most simplistic approach to vehicle arrival modeling is to assume that all vehicles are equally or uniformly spaced. This results in what is termed a deterministic, uniform arrival pattern. Under this assumption, if the traffic flow is 360 veh/h, the number of vehicles arriving in any 5-minute time interval is 30 and the headway between all vehicles is 10 seconds (because h will equal $3600/q$). However, actual observations show that such uniformity of traffic flow is not always realistic because some 5-minute intervals are likely to have more or less traffic flow than other 5-minute intervals. Thus a representation of vehicle arrivals that goes beyond the deterministic, uniform assumption is often warranted.

5.4.1 Poisson Model

Models that account for the nonuniformity of flow are derived by assuming that the pattern of vehicle arrivals (at a specified point) corresponds to some random process. The problem then becomes one of selecting a probability distribution that is a reasonable representation of observed traffic arrival patterns. An example of such a distribution is the Poisson distribution (the limitations of which will be discussed later), which is expressed as

$$P(n) = \frac{(\lambda t)^n e^{-\lambda t}}{n!} \tag{5.23}$$

where

$P(n)$ = probability of having n vehicles arrive in time t,
λ = average vehicle flow or arrival rate in vehicles per unit time,
t = duration of the time interval over which vehicles are counted, and
e = base of the natural logarithm ($e = 2.718$).

EXAMPLE 5.4

An observer counts 360 veh/h at a specific highway location. Assuming that the arrival of vehicles at this highway location is Poisson distributed, estimate the probabilities of having 0, 1, 2, 3, 4, and 5 or more vehicles arriving over a 20-second time interval.

5.4 Models of Traffic Flow

SOLUTION

The average arrival rate, λ, is 360 veh/h, or 0.1 vehicles per second (veh/s). Using this in Eq. 5.23 with $t = 20$ seconds, the probabilities of having exactly 0, 1, 2, 3, and 4 vehicles arrive are

$$P(0) = \frac{(0.1 \times 20)^0 e^{-0.1(20)}}{0!} = \underline{\underline{0.135}}$$

$$P(1) = \frac{(0.1 \times 20)^1 e^{-0.1(20)}}{1!} = \underline{\underline{0.271}}$$

$$P(2) = \frac{(0.1 \times 20)^2 e^{-0.1(20)}}{2!} = \underline{\underline{0.271}}$$

$$P(3) = \frac{(0.1 \times 20)^3 e^{-0.1(20)}}{3!} = \underline{\underline{0.180}}$$

$$P(4) = \frac{(0.1 \times 20)^4 e^{-0.1(20)}}{4!} = \underline{\underline{0.090}}$$

For five or more vehicles,

$$\begin{aligned} P(n \geq 5) &= 1 - P(n < 5) \\ &= 1 - 0.135 - 0.271 - 0.271 - 0.180 - 0.090 \\ &= \underline{\underline{0.053}} \end{aligned}$$

A histogram of these probabilities is shown in Fig. 5.5.

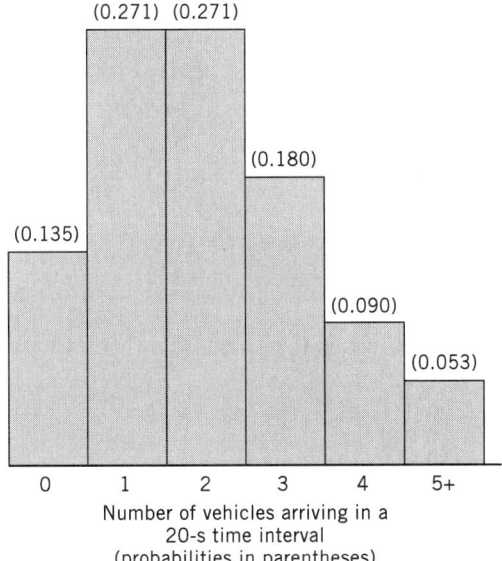

Figure 5.5 Histogram of the Poisson distribution for $\lambda = 0.1$ vehicles per second.

EXAMPLE 5.5

Traffic data are collected in 60-second intervals at a specific highway location as shown in Table 5.1. Assuming the traffic arrivals are Poisson distributed and continue at the same rate as that observed in the 15 time periods shown, what is the probability that six or more vehicles will arrive in each of the next three 60-second time intervals (12:15 P.M. to 12:16 P.M., 12:16 P.M. to 12:17 P.M., and 12:17 P.M. to 12:18 P.M.)?

SOLUTION

Table 5.1 shows that a total of 101 vehicles arrive in the 15-minute period from 12:00 P.M. to 12:15 P.M. Thus the average arrival rate, λ, is 0.112 veh/s (101/900). As in Example 5.4, Eq. 5.23 is applied to find the probabilities of exactly 0, 1, 2, 3, 4, and 5 vehicles arriving.

Applying Eq. 5.23, with $\lambda = 0.112$ veh/s and $t = 60$ seconds, the probabilities of having 0, 1, 2, 3, 4, and 5 vehicles arriving in a 60-second time interval are (using $\lambda t = 6.733$)

$$P(0) = \frac{(6.733)^0 e^{-6.733}}{0!} = \underline{\underline{0.0012}}$$

$$P(1) = \frac{(6.733)^1 e^{-6.733}}{1!} = \underline{\underline{0.008}}$$

$$P(2) = \frac{(6.733)^2 e^{-6.733}}{2!} = \underline{\underline{0.027}}$$

$$P(3) = \frac{(6.733)^3 e^{-6.733}}{3!} = \underline{\underline{0.0606}}$$

$$P(4) = \frac{(6.733)^4 e^{-6.733}}{4!} = \underline{\underline{0.102}}$$

$$P(5) = \frac{(6.733)^5 e^{-6.733}}{5!} = \underline{\underline{0.137}}$$

The summation of these probabilities is the probability that 0 to 5 vehicles will arrive in any given 60-second time interval, which is

$$P(n \leq 5) = \sum_{n=0}^{5} P(n)$$

$$= 0.0012 + 0.008 + 0.027 + 0.0606 + 0.102 + 0.137$$

$$= 0.3358$$

So 1 minus $P(n \leq 5)$ is the probability that 6 or more vehicles will arrive in any 60-second time interval, which is

5.4 Models of Traffic Flow

Table 5.1 Observed Traffic Data for Example 5.5

Time period	Observed number of vehicles
12:00 P.M. to 12:01 P.M.	3
12:01 P.M. to 12:02 P.M.	5
12:02 P.M. to 12:03 P.M.	4
12:03 P.M. to 12:04 P.M.	10
12:04 P.M. to 12:05 P.M.	7
12:05 P.M. to 12:06 P.M.	4
12:06 P.M. to 12:07 P.M.	8
12:07 P.M. to 12:08 P.M.	11
12:08 P.M. to 12:09 P.M.	9
12:09 P.M. to 12:10 P.M.	5
12:10 P.M. to 12:11 P.M.	3
12:11 P.M. to 12:12 P.M.	10
12:12 P.M. to 12:13 P.M.	9
12:13 P.M. to 12:14 P.M.	7
12:14 P.M. to 12:15 P.M.	6

$$P(n \geq 6) = 1 - P(n \leq 5)$$
$$= 1 - 0.3358$$
$$= 0.6642$$

The probability that 6 or more vehicles will arrive in three successive time intervals (t_1, t_2, and t_3) is simply the product of probabilities, which is

$$P(n \geq 6) \text{ for three successive time periods} = \prod_{t_i=1}^{3} P(n \geq 6)$$
$$= (0.6642)^3$$
$$= \underline{\underline{0.293}}$$

The assumption of Poisson vehicle arrivals also implies a distribution of the time intervals between the arrivals of successive vehicles (time headway). To show this, note that the average arrival rate is

$$\lambda = \frac{q}{3600} \quad (5.24)$$

where

λ = average vehicle arrival rate in veh/s,
q = flow in veh/h, and
3600 = number of seconds per hour.

Substituting Eq. 5.24 into Eq. 5.23 gives

$$P(n) = \frac{(qt/3600)^n e^{-qt/3600}}{n!} \qquad (5.25)$$

Note that the probability of having no vehicles arrive in a time interval of length t, $P(0)$, is equivalent to the probability of a vehicle headway, h, being greater than or equal to the time interval t. So from Eq. 5.25,

$$P(0) = P(h \geq t)$$

$$= e^{-qt/3600} \qquad (5.26)$$

This distribution of vehicle headways is known as the negative exponential distribution and is often simply referred to as the exponential distribution.

EXAMPLE 5.6

Consider the traffic situation in Example 5.4 (360 veh/h). Again assume that the vehicle arrivals are Poisson distributed. What is the probability that the gap between successive vehicles will be less than 8 seconds, and what is the probability that the gap between successive vehicles will be between 8 and 10 seconds?

SOLUTION

By definition, $P(h < t) = 1 - P(h \geq t)$. This expression gives the probability that the gap will be less than 8 seconds as

$$P(h < 8) = 1 - e^{-360(8)/3600}$$

$$= 1 - 0.449$$

$$= \underline{0.551}$$

To determine the probability that the gap will be between 8 and 10 seconds, compute the probability that the gap will be greater than or equal to 10 seconds:

$$P(h \geq 10) = e^{-360(10)/3600}$$

$$= 0.368$$

So the probability that the gap will be between 8 and 10 seconds is $\underline{0.081}$ ($1 - 0.551 - 0.368$).

To help in visualizing the shape of the exponential distribution, Fig. 5.6 shows the probability distribution implied by Eq. 5.26, with the flow, q, equal to 360 veh/h as in Example 5.4.

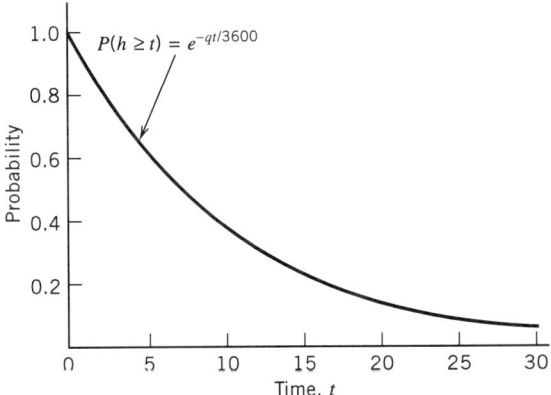

Figure 5.6 Exponentially distributed probabilities of headways greater than or equal to t, with $q = 360$ veh/h.

5.4.2 Limitations of the Poisson Model

Empirical observations have shown that the assumption of Poisson-distributed traffic arrivals is most realistic in lightly congested traffic conditions. As traffic flows become heavily congested or when traffic signals cause cyclical traffic stream disturbances, other distributions of traffic flow become more appropriate. The primary limitation of the Poisson model of vehicle arrivals is the constraint imposed by the Poisson distribution that the mean of period observations equals the variance. For example, the mean of period-observed traffic in Example 5.5 is 6.733 and the corresponding variance, σ^2, is 7.210. Because these two values are close, the Poisson model was appropriate for this example. If the variance is significantly greater than the mean, the data are said to be overdispersed, and if the variance is significantly less than the mean, the data are said to be underdispersed. In either case the Poisson distribution is no longer appropriate, and another distribution should be used. Such distributions are discussed in detail in more specialized sources [Transportation Research Board 1975; Poch and Mannering 1996].

5.5 QUEUING THEORY AND TRAFFIC FLOW ANALYSIS

The formation of traffic queues during congested periods is a source of considerable time delay and results in a loss of highway performance. Under extreme conditions, queuing delay can account for 90% or more of a motorist's total trip travel time. Given this, it is essential in traffic analysis that one develop a clear understanding of the characteristics of queue formation and dissipation along with mathematical formulations that can predict queuing-related elements.

As is well known, the problem of queuing is not unique to traffic analysis. Many non-transportation fields, such as the design and operation of industrial plants, retail stores, service-oriented industries, and computer networks, must also give serious consideration to the problem of queuing. The impact that queues have on performance and productivity in manufacturing, retailing, and other fields has led to numerous theories of queuing behavior (the process by which queues form and dissipate). As will be shown, the models of traffic flow presented earlier (uniform, deterministic arrivals and Poisson arrivals) will form the basis for studying traffic queues within the more general context of queuing theory.

5.5.1 Dimensions of Queuing Models

The purpose of traffic queuing models is to provide a means to estimate important measures of highway performance, including vehicle delay and traffic queue lengths. Such estimates are critical to roadway design (the required length of left-turn bays and the number of lanes at intersections) and traffic operations control, including the timing of traffic signals at intersections.

Queuing models are derived from underlying assumptions regarding arrival patterns, departure characteristics, and queue disciplines. Traffic arrival patterns were explored in Section 5.4, where, given an average vehicle arrival rate (λ), two possible distributions of the time between the arrival of successive vehicles were considered:

1. Equal time intervals (derived from the assumption of uniform, deterministic arrivals)

2. Exponentially distributed time intervals (derived from the assumption of Poisson-distributed arrivals)

In addition to vehicle arrival assumptions, the derivation of traffic queuing models requires assumptions relating to vehicle departure characteristics. Of particular interest is the distribution of the amount of time it takes a vehicle to depart—for example, the time to pass through an intersection at the beginning of a green signal, the time required to pay a toll at a toll booth, or the time a driver takes before deciding to proceed after stopping at a stop sign. As was the case for arrival patterns, given an average vehicle departure rate (denoted as μ, in vehicles per unit time), the assumption of a deterministic or exponential distribution of departure times is appropriate.

Another important aspect of queuing models is the number of available departure channels. For most traffic applications only one departure channel will exist, such as a highway lane or group of lanes passing through an intersection. However, multiple departure channels are encountered in some traffic applications, such as at toll booths on turnpikes and at entrances to bridges.

The final necessary assumption relates to the queue discipline. In this regard, two options have been popularized in the development of queuing models: first-in, first-out (FIFO), indicating that the first vehicle to arrive is the first to depart; and last-in, first-out (LIFO), indicating that the last vehicle to arrive is the first to depart. For virtually all traffic-oriented queues, the FIFO queuing discipline is the more appropriate of the two.

Queuing models are often identified by three alphanumeric values. The first value indicates the arrival rate assumption, the second value gives the departure rate assumption, and the third value indicates the number of departure channels. For traffic arrival and departure assumptions, the uniform, deterministic distribution is denoted D

5.5 Queuing Theory and Traffic Flow Analysis

and the exponential distribution is denoted *M*. Thus a *D/D/*1 queuing model assumes deterministic arrivals and departures with one departure channel. Similarly, an *M/D/*1 queuing model assumes exponentially distributed arrival times, deterministic departure times, and one departure channel.

5.5.2 *D/D/*1 Queuing

The case of deterministic arrivals and departures with one departure channel (*D/D/*1 queue) is an excellent starting point in understanding queuing models because of its simplicity. The *D/D/*1 queue lends itself to an intuitive graphical or mathematical solution that is best illustrated by an example.

EXAMPLE 5.7

Vehicles arrive at an entrance to a recreational park. There is a single gate (at which all vehicles must stop), where a park attendant distributes a free brochure. The park opens at 8:00 A.M., at which time vehicles begin to arrive at a rate of 480 veh/h. After 20 minutes the arrival flow rate declines to 120 veh/h, and it continues at that level for the remainder of the day. If the time required to distribute the brochure is 15 seconds, and assuming *D/D/*1 queuing, describe the operational characteristics of the queue.

SOLUTION

Begin by putting arrival and departure rates into common units of vehicles per minute:

$$\lambda = \frac{480 \text{ veh/h}}{60 \text{ min/h}} = 8 \text{ veh/min} \quad \text{for } t \leq 20 \text{ min}$$

$$\lambda = \frac{120 \text{ veh/h}}{60 \text{ min/h}} = 2 \text{ veh/min} \quad \text{for } t > 20 \text{ min}$$

$$\mu = \frac{60 \text{ s/min}}{15 \text{ s/veh}} = 4 \text{ veh/min} \quad \text{for all } t$$

Equations for the total number of vehicles that have arrived and departed up to a specified time, *t*, can now be written. Define *t* as the number of minutes after the start of the queuing process (in this case the number of minutes after 8:00 A.M.). The total number of vehicle arrivals at time *t* is equal to

$$8t \quad \text{for } t \leq 20 \text{ min}$$

and

$$160 + 2(t - 20) \quad \text{for } t > 20 \text{ min}$$

Similarly, the number of vehicle departures is

$$4t \quad \text{for all } t$$

154 Chapter 5 Fundamentals of Traffic Flow and Queuing Theory

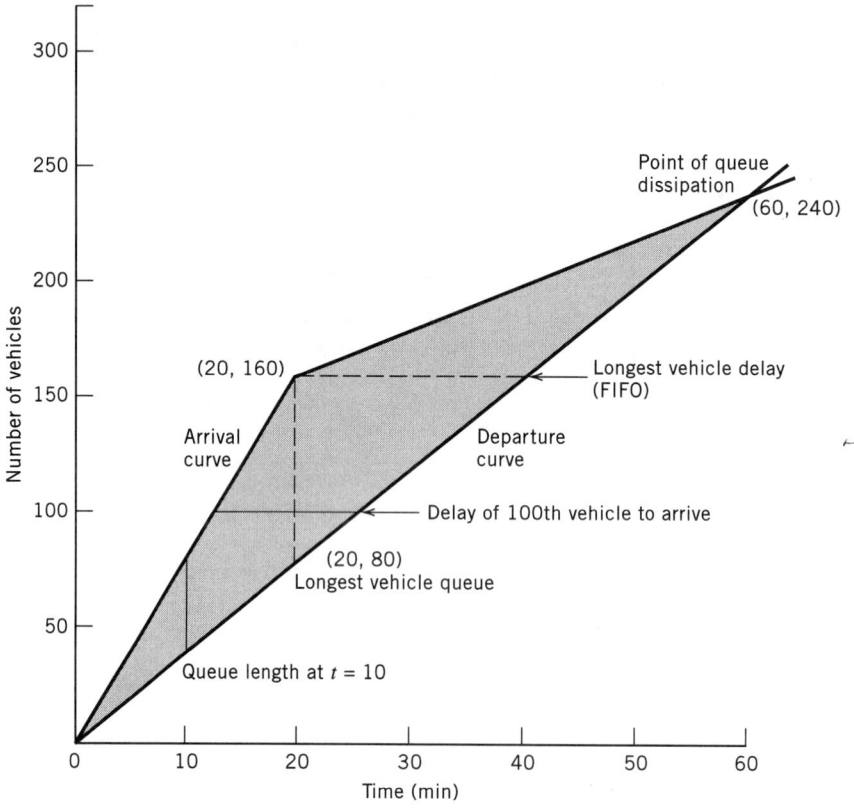

Figure 5.7 D/D/1 queuing diagram for Example 5.7.

The preceding equations can be illustrated graphically as shown in Fig. 5.7. When the arrival curve is above the departure curve, a queue condition exists. The point at which the arrival curve meets the departure curve is the moment when the queue dissipates (no more queue exists). In this example, the point of queue dissipation can be determined graphically by inspection of Fig. 5.7, or analytically by equating appropriate arrival and departure equations, that is,

$$160 + 2(t - 20) = 4t$$

Solving for t gives $t = 60$ minutes. Thus the queue that began to form at 8:00 A.M. will dissipate 60 minutes later (9:00 A.M.), at which time 240 vehicles will have arrived and departed (4 veh/min × 60 min).

Another aspect of interest is individual vehicle delay. Under the assumption of a FIFO queuing discipline, the delay of an individual vehicle is given by the horizontal distance between arrival and departure curves starting from the time of the vehicle's arrival in the queue. So, by inspection of Fig. 5.7, the 160th vehicle to arrive will have the longest delay, 20 minutes (the longest horizontal distance between arrival and departure curves), and vehicles arriving after the 239th vehicle will encounter no queue delay because the queue will have dissipated and the departure rate will continue to exceed the arrival rate. It follows that with the LIFO queuing discipline, the first vehicle to arrive would have to wait until the entire queue clears (60 minutes of delay).

The total length of the queue at a specified time, expressed as the number of vehicles, is given by the vertical distance between arrival and departure curves at that time. For example, at 10 minutes after the start of the queuing process (8:10 A.M.) the queue is 40 vehicles long, and the longest queue (longest vertical distance between arrival and departure curves) will occur at $t = 20$ minutes and is 80 vehicles long (see Fig. 5.7).

Total vehicle delay, defined as the summation of the delays for the individual vehicles, is given by the total area between the arrival and departure curves (see Fig. 5.7) and, in this case, is in units of vehicle-minutes. In this example, the area between the arrival and departure curves can be determined by summing triangular areas, giving total delay, D_t, as

$$D_t = \tfrac{1}{2}(80 \times 20) + \tfrac{1}{2}(80 \times 40)$$

$$= \underline{2400 \text{ veh-min}}$$

Finally, because 240 vehicles encounter queuing delay (as previously determined), the average delay per vehicle is 10 minutes (2400 veh-min/240 veh), and the average queue length is 40 vehicles (2400 veh-min/60 min).

EXAMPLE 5.8

After observing arrivals and departures at a highway toll booth over a 60-minute time period, the observer notes that the arrival and departure rates (or service rates) are deterministic, but instead of being uniform, they change over time according to a known function. The arrival rate is given by the function $\lambda(t) = 2.2 + 0.17t - 0.0032t^2$, and the departure rate is given by $\mu(t) = 1.2 + 0.07t$, where t is in minutes after the beginning of the observation period and $\lambda(t)$ and $\mu(t)$ are in vehicles per minute. Determine the total vehicle delay at the toll booth and the longest queue, assuming D/D/1 queuing.

SOLUTION

Note that this problem is an example of a time-dependent deterministic queue because the deterministic arrival and departure rates change over time. Begin by computing the time to queue dissipation by equating vehicle arrivals and departures:

$$\int_0^t 2.2 + 0.17t - 0.0032t^2 \, dt = \int_0^t 1.2 + 0.07t \, dt$$

$$2.2t + 0.085t^2 - 0.00107t^3 = 1.2t + 0.035t^2$$

$$-0.00107t^3 + 0.05t^2 + t = 0$$

which gives $t = 61.8$ minutes. Therefore, the total vehicle delay (the area between the arrival and departure functions) is

$$D_t = \int_0^{61.8} 2.2t + 0.085t^2 - 0.00107t^3 \, dt - \int_0^{61.8} 1.2t + 0.035t^2 \, dt$$

$$= 1.1t^2 + 0.0283t^3 - 0.0002675t^4 - 0.6t^2 - 0.0117t^3 \Big|_0^{61.8}$$

$$= -0.0002675(61.8)^4 + 0.0166(61.8)^3 + 0.5(61.8)^2$$

$$= \underline{1925.8 \text{ veh-min}}$$

The queue length (in vehicles) at any time t is given by the function

$$Q(t) = \int_0^t 2.2 + 0.17t - 0.0032t^2 \, dt - \int_0^t 1.2 + 0.07t \, dt$$

$$= -0.00107t^3 + 0.05t^2 + t$$

Solving for the time at which the maximum queue length occurs,

$$\frac{dQ(t)}{dt} = -0.00321t^2 + 0.1t + 1 = 0$$

$$t = \underline{39.12 \text{ min}}$$

Substituting with $t = 39.12$ minutes gives the maximum queue length:

$$Q(39.12) = -0.00107t^3 + 0.05t^2 + t \Big|_0^{39.12}$$

$$= -0.00107(39.12)^3 + 0.05(39.12)^2 + 39.12$$

$$= \underline{51.58 \text{ veh}}$$

5.5.3 M/D/1 Queuing

The assumption of exponentially distributed times between the arrivals of successive vehicles (Poisson arrivals) will, in some cases, give a more realistic representation of traffic flow than the assumption of uniformly distributed arrival times. Therefore, the M/D/1 queue (exponentially distributed arrivals, deterministic departures, and one departure channel) has some important applications within the traffic analysis field. Although a graphical solution to an M/D/1 queue is difficult, a mathematical solution is straightforward. Defining a new term (traffic intensity) for the ratio of average arrival to departure rates as

$$\rho = \frac{\lambda}{\mu} \qquad (5.27)$$

where

ρ = traffic intensity, unitless,
λ = average arrival rate in vehicles per unit time, and
μ = average departure rate in vehicles per unit time,

and assuming that ρ is less than 1, it can be shown that for an *M/D/*1 queue the following queuing performance equations apply:

$$\overline{Q} = \frac{\rho^2}{2(1-\rho)} \qquad (5.28)$$

$$\overline{w} = \frac{\rho}{2\mu(1-\rho)} \qquad (5.29)$$

$$\overline{t} = \frac{2-\rho}{2\mu(1-\rho)} \qquad (5.30)$$

where

$\overline{Q}$ = average length of queue in vehicles,
$\overline{w}$ = average waiting time in the queue, in unit time per vehicle,
$\overline{t}$ = average time spent in the system (the summation of average waiting time in the queue and average departure time), in unit time per vehicle, and
Other terms are as defined previously.

It is important to note that under the assumption that the traffic intensity is less than 1 ($\lambda < \mu$), the *D/D/*1 queue will predict no queue formation. However a queuing model that is derived based on random arrivals or departures, such as the *M/D/*1 queuing model, will predict queue formations under such conditions. Also, note that the *M/D/*1 queuing model presented here is based on steady-state conditions (constant average arrival and departure rates), with randomness arising from the assumed probability distribution of arrivals. This contrasts with the time-varying deterministic queuing case, presented in Example 5.8, in which arrival and departure rates changed over time, but randomness was not present.

EXAMPLE 5.9

Consider the entrance to the recreational park described in Example 5.7. However, let the average arrival rate be 180 veh/h and Poisson distributed (exponential times between arrivals) over the entire period from park opening time (8:00 A.M.) until closing at dusk. Compute the average length of queue (in vehicles), average waiting time in the queue, and average time spent in the system, assuming *M/D/*1 queuing.

SOLUTION

Putting arrival and departure rates into common units of vehicles per minute gives

$$\lambda = \frac{180 \text{ veh/h}}{60 \text{ min/h}} = 3 \text{ veh/min} \quad \text{for all } t$$

$$\mu = \frac{60 \text{ s/min}}{15 \text{ s/veh}} = 4 \text{ veh/min} \quad \text{for all } t$$

and

$$\rho = \frac{\lambda}{\mu} = \frac{3 \text{ veh/min}}{4 \text{ veh/min}} = 0.75$$

For the average length of queue (in vehicles), Eq. 5.28 is applied:

$$\overline{Q} = \frac{0.75^2}{2(1-0.75)}$$

$$= \underline{1.125 \text{ veh}}$$

For average waiting time in the queue, Eq. 5.29 gives

$$\overline{w} = \frac{0.75}{2(4)(1-0.75)}$$

$$= \underline{0.375 \text{ min/veh}}$$

For average time spent in the system [queue time plus departure (service) time], Eq. 5.30 is used:

$$\overline{t} = \frac{2-0.75}{2(4)(1-0.75)}$$

$$= \underline{0.625 \text{ min/veh}}$$

or, alternatively, because the departure (service) time is $1/\mu$ (the 0.25 minutes it takes the park attendant to distribute the brochure),

$$\overline{t} = \overline{w} + \frac{1}{\mu}$$

$$= 0.375 + \frac{1}{4}$$

$$= \underline{0.625 \text{ min/veh}}$$

5.5.4 M/M/1 Queuing

A queuing model that assumes one departure channel and exponentially distributed departure times in addition to exponentially distributed arrival times (an M/M/1 queue) is applicable in some traffic applications. For example, exponentially distributed departure patterns might be a reasonable assumption at a toll booth, where some arriving drivers have the correct toll and can be processed quickly, and others do not have the correct toll, producing a distribution of departures about some mean departure rate. Under standard *M/M/1* assumptions, it can be shown that the following queuing performance equations apply (again assuming that ρ is less than 1):

$$\overline{Q} = \frac{\rho^2}{1-\rho} \quad (5.31)$$

$$\overline{w} = \frac{\lambda}{\mu(\mu-\lambda)} \quad (5.32)$$

$$\overline{t} = \frac{1}{\mu-\lambda} \quad (5.33)$$

where

$\overline{Q}$ = average length of queue in vehicles,
$\overline{w}$ = average waiting time in the queue, in unit time per vehicle,
$\overline{t}$ = average time spent in the system ($\overline{w} + 1/\mu$), in unit time per vehicle, and
Other terms are as defined previously.

EXAMPLE 5.10

Assume that the park attendant in Examples 5.7 and 5.9 takes an average of 15 seconds to distribute brochures, but the distribution time varies depending on whether park patrons have questions relating to park operating policies. Given an average arrival rate of 180 veh/h as in Example 5.9, compute the average length of queue (in vehicles), average waiting time in the queue, and average time spent in the system, assuming *M/M/1* queuing.

SOLUTION

Using the average arrival rate, departure rate, and traffic intensity as determined in Example 5.9, the average length of queue is (from Eq. 5.31)

$$\overline{Q} = \frac{0.75^2}{1-0.75}$$

$$= \underline{2.25 \text{ veh}}$$

the average waiting time in the queue is (from Eq. 5.32)

$$\overline{w} = \frac{3}{4(4-3)}$$

$$= \underline{0.75 \text{ min/veh}}$$

and the average time spent in the system is (from Eq. 5.33)

$$\bar{t} = \frac{1}{4-3}$$

$$= 1 \text{ min/veh}$$

5.5.5 M/M/N Queuing

A more general formulation of the M/M/1 queue is the M/M/N queue, where N is the total number of departure channels. M/M/N queuing is a reasonable assumption at toll booths on turnpikes or at toll bridges, where there is often more than one departure channel available (more than one toll booth open). A parking lot is another example, with N being the number of parking stalls in the lot and the departure rate, μ, being the exponentially distributed times of parking duration. M/M/N queuing is also frequently encountered in non-transportation applications such as checkout lines at retail stores, security checks at airports, and so on.

The following equations describe the operational characteristics of M/M/N queuing. Note that unlike the equations for M/D/1 and M/M/1, which require that the traffic intensity, ρ, be less than 1, the following equations allow ρ to be greater than 1 but apply only when ρ/N (which is called the utilization factor) is less than 1.

$$P_0 = \frac{1}{\sum_{n_c=0}^{N-1} \frac{\rho^{n_c}}{n_c!} + \frac{\rho^N}{N!(1-\rho/N)}} \tag{5.34}$$

$$P_n = \frac{\rho^n P_0}{n!} \quad \text{for } n \leq N \tag{5.35}$$

$$P_n = \frac{\rho^n P_0}{N^{n-N} N!} \quad \text{for } n \geq N \tag{5.36}$$

$$P_{n>N} = \frac{P_0 \rho^{N+1}}{N! N (1-\rho/N)} \tag{5.37}$$

where

P_0 = probability of having no vehicles in the system,
P_n = probability of having n vehicles in the system,
$P_{n>N}$ = probability of waiting in a queue (the probability that the number of vehicles in the system is greater than the number of departure channels),
n = number of vehicles in the system,
N = number of departure channels,
n_c = departure channel number, and
ρ = traffic intensity (λ/μ).

$$\overline{Q} = \frac{P_0 \rho^{N+1}}{N!N} \left[\frac{1}{(1-\rho/N)^2} \right] \tag{5.38}$$

$$\overline{w} = \frac{\rho + \overline{Q}}{\lambda} - \frac{1}{\mu} \tag{5.39}$$

$$\overline{t} = \frac{\rho + \overline{Q}}{\lambda} \tag{5.40}$$

where

$\overline{Q}$ = average length of queue (in vehicles),
$\overline{w}$ = average waiting time in the queue, in unit time per vehicle,
$\overline{t}$ = average time spent in the system, in unit time per vehicle, and
Other terms are as defined previously.

EXAMPLE 5.11

At an entrance to a toll bridge, four toll booths are open. Vehicles arrive at the bridge at an average rate of 1200 veh/h, and at the booths, drivers take an average of 10 seconds to pay their tolls. Both the arrival and departure rates can be assumed to be exponentially distributed. How would the average queue length, time in the system, and probability of waiting in a queue change if a fifth toll booth were opened?

SOLUTION

Using the equations for M/M/N queuing, we first compute the four-booth case. Note that $\mu = 6$ veh/min and $\lambda = 20$ veh/min, and therefore $\rho = 3.333$. Also, because $\rho/N = 0.833$ (which is less than 1), Eqs. 5.34 to 5.40 can be used. The probability of having no vehicles in the system with four booths open (using Eq. 5.34) is

$$P_0 = \frac{1}{1 + \dfrac{3.333}{1!} + \dfrac{3.333^2}{2!} + \dfrac{3.333^3}{3!} + \dfrac{3.333^4}{4!(0.1667)}}$$

$$= 0.0213$$

The average queue length is (from Eq. 5.38)

$$\bar{Q} = \frac{0.0213(3.333)^5}{4!4}\left[\frac{1}{(0.1667)^2}\right]$$

$$= 3.287 \text{ veh}$$

The average time spent in the system is (from Eq. 5.40)

$$\bar{t} = \frac{3.333 + 3.287}{20}$$

$$= 0.331 \text{ min/veh}$$

And the probability of having to wait in a queue is (from Eq. 5.37)

$$P_{n>N} = \frac{0.0213(3.333)^5}{4!4(0.1667)}$$

$$= 0.548$$

With a fifth booth open, the probability of having no vehicles in the system is (from Eq. 5.34)

$$P_0 = \frac{1}{1 + \dfrac{3.333}{1!} + \dfrac{3.333^2}{2!} + \dfrac{3.333^3}{3!} + \dfrac{3.333^4}{4!} + \dfrac{3.333^5}{5!(0.3333)}}$$

$$= 0.0318$$

The average queue length is (from Eq. 5.38)

$$\bar{Q} = \frac{0.0318(3.333)^6}{5!5}\left[\frac{1}{(0.3333)^2}\right]$$

$$= 0.654 \text{ veh}$$

The average time spent in the system is (from Eq. 5.40)

$$\bar{t} = \frac{3.333 + 0.654}{20}$$

$$= 0.199 \text{ min/veh}$$

And the probability of having to wait in a queue is (from Eq. 5.37)

$$P_{n>N} = \frac{0.0318(3.333)^6}{5!5(0.3333)}$$

$$= 0.218$$

So opening a fifth booth would reduce the average queue length by 2.633 veh (3.287 − 0.654), the average time in the system by 0.132 min/veh (0.331 − 0.199), and the probability of waiting in a queue by 0.330 (0.548 − 0.218).

EXAMPLE 5.12

A convenience store has four available parking spaces. The owner predicts that the duration of customer shopping (the time that a customer's vehicle will occupy a parking space) is exponentially distributed with a mean of 6 minutes. The owner knows that in the busiest hour customer arrivals are exponentially distributed with a mean arrival rate of 20 customers per hour. What is the probability that a customer will not find an open parking space when arriving at the store?

SOLUTION

Putting mean arrival and departure rates in common units gives $\mu = 10$ veh/h and $\lambda = 20$ veh/h. So $\rho = 2.0$, and because $\rho/N = 0.5$ (which is less than 1), Eqs. 5.34 to 5.40 can be used. The probability of having no vehicles in the system with four parking spaces available (using Eq. 5.34) is

$$P_0 = \frac{1}{1 + \frac{2}{1!} + \frac{2^2}{2!} + \frac{2^3}{3!} + \frac{2^4}{4!(0.5)}}$$

$$= 0.1304$$

Thus the probability of not finding an open parking space upon arrival is (from Eq. 5.37)

$$P_{n>N} = \frac{0.1304(2)^5}{4!4(0.5)}$$

$$= \underline{0.087}$$

5.6 TRAFFIC ANALYSIS AT HIGHWAY BOTTLENECKS

Some of the most severe congestion problems occur at highway bottlenecks, which are defined as a portion of highway with a lower capacity (q_{cap}) than the incoming section of highway. This reduction in capacity can originate from a number of sources, including a decrease in the number of highway lanes and reduced shoulder widths (which tend to cause drivers to slow and thus effectively reduce highway capacity, as will be discussed in Chapter 6). There are two general types of traffic bottlenecks—those that are recurring and those that are incident induced. Recurring bottlenecks exist where the highway itself limits capacity—for example, by a physical reduction in the number of lanes. Traffic congestion at such bottlenecks results from recurring traffic flows that exceed the vehicle capacity of the highway in the bottleneck area. In contrast, incident-induced bottlenecks occur as a result of vehicle breakdowns or accidents that effectively reduce highway capacity by restricting the through movement of traffic. Because incident-induced bottlenecks are unanticipated and temporary in nature, they have features that distinguish them from recurring bottlenecks, such as the possibility that the capacity resulting from an incident-induced bottleneck may change over time. For example, an accident may initially stop traffic flow completely, but as the wreckage is cleared, partial capacity (one lane open) may be provided for a period of time before full capacity is eventually restored. A feature shared by recurring and incident-induced bottlenecks is the adjustment in traffic flow that may occur as travelers choose other routes and/or different trip departure times, to avoid the bottleneck area, in response to visual information or traffic advisories.

The analysis of traffic flow at bottlenecks can be undertaken using the queuing models discussed in Section 5.5. The most intuitive approach to analyzing traffic congestion at bottlenecks is to assume $D/D/1$ queuing.

EXAMPLE 5.13

An incident occurs on a freeway that has a capacity in the northbound direction, before the incident, of 4000 veh/h and a constant flow of 2900 veh/h during the morning commute (no adjustments to traffic flow result from the incident). At 8:00 A.M. a traffic accident closes the freeway to all traffic. At 8:12 A.M. the freeway is partially opened with a capacity of 2000 veh/h. Finally, the wreckage is removed, and the freeway is restored to full capacity (4000 veh/h) at 8:31 A.M. Assume $D/D/1$ queuing to determine time of queue dissipation, longest queue length, total delay, average delay per vehicle, and longest wait of any vehicle (assuming FIFO).

SOLUTION

Let μ be the full-capacity departure rate and μ_r be the restrictive partial-capacity departure rate. Putting arrival and departure rates in common units of vehicles per minute,

$$\mu = \frac{4000 \text{ veh/h}}{60 \text{ min/h}} = 66.67 \text{ veh/min}$$

$$\mu_r = \frac{2000 \text{ veh/h}}{60 \text{ min/h}} = 33.33 \text{ veh/min}$$

$$\lambda = \frac{2900 \text{ veh/h}}{60 \text{ min/h}} = 48.33 \text{ veh/min}$$

The arrival rate is constant over the entire time period, and the total number of vehicles is equal to λt, where t is the number of minutes after 8:00 A.M. The total number of departing vehicles is

$$\begin{cases} 0 & \text{for } t \leq 12 \text{ min} \\ \mu_r(t - 12) & \text{for } 12 \text{ min} < t \leq 31 \text{ min} \\ 633.33 + \mu(t - 31) & \text{for } t > 31 \text{ min} \end{cases}$$

Note that the value of 633.33 in the departure rate function for $t > 31$ is based on the preceding departure rate function [$33\frac{1}{3}(31 - 12)$]. These arrival and departure rates can be represented graphically as shown in Fig. 5.8. As discussed in Section 5.5, for $D/D/1$ queuing, the queue will dissipate at the intersection point of the arrival and departure curves, which can be determined as

$$\lambda t = 633.33 + \mu(t - 31) \quad \text{or} \quad t = \underline{78.16 \text{ min}} \text{ (just after 9:18 A.M.)}$$

At this time a total of 3777.5 vehicles (48.33×78.16) will have arrived and departed (for the sake of clarity, fractions of vehicles are used). The longest queue (longest vertical distance between arrival and departure curves) occurs at 8:31 A.M. and is

$$\begin{aligned} Q_{max} &= \lambda t - \mu_r(t - 12) \\ &= 48.33(31) - 33.33(19) \\ &= \underline{865 \text{ veh}} \end{aligned}$$

Total vehicle delay is (using equations for triangular and trapezoidal areas to calculate the total area between the arrival and departure curves)

$$\begin{aligned} D_t &= \tfrac{1}{2}(12)(580) + \tfrac{1}{2}(580 + 1498.33)(19) - \tfrac{1}{2}(19)(633.33) \\ &\quad + \tfrac{1}{2}(1498.33 - 633.33)(78.16 - 31) \\ &= \underline{37{,}604.2 \text{ veh-min}} \end{aligned}$$

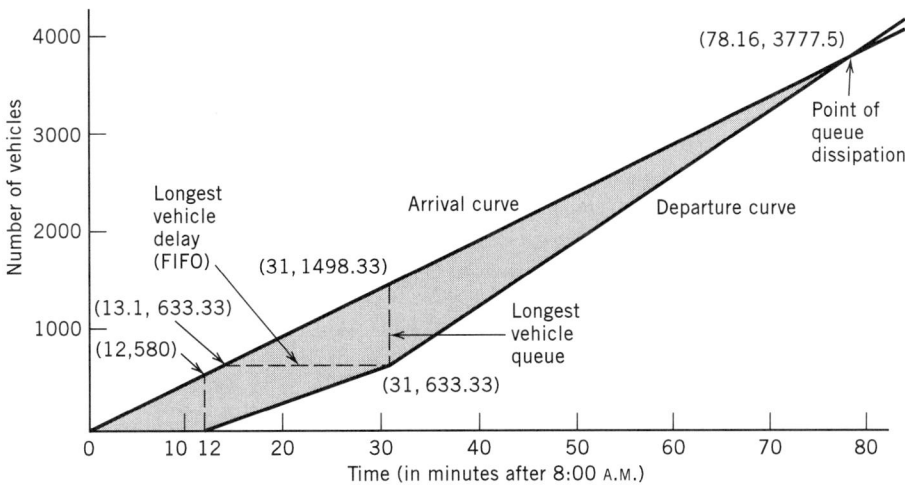

Figure 5.8 $D/D/1$ queuing diagram for Example 5.13.

The average delay per vehicle is 9.95 min (37,604.2/3777.5). The longest wait of any vehicle (the longest horizontal distance between the arrival and departure curves), assuming a FIFO queuing discipline, will be the delay time of the 633.33rd vehicle to arrive. This vehicle will arrive 13.1 minutes (633.33/48.33) after 8:00 A.M. and will depart at 8:31 A.M., being delayed a total of 17.9 min.

NOMENCLATURE FOR CHAPTER 5

D	deterministic arrivals or departures	Q_{max}	maximum length of queue
D_t	total vehicle delay	s	vehicle spacing
h	vehicle time headway	t	time
k	traffic density	$\bar{t}$	average time spent in the system
k_j	traffic jam density	u	space-mean speed (also denoted $\bar{u}_s$)
k_{cap}	traffic density at capacity	u_i	spot speed for vehicle i
l	roadway length	u_f	free-flow speed
M	exponentially distributed arrivals or departures	u_{cap}	speed at capacity
n	number of vehicles	$\bar{u}_s$	space-mean speed (also denoted simply as u)
n_c	departure channel number	$\bar{u}_t$	time-mean speed
N	total number of departure channels	$\bar{w}$	average time waiting in the queue
q	traffic flow	λ	arrival rate
q_{cap}	traffic flow at capacity (maximum traffic flow)	μ	departure rate
Q	length of queue	ρ	traffic intensity
$\bar{Q}$	average length of queue		

REFERENCES

Drew, D. R. "Deterministic Aspects of Freeway Operations and Control." *Highway Research Record*, 99, 1965.

Pipes, L. A. "Car Following Models and the Fundamental Diagram of Road Traffic." *Transportation Research*, vol. 1, no. 1, 1967.

Poch, M., and F. Mannering. "Negative Binomial Analysis of Intersection-Accident Frequencies," *Journal of Transportation Engineering*, vol. 122, no. 2, March/April 1996.

Transportation Research Board. *Traffic Flow Theory: A Monograph*. Special Report 165. Washington, DC: National Research Council, 1975.

Transportation Research Board. *Highway Capacity Manual*. Washington, DC: National Research Council, 2000.

PROBLEMS

5.1 On a specific westbound section of highway, studies show that the speed-density relationship is

$$u = u_f \left[1 - \left(\frac{k}{k_j}\right)^{3.5}\right]$$

It is known that the capacity is 3800 veh/h and the jam density is 225 veh/mi. What is the space-mean speed of the traffic at capacity, and what is the free-flow speed?

5.2 A section of highway has a speed-flow relationship of the form

$$q = au^2 + bu$$

It is known that at capacity (which is 2900 veh/h) the space-mean speed of traffic is 30 mi/h. Determine the speed when the flow is 1400 veh/h and the free-flow speed.

5.3 A section of highway has the following flow-density relationship:

$$q = 50k - 0.156k^2$$

What is the capacity of the highway section, the speed at capacity, and the density when the highway is at one-quarter of its capacity?

5.4 Assume you are observing traffic in a single lane of a highway at a specific location. You measure the average headway and average spacing of passing vehicles as 3 seconds and 150 ft, respectively. Calculate the flow, average speed, and density of the traffic stream in this lane.

5.5 Assume you are an observer standing at a point along a three-lane roadway. All vehicles in lane 1 are traveling at 30 mi/h, all vehicles in lane 2 are traveling at 45 mi/h, and all vehicles in lane 3 are traveling at 60 mi/h. There is also a constant spacing of 0.5 mile between vehicles. If you collect spot speed data for all vehicles as they cross your observation point, for 30 minutes, what will be the time-mean speed and space-mean speed for this traffic stream?

5.6 Four race cars are traveling on a 2.5-mile tri-oval track. The four cars are traveling at constant speeds of 195 mi/h, 190 mi/h, 185 mi/h, and 180 mi/h, respectively. Assume you are an observer standing at a point on the track for a period of 30 minutes and are recording the instantaneous speed of each vehicle as it crosses your point. What is the time-mean speed and space-mean speed for these vehicles for this time period? (*Note:* Be careful with rounding.)

5.7 For Problem 5.6, calculate the space-mean speed assuming you were provided with only an aerial photo of the circling race cars and the constant travel speed of each of the vehicles.

5.8 An observer has determined that the time headways between successive vehicles on a section of highway are exponentially distributed and that 60% of the headways between vehicles are 13 seconds or greater. If the observer decides to count traffic in 30-second time intervals, estimate the probability of the observer counting exactly four vehicles in an interval.

5.9 At a specified point on a highway, vehicles are known to arrive according to a Poisson process. Vehicles are counted in 20-second intervals, and vehicle counts are taken in 120 of these time intervals. It is noted that no cars arrive in 18 of these 120 intervals. Approximate the number of these 120 intervals in which exactly three cars arrive.

5.10 For the data collected in Problem 5.9, estimate the percentage of time headways that will be 10 seconds or greater and those that will be less than 6 seconds.

5.11 A vehicle pulls out onto a single-lane highway that has a flow rate of 280 veh/h (Poisson distributed). The driver of the vehicle does not look for oncoming traffic. Road conditions and vehicle speeds on the highway are such that it takes 1.5 seconds for an oncoming vehicle to stop once the brakes are applied. Assuming a standard driver reaction time of 2.5 seconds, what is the probability that the vehicle pulling out will get in an accident with oncoming traffic?

5.12 Consider the conditions in Problem 5.11. How short would the driver reaction times of oncoming vehicles have to be for the probability of an accident to equal 0.15?

5.13 A toll booth on a turnpike is open from 8:00 A.M. to 12 midnight. Vehicles start arriving at 7:45 A.M. at a uniform deterministic rate of six per minute until 8:15 A.M. and from then on at two per minute. If vehicles are processed at a uniform deterministic rate of six per minute, determine when the queue will dissipate, the total delay, the maximum queue length (in vehicles), the longest vehicle delay under FIFO, and the longest vehicle delay under LIFO.

5.14 Vehicles begin to arrive at a parking lot at 6:00 A.M. at a rate of eight per minute. Due to an accident on the access highway, no vehicles arrive from 6:20 to 6:30 A.M. From 6:30 A.M. on, vehicles arrive at a rate of two per minute. The parking lot attendant processes incoming vehicles (collects parking fees) at a rate of four per minute throughout the day. Assuming D/D/1 queuing, determine total vehicle delay.

5.15 The arrival rate at a parking lot is 6 veh/min. Vehicles start arriving at 6:00 P.M., and when the queue reaches 36 vehicles, service begins. If company policy is that total vehicle delay should be equal to 500 veh-min, what is the departure rate? (Assume D/D/1 queuing and a constant service rate.)

5.16 Vehicles begin to arrive at a toll booth at 8:50 A.M. with an arrival rate of $\lambda(t) = 4.1 + 0.01t$ [with t in minutes and $\lambda(t)$ in vehicles per minute]. The toll booth opens at 9:00 A.M. and processes vehicles at a rate of 12 per minute throughout the day. Assuming D/D/1 queuing, when will the queue dissipate and what will be the total vehicle delay?

5.17 Vehicles begin to arrive at a toll booth at 7:50 A.M. with an arrival rate of $\lambda(t) = 5.2 - 0.01t$ (with t in minutes after 7:50 A.M. and λ in vehicles per minute). The toll booth opens at 8:00 A.M. and serves vehicles at a rate of $\mu(t) = 3.3 + 2.4t$ (with t in minutes after 8:00 A.M. and μ in vehicles per minute). Once the service rate reaches 10 veh/min, it stays at that level for the rest of the day. If queuing is D/D/1, when will the queue that formed at 7:50 A.M. be cleared?

5.18 Vehicles arrive at a freeway on-ramp meter at a constant rate of six per minute starting at 6:00 A.M. Service begins at 6:00 A.M. such that $\mu(t) = 2 + 0.5t$, where $\mu(t)$ is in veh/min and t is in minutes after 6:00 A.M. What is the total delay and the maximum queue length (in vehicles)?

5.19 Vehicles arrive at a tollbooth according to the function $\lambda(t) = 5.2 - 0.20t$, where $\lambda(t)$ is in vehicles per minute and t is in minutes. The toll booth operator processes one vehicle every 20 seconds. Determine total delay, maximum queue length, and the time that the 20th vehicle to arrive waits from its arrival to its departure.

5.20 There are 10 vehicles in a queue when an attendant opens a toll booth. Vehicles arrive at the booth at a rate of 4 per minute. The attendant opens the booth and improves the service rate over time following the function $\mu(t) = 1.1 + 0.30t$, where $\mu(t)$ is in vehicles per minute and t is in minutes. When will the queue clear, what is the total delay, and what is the maximum queue length?

5.21 Vehicles begin to arrive at a parking lot at 6:00 A.M. with an arrival rate function (in vehicles per minute) of $\lambda(t) = 1.2 + 0.3t$, where t is in minutes. At 6:10 A.M. the parking lot opens and processes vehicles at a rate of 12 per minute. What is the total delay and the maximum queue length?

5.22 At a parking lot, vehicles arrive according to a Poisson process and are processed (parking fee collected) at a uniform deterministic rate at a single station. The mean arrival rate is 4 veh/min and the processing rate is 5 veh/min. Determine the average length of queue, the average time spent in the system, and the average waiting time in the queue.

5.23 Consider the parking lot and conditions described in Problem 5.22. If the rate at which vehicles are processed became exponentially distributed (instead of deterministic) with a mean processing rate of 5 veh/min, what would be the average length of queue, the average time spent in the system, and the average waiting time in the queue?

5.24 Vehicles arrive at a toll booth with a mean arrival rate of 2 veh/min (the time between arrivals is exponentially distributed). The toll booth operator processes vehicles (collects tolls) at a uniform deterministic rate of one every 20 seconds. What is the average length of queue, the average time spent in the system, and the average waiting time in the queue?

5.25 A business owner decides to pass out free transistor radios (along with a promotional brochure) at a booth in a parking lot. The owner begins giving the radios away at 9:15 A.M. and continues until 10:00 A.M. Vehicles start arriving for the radios at 8:45 A.M. at a uniform deterministic rate of 4 per minute and continue to arrive at this rate until 9:15 A.M. From 9:15 to 10:00 A.M. the arrival rate becomes 8 per minute. The radios and brochures are distributed at a uniform deterministic rate of 11 cars per minute over the 45-minute time period. Determine total delay, maximum queue length, and longest vehicle delay assuming FIFO and LIFO.

5.26 Consider the conditions described in Problem 5.25. Suppose the owner decides to accelerate the radio-brochure distribution rate (in veh/min) so that the queue that forms will be cleared by 9:45 A.M. What would this new distribution rate be?

5.27 A ferryboat queuing lane holds 30 vehicles. If vehicles are processed (tolls collected) at a uniform deterministic rate of 4 vehicles per minute and processing begins when the lane reaches capacity, what is the uniform deterministic arrival rate if the vehicle queue is cleared 30 minutes after vehicles begin to arrive?

5.28 At a toll booth, vehicles arrive and are processed (tolls collected) at uniform deterministic rates λ and μ, respectively. The arrival rate is 2 veh/min. Processing begins 13 minutes after the arrival of the first vehicle, and the queue dissipates t minutes after the arrival of the first vehicle. Letting the number of vehicles that must actually wait in a queue be x, develop an expression for determining processing rates in terms of x.

5.29 Vehicles arrive at a recreational park booth at a uniform deterministic rate of 4 veh/min. If uniform deterministic processing of vehicles (collecting of fees) begins 30 minutes after the first arrival and the total delay is 3600 veh-min, how long after the arrival of the first vehicle will it take for the queue to be cleared?

5.30 Trucks begin to arrive at a truck weigh station (with a single scale) at 6:00 A.M. at a deterministic but time-varying rate of $\lambda(t) = 4.3 - 0.22t$ [$\lambda(t)$ is in veh/min and t is in minutes]. The departure rate is a constant 2 veh/min (time to weigh a truck is 30 seconds). When will the queue that forms be cleared, what will be the total delay, and what will be the maximum queue length?

5.31 Vehicles begin to arrive at a remote parking lot after the start of a major sporting event. They are arriving at a deterministic but time-varying rate of $\lambda(t) = 3.3 - 0.1t$ [$\lambda(t)$ is in veh/min and t is in minutes]. The parking lot attendant processes vehicles (assigns spaces and collects fees) at a deterministic rate at a single station. A queue exceeding four vehicles will back up onto a congested street, and is to be avoided. How many vehicles per minute must the attendant process to ensure that the queue does not exceed four vehicles?

5.32 A truck weighing station has a single scale. The time between truck arrivals at the station is exponentially distributed with a mean arrival rate of 1.5 veh/min. The time it takes vehicles to be weighed is exponentially distributed with a mean rate of 2 veh/min. When more than 5 trucks are in the system, the queue backs up onto the highway and interferes with through traffic. What is the probability that the number of trucks in the system will exceed 5?

5.33 Consider the convenience store described in Example 5.12. The owner is concerned about customers not finding an available parking space when they arrive during the busiest hour. How many spaces must be provided for there to be less than a 1% chance of an arriving customer not finding an open parking space?

5.34 Vehicles arrive at a toll bridge at a rate of 430 veh/h (the time between arrivals is exponentially distributed). Two toll booths are open and each can process arrivals (collect tolls) at a mean rate of 10 seconds per vehicle (the processing time is also exponentially distributed). What is the total time spent in the system by all vehicles in a 1-hour period?

5.35 Vehicles leave an airport parking facility (arrive at parking fee collection booths) at a rate of 500 veh/h (the time between arrivals is exponentially distributed). The parking facility has a policy that the average time a patron spends in a queue waiting to pay for parking is not to exceed 5 seconds. If the time required to pay for parking is exponentially distributed with a mean of 15 seconds, what is the fewest number of payment processing booths that must be open to keep the average time spent in a queue below 5 seconds?

Chapter 6

Highway Capacity and Level-of-Service Analysis

6.1 INTRODUCTION

The underlying objective of traffic analysis is to quantify a roadway's performance with regard to specified traffic volumes. This performance can be measured in terms of travel delay (as the roadway becomes increasingly congested) as well as other factors. The comparative performance of various roadway segments (which is determined from an analysis of traffic) is important because it can be used as a basis to allocate limited roadway construction and improvement funds. The purpose of this chapter is to apply the elements of uninterrupted traffic flow theory covered in Chapter 5 to the practical field analysis of traffic flow and capacity on freeways, multilane highways, and two-lane highways.

The main challenge of such a process is to adapt the theoretical formulations to the wide range of conditions that occur in the field. These diverse field conditions must be accounted for in a traffic analysis methodology, yet the methodology must remain theoretically consistent. For example, in Chapter 5, capacity (q_{cap}) is simply defined as the highest traffic flow rate that the roadway is capable of supporting. For applied traffic analysis, a consistent and reasonably precise method of determining capacity must be developed within this definition. Because it can readily be shown that the capacity of a roadway segment is a function of factors such as roadway type (freeway, multilane highway, or two-lane highway), free-flow speed, number of lanes, and widths of lanes and shoulders, the method of capacity determination clearly must account for a wide variety of physical and operational roadway characteristics. Additionally, recall that Chapter 5 defines traffic flow on the basis of units of vehicles per hour. Two practical issues arise concerning this unit of measure. First, in many cases vehicular traffic consists of a variety of vehicle types with substantially different performance characteristics. These performance differentials are likely to be magnified by changing roadway geometrics, such as upgrades or downgrades, which have a differential effect on the acceleration and deceleration capabilities of the various types of vehicles; for example, grades have a greater impact on the performance of large trucks relative to automobiles. As a result, traffic must be defined not only in terms of vehicles per unit time but also in terms of vehicle composition, because it is clear that a 1500-veh/h traffic flow consisting of 100% automobiles will differ significantly with regard to operating speed and traffic density from a 1500-veh/h traffic flow that consists of 50% automobiles and 50% heavy trucks.

The other flow-related concern is the temporal distribution of traffic. In practice, the analysis of roadway traffic usually focuses on the most critical condition, which is the most congested hour within a 24-hour daily period (the temporal distribution of traffic will be discussed in more detail in Section 6.7). However, within this most congested peak hour, traffic flow is likely to be nonuniform. It is therefore necessary to arrive at some method of defining and measuring the nonuniformity of flow within the peak hour.

To summarize, the objective of applied traffic analysis is to provide a practical method of quantifying the degree of traffic congestion and to relate this to the overall traffic-related performance of the roadway. The following sections of this chapter discuss and demonstrate accepted standards for applied traffic analysis for the three major types of uninterrupted-flow roadways: freeways, multilane highways, and two-lane highways (one lane in each direction).

6.2 LEVEL-OF-SERVICE CONCEPT

The *Highway Capacity Manual* (HCM), produced by the Transportation Research Board, is a synthesis of the state of the art in methodologies for quantifying traffic operational performance and capacity utilization (congestion level) for a variety of transportation facilities. One of the foundations of the HCM is the concept of level of service (LOS). The level of service represents a qualitative ranking of the traffic operational conditions experienced by users of a facility under specified roadway, traffic, and traffic control (if present) conditions. Current practice designates six levels of service ranging from A to F, with level of service A representing the best operating conditions and level of service F the worst.

A number of operational performance measures, such as speed, flow, and density, can be measured or calculated for any transportation facility. For the level-of-service concept to be applied to traffic analysis, it is necessary to select a performance measure that is representative of how motorists actually perceive the quality of service they are receiving on a facility. Motorists tend to evaluate their received quality of service in terms of factors such as speed and travel time, freedom to maneuver, traffic interruptions, and comfort and convenience. Thus, it is important to select a measure that encompasses some or all of these factors. The performance measure that is selected for level-of-service (LOS) analysis for a particular transportation facility is referred to as the service measure.

The HCM [Transportation Research Board 2000] defines the LOS categories for freeways and multilane highways as follows:

Level of service A. LOS A represents free-flow conditions (traffic operating at free-flow speeds, as defined in Chapter 5). Individual users are virtually unaffected by the presence of others in the traffic stream. Freedom to select speeds and to maneuver within the traffic stream is extremely high. The general level of comfort and convenience provided to drivers is excellent.

Level of service B. LOS B also allows speeds at or near free-flow speeds, but the presence of other users in the traffic stream begins to be noticeable. Freedom to select speeds is relatively unaffected, but there is a slight decline in the freedom to maneuver within the traffic stream relative to LOS A.

Level of service C. LOS C has speeds at or near free-flow speeds, but the freedom to maneuver is noticeably restricted (lane changes require careful attention on the part of drivers). The general level of comfort and convenience declines significantly at this level. Disruptions in the traffic stream, such as an incident (for example, vehicular accident or disablement), can result in significant queue formation and vehicular delay. In contrast, the effects of incidents at LOS A or LOS B are minimal, with only minor delay in the immediate vicinity of the event.

Level of service D. LOS D represents the conditions where speeds begin to decline slightly with increasing flow. The freedom to maneuver becomes more restricted, and drivers experience reductions in physical and psychological comfort. Incidents can generate lengthy queues because the higher density associated with this LOS provides little space to absorb disruptions in the traffic flow.

Level of service E. LOS E represents operating conditions at or near the roadway's capacity. Even minor disruptions to the traffic stream, such as vehicles entering from a ramp or vehicles changing lanes, can cause delays as other vehicles give way to allow such maneuvers. In general, maneuverability is extremely limited, and drivers experience considerable physical and psychological discomfort.

Level of service F. LOS F describes a breakdown in vehicular flow. Queues form quickly behind points in the roadway where the arrival flow rate temporarily exceeds the departure rate, as determined by the roadway's capacity (see Chapter 5). Such points occur at incidents and on- and off-ramps, where incoming traffic results in capacity being exceeded. Vehicles typically operate at low speeds under these conditions and are often required to come to a complete stop, usually in a cyclic fashion. The cyclic formation and dissipation of queues is a key characterization of LOS F.

A visual perspective of the level-of-service definitions for freeways is provided in Fig. 6.1. In dealing with level of service it is important to remember that when the traffic volume is at or near the roadway capacity (which will be shown as a function of the prevailing traffic and physical characteristics of the roadway), the roadway is operating at LOS E. This, however, is not a desirable condition because under LOS E conditions there is considerable driver discomfort, which could increase the likelihood of vehicular crashes and overall delay. In roadway design, the possibility of degradation of level of service to LOS E should be avoided, although this is not always possible due to financial and environmental constraints that may limit the design speed, number of lanes, and other factors affecting roadway capacity.

6.2 Level-of-Service Concept   **173**

Figure 6.1 Illustration of freeway level of service (A to F).
Reproduced by permission from Transportation Research Board, *Highway Capacity Manual,* National Research Council, Washington, DC, 2000.

6.3 LEVEL-OF-SERVICE DETERMINATION

There are several steps in a basic level-of-service determination for an uninterrupted-flow facility. The remainder of this section describes the general details of each step, as applicable to uninterrupted-flow facility analyses. Facility-specific details of these steps are described in the sections that follow.

6.3.1 Base Conditions and Capacity

The determination of a roadway's level of service begins with the specification of base roadway conditions. Recall that in the introduction of this chapter, the effects of vehicle performance and roadway design characteristics on traffic flow were discussed qualitatively. In practice, the effects of such factors on traffic flow are measured quantitatively, relative to the base traffic and roadway design conditions. For uninterrupted-flow roadways, base conditions can be categorized as those relating to roadway conditions, such as lane widths, lateral clearances, access frequency, and terrain; and traffic stream conditions such as the effects of heavy vehicles (large trucks, buses, and RVs) and driver population characteristics. Base conditions are defined as those conditions that represent unrestrictive geometric and traffic conditions. Additionally, base conditions are assumed to consist of favorable environmental conditions (such as dry roadways).

The capacity of a particular roadway segment will be greatest when all roadway and traffic conditions meet or exceed their base values. Empirical studies have identified the values of these base conditions for which the capacity of a roadway segment is maximized. Values in excess of the base conditions will not increase the capacity of the roadway, but values more restrictive than the base conditions will result in a lower capacity. For example, studies have identified a base lane width of 12 ft (3.6 m). That is, lane widths in excess of 12 ft (3.6 m) will not result in increased capacity; however, lane widths less than 12 ft (3.6 m) will result in a reduction in capacity. Capacity values for base conditions have been determined for all uninterrupted-flow facility types from field studies. It should be noted that for purposes of level-of-service analysis, capacity is defined not as the absolute maximum flow rate ever observed for a particular facility type, but rather as the maximum flow rate that can be reasonably expected on a recurring basis.

Because all base conditions for a particular roadway type are seldom realized in practice, methods for converting the measured flow rate into an equivalent analysis flow rate in terms of passenger cars for the given traffic conditions and estimating the actual free-flow speed for the given roadway conditions are needed. The following sections describe the procedures for arriving at flow and speed values for given roadway and traffic conditions.

6.3.2 Determine Free-Flow Speed

Free-flow speed (*FFS*) is a term that was introduced in Chapter 5 as the speed of traffic as the traffic density approaches zero. In practice, *FFS* is governed by roadway design characteristics (horizontal and vertical curves, lane and shoulder widths, and median design), the frequency of access points, the complexity of the driving environment (possible distractions from roadway signs and so on), and posted speed limits.

The free-flow speed must be determined given the characteristics of the roadway segment. *FFS* is the mean speed of traffic as measured when flow rates are low to moderate (specific values are given under the individual sections for each roadway type). Ideally, *FFS* should be measured directly in the field at the site of interest. However, if this is not possible or feasible, an alternative method can be employed to arrive at an estimate of *FFS* under the prevailing conditions. This method makes adjustments to a base *FFS* (*BFFS*) depending on the physical characteristics of the roadway segment, such as lane width, shoulder width, and access frequency. This method has the same basic structure for the various roadway types, but contains adjustment factors and values appropriate for each roadway type.

6.3.3 Determine Analysis Flow Rate

One of the fundamental inputs to a traffic analysis is the actual traffic volume on the roadway, in vehicles per hour, which is given the symbol V. Generally, the highest volume in a 24-hour period (the peak-hour volume) is used for V in traffic analysis computations. However, this hourly volume needs to be adjusted to reflect the temporal variation of traffic demand within the analysis hour, the impacts due to heavy vehicles, and, in the case of freeway and multilane roadways, the characteristics of the driving population. To account for these effects, the hourly volume is divided by adjustment factors to obtain an equivalent flow rate in terms of passenger cars per hour (pc/h). Additionally, the flow rate is expressed on a per-lane basis (pc/h/ln) by dividing by the number of lanes in the analysis segment.

6.3.4 Calculate Service Measure(s) and Determine LOS

Once the previous steps have been completed, all that remains is to calculate the value of the service measure and then determine the LOS from the service measure value. For freeways and multilane highways, this is a relatively straightforward task. However, for two-lane highways, there are actually two service measures, and the calculation of these and the subsequent LOS determination are more involved.

6.4 BASIC FREEWAY SEGMENTS

A basic freeway segment is defined as a section of a divided roadway having two or more lanes in each direction, full access control, and traffic that is unaffected by merging or diverging movements near ramps. It is important to note that capacity analysis for divided roadways focuses on the traffic flow in one direction only. This is reasonable because the objective is to measure the highest level of congestion.

Due to directional imbalance of traffic flows—for example, morning rush hours having higher volumes going toward the central city and evening rush hours having higher volumes going away from the central city—consideration of traffic volumes in both directions is likely to seriously understate the true level of traffic congestion.

Table 6.1 provides the level-of-service criteria corresponding to traffic density, speed, volume-to-capacity ratio, and maximum service flow rate. A graphical representation of this table is provided in Fig. 6.2. The maximum service flow rate is simply the maximum flow rate, under base conditions, that can be sustained for a given level of service. This value is related to speed and density as discussed in Chapter 5. This speed-flow-density relationship is central to the analysis of basic freeway segments, as will be outlined in the remainder of this section.

6.4.1 Base Conditions and Capacity

The base conditions for a basic freeway segment are defined as [Transportation Research Board 2000]

- 12-ft (3.6-m) minimum lane widths
- 6-ft (1.8-m) minimum right-shoulder clearance between the edge of the travel lane and objects (utility poles, retaining walls, etc.) that influence driver behavior
- 2-ft (0.6-m) minimum median lateral clearance
- Only passenger cars in the traffic stream
- Five or more lanes in each travel direction (urban areas only)
- 2-mi (3.2-km) or greater interchange spacing
- Level terrain (no grades greater than 2%)
- A driver population of mostly familiar roadway users

These conditions represent a high operating level, with a free-flow speed of 70 mi/h (110 km/h) or higher.

The capacity, c, for basic freeway segments, in passenger cars per hour per lane (pc/h/ln), is given in Table 6.2. From Table 6.1, note that, by definition, the upper boundary of LOS E corresponds to the value of capacity and a v/c of 1.0. Other values of v/c for a specific level of service are obtained by simply dividing the maximum service flow rate for that level of service by capacity (the maximum service flow rate at LOS E).

Table 6.1 LOS Criteria for Basic Freeway Segments

Criterion	LOS					Criterion	LOS				
	A	B	C	D	E		A	B	C	D	E
FFS = 75 mi/h						*FFS = 120 km/h*					
Maximum density (pc/mi/ln)	11	18	26	35	45	Maximum density (pc/km/ln)	7	11	16	22	28
Average speed (mi/h)	75.0	74.8	70.6	62.2	53.3	Average speed (km/h)	120.0	120.0	114.6	99.6	85.7
Maximum v/c	0.34	0.56	0.76	0.90	1.00	Maximum v/c	0.35	0.55	0.77	0.92	1.00
Maximum service flow rate (pc/h/ln)	820	1350	1830	2170	2400	Maximum service flow rate (pc/h/ln)	840	1320	1840	2200	2400
FFS = 70 mi/h						*FFS = 110 km/h*					
Maximum density (pc/mi/ln)	11	18	26	35	45	Maximum density (pc/km/ln)	7	11	16	22	28
Average speed (mi/h)	70.0	70.0	68.2	61.5	53.3	Average speed (km/h)	110.0	110.0	108.5	97.2	83.9
Maximum v/c	0.32	0.53	0.74	0.90	1.00	Maximum v/c	0.33	0.51	0.74	0.91	1.00
Maximum service flow rate (pc/h/ln)	770	1260	1770	2150	2400	Maximum service flow rate (pc/h/ln)	770	1210	1740	2135	2350
FFS = 65 mi/h						*FFS = 100 km/h*					
Maximum density (pc/mi/ln)	11	18	26	35	45	Maximum density (pc/km/ln)	7	11	16	22	28
Average speed (mi/h)	65.0	65.0	64.6	59.7	52.2	Average speed (km/h)	100.0	100.0	100.0	93.8	82.1
Maximum v/c	0.30	0.50	0.71	0.89	1.00	Maximum v/c	0.30	0.48	0.70	0.90	1.00
Maximum service flow rate (pc/h/ln)	710	1170	1680	2090	2350	Maximum service flow rate (pc/h/ln)	700	1100	1600	2065	2300
FFS = 60 mi/h						*FFS = 90 km/h*					
Maximum density (pc/mi/ln)	11	18	26	35	45	Maximum density (pc/km/ln)	7	11	16	22	28
Average speed (mi/h)	60.0	60.0	60.0	57.6	51.1	Average speed (km/h)	90.0	90.0	90.0	89.1	80.4
Maximum v/c	0.29	0.47	0.68	0.88	1.00	Maximum v/c	0.28	0.44	0.64	0.87	1.00
Maximum service flow rate (pc/h/ln)	660	1080	1560	2020	2300	Maximum service flow rate (pc/h/ln)	630	990	1440	1955	2250
FFS = 55 mi/h											
Maximum density (pc/mi/ln)	11	18	26	35	45						
Average speed (mi/h)	55.0	55.0	55.0	54.7	50.0						
Maximum v/c	0.27	0.44	0.64	0.85	1.00						
Maximum service flow rate (pc/h/ln)	600	990	1430	1910	2250						

Note: The exact mathematical relationship between density and v/c has not always been maintained at LOS boundaries because of the use of rounded values. Density is the primary determinant of LOS. The speed criterion is the speed at maximum density for a given LOS. LOS F is characterized by highly unstable and variable traffic flow. Prediction of accurate flow rate, density, and speed at LOS F is difficult.

Source: Transportation Research Board. *Highway Capacity Manual*. Washington, DC: National Research Council, 2000

178 Chapter 6 Highway Capacity and Level-of-Service Analysis

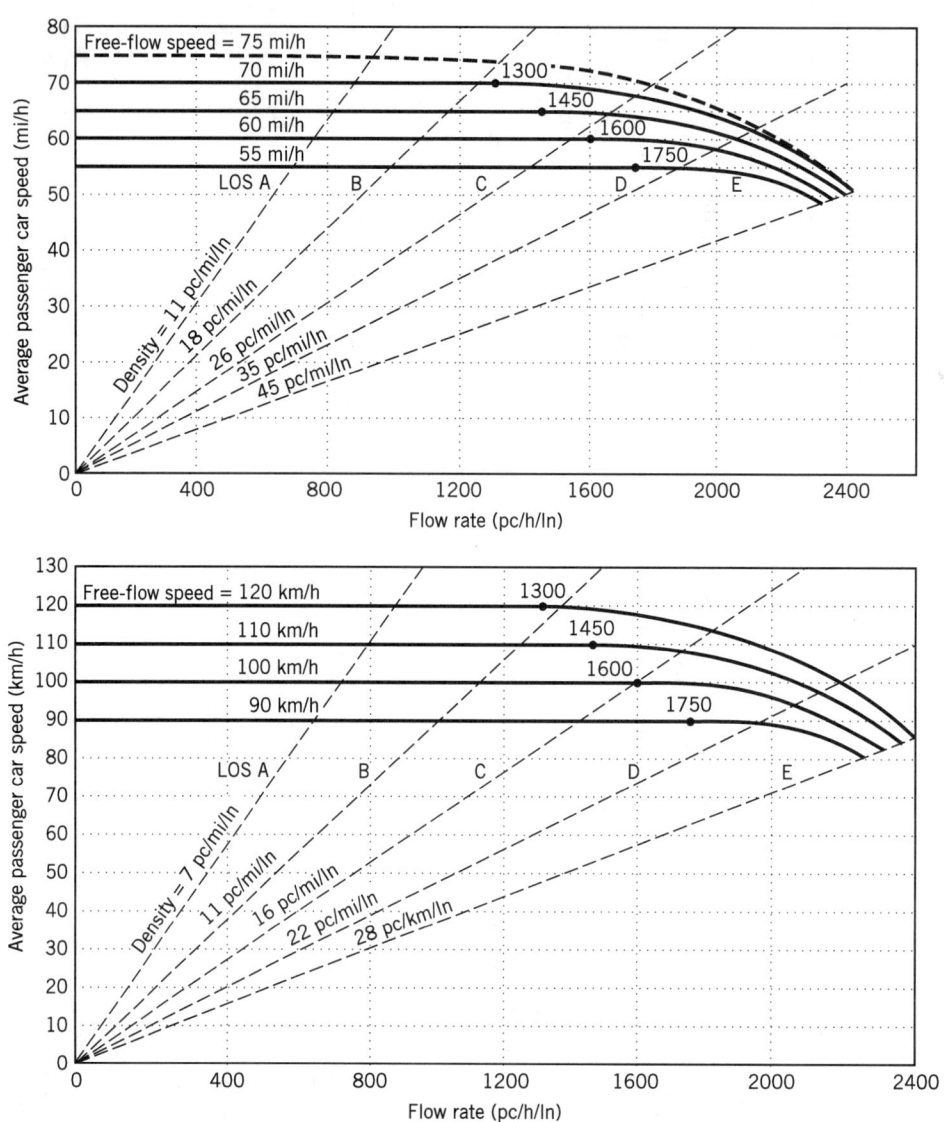

Figure 6.2 Basic freeway segment speed-flow curves and level-of-service criteria.
Reproduced by permission from Transportation Research Board, *Highway Capacity Manual*, National Research Council, Washington, DC, 2000.

Table 6.2 Relationship between Free-Flow Speed and Capacity on Basic Freeway Segments

Free-flow speed (mi/h)	Capacity (pc/h/ln)	Free-flow speed (km/h)	Capacity (pc/h/ln)
75	2400	120	2400
70	2400	110	2350
65	2350	100	2300
60	2300	90	2250
55	2250		

Source: Transportation Research Board, *Highway Capacity Manual.* Washington, DC: National Research Council, 2000.

6.4.2 Service Measure

The service measure for basic freeway segments is density. Density, as discussed in Chapter 5, is typically measured in terms of passenger cars per mile (kilometer) per lane (pc/mi/ln, pc/km/ln) and therefore provides a good measure of the relative mobility of individual vehicles in the traffic stream. A low traffic stream density gives individual vehicles the ability to change lanes and speeds with relative ease, while a high density makes it very difficult for individual vehicles to maneuver within the traffic stream. Thus, traffic density is the primary determinant of freeway level of service.

Recall Eq. 5.14 from Chapter 5:

$$q = uk \qquad (6.1)$$

where

q = flow in veh/h,
u = speed in mi/h (km/h), and
k = density in veh/mi (veh/km).

Density is therefore calculated as flow divided by speed. The following sections will describe how to arrive at flow and speed values for the given roadway and traffic conditions. Once the flow and speed values have been determined according to the given conditions, a density can be calculated and then referenced in Table 6.1 or Fig. 6.2 to arrive at a level of service for the freeway segment.

6.4.3 Determining Free-Flow Speed

For basic freeway segments *FFS* is the mean speed of passenger cars operating in flow rates up to 1300 passenger cars per hour per lane (pc/h/ln). If *FFS* is to be estimated rather than measured, the following equation can be used. It accounts for the roadway characteristics of lane width, right-shoulder lateral clearance, number of lanes, and interchange density.

$$FFS = BFFS - f_{LW} - f_{LC} - f_N - f_{ID} \qquad (6.2)$$

where

FFS = estimated free-flow speed in mi/h (km/h),
$BFFS$ = estimated free-flow speed, in mi/h (km/h), for base conditions,
f_{LW} = adjustment for lane width in mi/h (km/h),
f_{LC} = adjustment for lateral clearance in mi/h (km/h),
f_N = adjustment for number of lanes in mi/h (km/h), and
f_{ID} = adjustment for interchange density in mi/h (km/h).

$BFFS$ is assumed to be 70 mi/h (110 km/h) for freeways in urban areas and 75 mi/h (120 km/h) for freeways in rural areas. The following sections describe the procedures for estimating the adjustment factor values.

Lane Width Adjustment

When lane widths are narrower than the base 12 ft (3.6 m), the adjustment factor f_{LW} is used to reflect the impact on free-flow speed. Such an adjustment is needed because narrow lanes cause traffic to slow as a result of reduced psychological comfort and limits on driver maneuvering and accident avoidance options. Thus, FFS under these conditions is less than the value that would be observed if base lane widths were provided. The adjustment factors used in current practice are presented in Table 6.3.

Lateral Clearance Adjustment

When obstructions are closer than 6 ft (1.8 m) (at the roadside) from the traveled pavement, the adjustment factor f_{LC} is used to reflect the impact on FFS. Again, these conditions lead to reduced psychological comfort for the driver, and consequently, reduced speeds. An obstruction is a right-side object that can either be continuous (such as a retaining wall or barrier) or periodic (such as light posts or utility poles). Table 6.4 provides corrections for obstructions on the right side of the roadway.

Table 6.3 Adjustment for Lane Width

Lane width (ft)	Reduction in free-flow speed, f_{LW} (mi/h)
12	0.0
11	1.9
10	6.6

Lane width (m)	Reduction in free-flow speed, f_{LW} (km/h)
3.6	0.0
3.5	1.0
3.4	2.1
3.3	3.1
3.2	5.6
3.1	8.1
3.0	10.6

Source: Transportation Research Board. *Highway Capacity Manual.* Washington, DC: National Research Council, 2000.

Table 6.4 Adjustment for Right-Shoulder Lateral Clearance

Right-shoulder lateral clearance (ft)	Reduction in free-flow speed, f_{LC} (mi/h), lanes in one direction			
	2	3	4	≥5
≥ 6	0.0	0.0	0.0	0.0
5	0.6	0.4	0.2	0.1
4	1.2	0.8	0.4	0.2
3	1.8	1.2	0.6	0.3
2	2.4	1.6	0.8	0.4
1	3.0	2.0	1.0	0.5
0	3.6	2.4	1.2	0.6

Right-shoulder lateral clearance (m)	Reduction in free-flow speed, f_{LC} (km/h), lanes in one direction			
	2	3	4	≥5
≥ 1.8	0.0	0.0	0.0	0.0
1.5	1.0	0.7	0.3	0.2
1.2	1.9	1.3	0.7	0.4
0.9	2.9	1.9	1.0	0.6
0.6	3.9	2.6	1.3	0.8
0.3	4.8	3.2	1.6	1.1
0.0	5.8	3.9	1.9	1.3

Source: Transportation Research Board. *Highway Capacity Manual.* Washington, DC: National Research Council, 2000.

Number of Lanes Adjustment

A freeway segment with five or more lanes in a single direction is the base condition with respect to free-flow speed. Adjustments in *FFS* for fewer directional lanes are shown in Table 6.5. These adjustments apply to freeways in urban and suburban areas. Rural freeways typically have only two lanes in each direction; thus, no *FFS* adjustment is made for number of lanes for rural freeways.

Table 6.5 Adjustment for Number of Lanes on Urban Freeways

Number of lanes (one direction)	Reduction in free-flow speed, f_N	
	mi/h	km/h
≥ 5	0.0	0.0
4	1.5	2.4
3	3.0	4.8
2	4.5	7.3

Note: For all rural freeway segments, f_N is 0.0.
Source: Transportation Research Board. *Highway Capacity Manual.* Washington, DC: National Research Council, 2000.

Table 6.6 Adjustment for Interchange Density

Interchanges per mile	Reduction in free-flow speed, f_{ID} (mi/h)	Interchanges per kilometer	Reduction in free-flow speed, f_{ID} (km/h)
≤ 0.50	0.0	≤ 0.3	0.0
0.75	1.3	0.4	1.1
1.00	2.5	0.5	2.1
1.25	3.7	0.6	3.9
1.50	5.0	0.7	5.0
1.75	6.3	0.8	6.0
≥ 2.00	7.5	0.9	8.1
		1.0	9.2
		1.1	10.2
		≥ 1.2	12.1

Source: Transportation Research Board. *Highway Capacity Manual.* Washington, DC: National Research Council, 2000.

Interchange Density Adjustment

When interchanges are spaced more closely than one every 2 miles (one every 3.2 km), the traffic disturbances generated from merging and diverging movements will reduce *FFS*. Interchange density should be measured over a length extending 3 miles (5 km) upstream and 3 miles (5 km) downstream of the analysis segment. To be counted as an interchange, it should have at least one on-ramp. Thus, interchanges with only off-ramps would not be counted toward the interchange density for purposes of this analysis. Adjustments for interchange density are shown in Table 6.6.

6.4.4 Determining Analysis Flow Rate

The analysis flow rate is calculated using the following equation:

$$v_p = \frac{V}{PHF \times N \times f_{HV} \times f_p} \tag{6.3}$$

where

v_p = 15-min passenger car equivalent flow rate (pc/h/ln),
V = hourly volume (veh/h),
PHF = peak-hour factor,
N = number of lanes,
f_{HV} = heavy-vehicle adjustment factor, and
f_p = driver population factor.

The adjustment factors PHF, f_{HV}, and f_p are described next.

Peak-Hour Factor

As previously mentioned, vehicle arrivals during the period of analysis [typically the highest hourly volume within a 24-h period (peak hour)] will likely be nonuniform. To account for this varying arrival rate, the peak 15-min vehicle arrival rate within the analysis hour is usually used for practical traffic analysis purposes. The peak-hour factor has been developed for this purpose, and is defined as the ratio of the hourly volume to the maximum 15-min flow rate expanded to an hourly volume, as follows:

$$PHF = \frac{V}{V_{15} \times 4} \qquad (6.4)$$

where

PHF = peak-hour factor,
V = hourly volume for hour of analysis,
V_{15} = maximum 15-min flow rate within hour of analysis, and
4 = number of 15-min periods per hour.

Equation 6.4 indicates that the further the *PHF* is from unity, the more *peaked* or nonuniform the traffic flow is during the hour. For example, consider two roads both of which have a peak-hour volume, V, of 1800 veh/h. The first road has 600 vehicles arriving in the highest 15-min interval, and the second road has 500 vehicles arriving in the highest 15-min interval. The first road has a more nonuniform flow, as indicated by its *PHF* of 0.75 [1800/(600 × 4)], which is further from unity than the second road's *PHF* of 0.90 [1800/(500 × 4)].

Heavy-Vehicle Adjustment

Large trucks, buses, and recreational vehicles have performance characteristics (slow acceleration and inferior braking) and dimensions (length, height, and width) that have an adverse effect on roadway capacity. Recall that base conditions stipulate that no heavy vehicles are present in the traffic stream, and when prevailing conditions indicate the presence of such vehicles, the adjustment factor f_{HV} is used to translate the traffic stream from base to prevailing conditions. The f_{HV} correction term is found using a two-step process. The first step is to determine the passenger car equivalent (PCE) for each large truck, bus, and recreational vehicle in the traffic stream. These values represent the number of passenger cars that would consume the same amount of roadway capacity as a single large truck, bus, or recreational vehicle. These passenger car equivalents are denoted E_T for large trucks and buses, and E_R for recreational vehicles, and are a function of roadway grades because steep grades will tend to magnify the poor performance of heavy vehicles as well as the sight distance problems caused by their larger dimensions (the visibility afforded to drivers in vehicles following heavy vehicles). For segments of freeway that contain a mix of grades, an extended segment analysis can be used as long as no single grade is steep enough or long enough to significantly impact the overall operations of the segment. As a guideline, an extended segment analysis can be used for freeway segments where no single grade that is less than 3% is more than 0.5 mi (0.8 km) long, or no single grade that is 3% or greater is longer than 0.25 mi (0.4 km). If an extended segment analysis is used, the terrain must be generally classified according to the following definitions [Transportation Research Board 2000]:

Level terrain. Any combination of horizontal and vertical alignment permitting heavy vehicles to maintain approximately the same speed as passenger cars. This generally includes short grades of no more than 2%.

Rolling terrain. Any combination of horizontal and vertical alignment that causes heavy vehicles to reduce their speed substantially below those of passenger cars but does not cause heavy vehicles to operate at their limiting speed for the given terrain for any significant length of time or at frequent intervals [not having $F_{net}(V) = 0$ due to high grade resistance as illustrated in Fig. 2.6].

Mountainous terrain. Any combination of horizontal and vertical alignment that causes heavy vehicles to operate at their limiting speed for significant distances or at frequent intervals.

The passenger car equivalency factors for an extended segment analysis can be obtained from Table 6.7.

Any grade that does not meet the conditions for an extended segment analysis must be analyzed as a separate segment because of its significant impact on traffic operations. In these cases, grade-specific PCE values must be used. Tables 6.8 and 6.9 provide these values for positive grades (upgrades). These tables assume typical large trucks [with average weight-to-horsepower ratios between 125 and 150 lb/hp (75 and 90 kg/kW)] and recreational vehicles [with average weight-to-horsepower ratios between 30 and 60 lb/hp (20 and 40 kg/kW)]. Note that the equivalency factors presented in these tables increase with increasing grade and length of grade, but decrease with increasing heavy vehicle percentage. This decrease with increasing percentage is due to the fact that heavy vehicles tend to group together as their percentages increase on steep, extended grades, thus decreasing their adverse impact on the traffic stream.

Sometimes it is necessary to determine the cumulative effect on traffic operations of several significant grades in succession. For this situation, a distance-weighted average may be used if all grades are less than 4% or the total combined length of the grades is less than 4000 ft (1220 m). For example, a 2% upgrade for 1000 ft (305 m) followed immediately by a 3% upgrade for 2000 ft (610 m) would use the equivalency factor for a 2.67% upgrade [(2 × 1000 + 3 × 2000)/3000] for 3000 ft (914 m) or

Table 6.7 Passenger Car Equivalents (PCEs) for Extended Freeway Segments

	Type of terrain		
Factor	Level	Rolling	Mountainous
E_T (trucks and buses)	1.5	2.5	4.5
E_R (RVs)	1.2	2.0	4.0

Source: Transportation Research Board. *Highway Capacity Manual.* Washington, DC: National Research Council, 2000.

0.568 mi. For information on additional analysis situations involving composite grades, refer to the *Highway Capacity Manual* [Transportation Research Board 2000]. These situations include combining two or more successive grades when the grades exceed 4% or the combined length is greater than 4000 ft (1220 m), determining the length of a grade that starts or ends on a vertical curve, and determining the point of greatest traffic impact in a series of grades (for example, if a long 5% grade were immediately followed by a 2% grade, the end of the 5% grade would be used, as this would be the point of minimum vehicle speed).

Table 6.8 Passenger Car Equivalents (E_T) for Trucks and Buses on Specific Upgrades

Upgrade (%)	Length (mi)	Length (km)	Percentage of trucks and buses								
			2	4	5	6	8	10	15	20	25
< 2	All	All	1.5	1.5	1.5	1.5	1.5	1.5	1.5	1.5	1.5
≥ 2–3	0.0–0.25	0.0–0.4	1.5	1.5	1.5	1.5	1.5	1.5	1.5	1.5	1.5
	>0.25–0.50	>0.4–0.8	1.5	1.5	1.5	1.5	1.5	1.5	1.5	1.5	1.5
	>0.50–0.75	>0.8–1.2	1.5	1.5	1.5	1.5	1.5	1.5	1.5	1.5	1.5
	>0.75–1.00	>1.2–1.6	2.0	2.0	2.0	2.0	1.5	1.5	1.5	1.5	1.5
	>1.00–1.50	>1.6–2.4	2.5	2.5	2.5	2.5	2.0	2.0	2.0	2.0	2.0
	>1.50	>2.4	3.0	3.0	2.5	2.5	2.0	2.0	2.0	2.0	2.0
>3–4	0.00–0.25	0.0–0.4	1.5	1.5	1.5	1.5	1.5	1.5	1.5	1.5	1.5
	>0.25–0.50	>0.4–0.8	2.0	2.0	2.0	2.0	2.0	2.0	1.5	1.5	1.5
	>0.50–0.75	>0.8–1.2	2.5	2.5	2.0	2.0	2.0	2.0	2.0	2.0	2.0
	>0.75–1.00	>1.2–1.6	3.0	3.0	2.5	2.5	2.5	2.5	2.0	2.0	2.0
	>1.00–1.50	>1.6–2.4	3.5	3.5	3.0	3.0	3.0	3.0	2.5	2.5	2.5
	>1.50	>2.4	4.0	3.5	3.0	3.0	3.0	3.0	2.5	2.5	2.5
>4–5	0.0–0.25	0.0–0.4	1.5	1.5	1.5	1.5	1.5	1.5	1.5	1.5	1.5
	>0.25–0.50	>0.4–0.8	3.0	2.5	2.5	2.5	2.0	2.0	2.0	2.0	2.0
	>0.50–0.75	>0.8–1.2	3.5	3.0	3.0	3.0	2.5	2.5	2.5	2.5	2.5
	>0.75–1.00	>1.2–1.6	4.0	3.5	3.5	3.5	3.0	3.0	3.0	3.0	3.0
	>1.00	>1.6	5.0	4.0	4.0	4.0	3.5	3.5	3.0	3.0	3.0
>5–6	0.00–0.25	0.0–0.4	2.0	2.0	1.5	1.5	1.5	1.5	1.5	1.5	1.5
	>0.35–0.30	>0.4–0.5	4.0	3.0	2.5	2.5	2.0	2.0	2.0	2.0	2.0
	>0.30–0.50	>0.5–0.8	4.5	4.0	3.5	3.0	2.5	2.5	2.5	2.5	2.5
	>0.50–0.75	>0.8–1.2	5.0	4.5	4.0	3.5	3.0	3.0	3.0	3.0	3.0
	>0.75–1.00	>1.2–1.5	5.5	5.0	4.5	4.0	3.0	3.0	3.0	3.0	3.0
	>1.00	>1.6	6.0	5.0	5.0	4.5	3.5	3.5	3.5	3.5	3.5
>6	0.00–0.25	0.0–0.4	4.0	3.0	2.5	2.5	2.5	2.5	2.0	2.0	2.0
	>0.25–0.30	>0.4–0.5	4.5	4.0	3.5	3.5	3.5	3.0	2.5	2.5	2.5
	>0.30–0.50	>0.5–0.8	5.0	4.5	4.0	4.0	3.5	3.0	2.5	2.5	2.5
	>0.50–0.75	>0.8–1.2	5.5	5.0	4.5	4.5	4.0	3.5	3.0	3.0	3.0
	>0.75–1.00	>1.2–1.6	6.0	5.5	5.0	5.0	4.5	4.0	3.5	3.5	3.5
	>1.00	>1.6	7.0	6.0	5.5	5.5	5.0	4.5	4.0	4.0	4.0

Source: Transportation Research Board. *Highway Capacity Manual.* Washington, DC: National Research Council, 2000.

Negative grades (downgrades) also have an impact on equivalency factors because the comparatively poor braking characteristics of heavy vehicles have a more deleterious effect on the traffic stream compared with the level-terrain case. Table 6.10 gives the passenger car equivalents for trucks and buses on downgrades. It is assumed that recreational vehicles are not significantly impacted by downgrades, and therefore downgrade values for E_R are drawn from the level-terrain column in Table 6.7.

Table 6.9 Passenger Car Equivalents (E_R) for RVs on Specific Upgrades

Upgrade (%)	Length (mi)	Length (km)	Percentage of RVs								
			2	4	5	6	8	10	15	20	25
≤ 2	All	All	1.2	1.2	1.2	1.2	1.2	1.2	1.2	1.2	1.2
> 2–3	0.00–0.50	0.0–0.8	1.2	1.2	1.2	1.2	1.2	1.2	1.2	1.2	1.2
	> 0.50	> 0.8	3.0	1.5	1.5	1.5	1.5	1.5	1.2	1.2	1.2
> 3–4	0.00–0.25	0.0–0.4	1.2	1.2	1.2	1.2	1.2	1.2	1.2	1.2	1.2
	> 0.25–0.50	> 0.4–0.8	2.5	2.5	2.0	2.0	2.0	2.0	1.5	1.5	1.5
	> 0.50	> 0.8	3.0	2.5	2.5	2.5	2.0	2.0	2.0	1.5	1.5
> 4–5	0.00–0.25	0.0–0.4	2.5	2.0	2.0	2.0	1.5	1.5	1.5	1.5	1.5
	> 0.25–0.50	> 0.4–0.8	4.0	3.0	3.0	3.0	2.5	2.5	2.0	2.0	2.0
	> 0.50	> 0.8	4.5	3.5	3.0	3.0	3.0	2.5	2.5	2.0	2.0
> 5	0.00–0.25	0.0–0.4	4.0	3.0	2.5	2.5	2.5	2.0	2.0	2.0	1.5
	> 0.25–0.50	> 0.4–0.8	6.0	4.0	4.0	3.5	3.0	3.0	2.5	2.5	2.0
	> 0.50	> 0.8	6.0	4.5	4.0	4.5	3.5	3.0	3.0	2.5	2.0

Source: Transportation Research Board. *Highway Capacity Manual.* Washington, DC: National Research Council, 2000.

Table 6.10 Passenger Car Equivalents (E_T) for Trucks and Buses on Specific Downgrades

Downgrade (%)	Length (mi)	Length (km)	Percentage of trucks			
			5	10	15	20
< 4	All	All	1.5	1.5	1.5	1.5
4–5	≤ 4	≤ 6.4	1.5	1.5	1.5	1.5
4–5	> 4	> 6.4	2.0	2.0	2.0	1.5
> 5–6	≤ 4	≤ 6.4	1.5	1.5	1.5	1.5
> 5–6	> 4	> 6.4	5.5	4.0	4.0	3.0
> 6	≤ 4	≤ 6.4	1.5	1.5	1.5	1.5
> 6	> 4	> 6.4	7.5	6.0	5.5	4.5

Source: Transportation Research Board. *Highway Capacity Manual.* Washington, DC: National Research Council, 2000.

Once the appropriate equivalency factors have been obtained, the following equation is applied to arrive at the heavy-vehicle adjustment factor f_{HV}:

$$f_{HV} = \frac{1}{1 + P_T(E_T - 1) + P_R(E_R - 1)} \tag{6.5}$$

where

f_{HV} = heavy-vehicle adjustment factor,
P_T = proportion of trucks and buses in the traffic stream,
P_R = proportion of recreational vehicles in the traffic stream,
E_T = passenger car equivalent for trucks and buses, from Table 6.7, 6.8, or 6.10, and
E_R = passenger car equivalent for recreational vehicles, from Table 6.7 or 6.9.

As an example of how the heavy-vehicle adjustment factor is computed, consider a freeway with a 1.0-mi (1.6-km) 4% upgrade with a traffic stream having 8% trucks, 2% buses, and 2% recreational vehicles. Tables 6.8 and 6.9 must be used because the grade is too steep and long for Table 6.7 to apply. The corresponding equivalency factors for this roadway are $E_T = 2.5$ (for a combined truck and bus percentage of 10) and $E_R = 3.0$, as obtained from Tables 6.8 and 6.9, respectively. Also, from the given percentages of heavy vehicles in the traffic stream, $P_T = 0.1$ and $P_R = 0.02$. Substituting these values into Eq. 6.5 gives $f_{HV} = 0.84$, or a 16% reduction in effective roadway capacity relative to the base condition of having no heavy vehicles in the traffic stream.

Driver Population Adjustment

Under base conditions, the traffic stream is assumed to consist of regular weekday drivers and commuters. Such drivers have a high familiarity with the roadway and generally maneuver and respond to the maneuvers of other drivers in a safe and predictable fashion. There are times, however, when the traffic stream has a driver population that is less familiar with the roadway in question (such as weekend drivers or recreational drivers). Such drivers can cause a significant reduction in roadway capacity relative to the base condition of having only familiar drivers.

To account for the composition of the driver population, the adjustment factor f_p is used, and its recommended range is 0.85–1.00. Normally, the analyst should select a value of 1.00 for primarily commuter (or familiar-driver) traffic streams. But for other driver populations (for example, a large percentage of tourists), the loss in roadway capacity can vary from 1% to 15%. The exact value of the driver population adjustment factor is dependent on local conditions such as roadway characteristics and the surrounding environment (possible driver distractions such as scenic views, and so on). When the driver population consists of a significant percentage of nonfamiliar users, judgment is necessary to determine the exact value of this factor. This usually involves collection of data on local conditions (for further information, see [Transportation Research Board 2000]).

188 Chapter 6 Highway Capacity and Level-of-Service Analysis

6.4.5 Calculating Density and Determining LOS

With all the terms in the previous equations defined, these equations can now be applied to determine freeway level of service and freeway capacity. The final step before level of service can be determined is to calculate the density of the traffic stream. The alternative notation to Eq. 6.1 is shown in Eq. 6.6, which will be used in subsequent example problems (for consistency with the *Highway Capacity Manual*).

$$D = \frac{v_p}{S} \tag{6.6}$$

where

D = density in pc/mi/ln (pc/km/ln),
v_p = flow rate in pc/h/ln, and
S = average passenger car speed in mi/h (km/h).

The average passenger car speed is found by reading it from the y-axis of Fig. 6.2 for the corresponding flow rate (v_p) and free-flow speed. Once the density value is calculated, the level of service can be read from Table 6.1 or Fig. 6.2.

Application of the process for determining basic freeway segment capacity and level of service will now be demonstrated by example.

EXAMPLE 6.1

A six-lane urban freeway (three lanes in each direction) is on rolling terrain with 11-ft (3.4-m) lanes, obstructions 2 ft (0.6 m) from the right edge of the traveled pavement, and 1.5 interchanges per mile (0.93 per km). The traffic stream consists of primarily commuters. A directional weekday peak-hour volume of 2200 vehicles is observed, with 700 vehicles arriving in the most congested 15-min period. If the traffic stream has 15% large trucks and buses and no recreational vehicles, determine the level of service.

SOLUTION

Determine the free-flow speed according to Eq. 6.2.

$$FFS = BFFS - f_{LW} - f_{LC} - f_N - f_{ID}$$

with

$BFFS$ = 70 mi/h (urban freeway)
f_{LW} = 1.9 mi/h (Table 6.3)
f_{LC} = 1.6 mi/h (Table 6.4)
f_N = 3.0 mi/h (Table 6.5)
f_{ID} = 5.0 mi/h (Table 6.6)

$$FFS = 70 - 1.9 - 1.6 - 3.0 - 5.0 = 58.5 \text{ mi/h}$$

Determine the flow rate according to Eq. 6.3:

$$v_p = \frac{V}{PHF \times N \times f_{HV} \times f_p}$$

with

$$PHF = \frac{2200}{700 \times 4} = 0.786$$

$N = 3$ (given)
$f_p = 1.0$ (commuters)
$E_T = 2.5$ (rolling terrain, Table 6.7)

From Eq. 6.5 we obtain

$$f_{HV} = \frac{1}{1 + 0.15(2.5 - 1)} = 0.816$$

So,

$$v_p = \frac{2200}{0.786 \times 3 \times 0.816 \times 1.0} = 1143.4 \rightarrow 1144 \text{ pc/h/ln}$$

Obtaining average passenger car speed from Fig. 6.2 for a flow rate of 1144 and *FFS* of 58.5 mi/h yields 58.5 mi/h. In this case, the average speed is the same as *FFS* because the flow rate is low enough that it is still on the linear/flat part of the speed-flow curve.

Now density can be calculated with Eq. 6.6:

$$D = \frac{1144}{58.5} = 19.6 \text{ pc/mi/ln}$$

From Table 6.1, it can be seen that this corresponds to LOS C [18.0 (maximum density for LOS B) < 19.6 < 26.0 (maximum density for LOS C)]. Thus, this freeway segment operates at level of service C.

This problem can also be solved graphically by applying Fig. 6.2. Using this figure, draw a vertical line up from 1144 pc/h/ln (on the figure's *x*-axis) and find that this line intersects the 58.5-mi/h free-flow speed curve (between the 55 mi/h and 60 mi/h curves) in the LOS C density region (the dashed diagonal lines).

EXAMPLE 6.2

Consider the freeway and traffic conditions in Example 6.1. At some point further along the roadway there is a 6% upgrade that is 1.5 mi (2.4 km) long. All other characteristics are the same as in Example 6.1. What is the level of service of this portion of the roadway, and how many vehicles can be added before the roadway reaches capacity (assuming that the proportion of vehicle types and the peak-hour factor remain constant)?

SOLUTION

To determine the LOS of this segment of the freeway, we note that all adjustment factors are the same as those in Example 6.1 except f_{HV}, which must now be determined using an equivalency factor, E_T, drawn from the specific-upgrade tables (in this case Table 6.8). From Table 6.8, $E_T = 3.5$, which gives

$$f_{HV} = \frac{1}{1 + 0.15(3.5 - 1)} = 0.727$$

So,

$$v_p = \frac{2200}{0.786 \times 3 \times 0.727 \times 1.0} = 1283.3 \rightarrow 1284 \text{ pc/h/ln}$$

The average passenger car speed remains 58.5 mi/h; thus

$$D = \frac{1284}{58.5} = 22.0 \text{ pc/mi/ln}$$

which still gives <u>LOS C</u> from Table 6.1.

To determine how many vehicles can be added before capacity is reached, the hourly volume at capacity must be computed. Recall that capacity corresponds to a volume-to-capacity ratio of 1.0 (the threshold between LOS E and LOS F). The maximum service flow rate that can be accommodated for a free-flow speed of 58.5 mi/h is 2285 pc/h/ln (by linear interpolation). Equation 6.3 is rearranged and used to solve for the hourly volume based upon the maximum service flow rate:

$$V = v_p \times PHF \times N \times f_{HV} \times f_p \quad \Rightarrow \quad V = 2285 \times 0.786 \times 3 \times 0.727 \times 1.0$$

which gives $V = 3917$ veh/h. This means that about <u>1717 vehicles</u> (3917 − 2200) can be added during the peak hour before capacity is reached. It should be noted that the assumption that the peak-hour factor will remain constant as the roadway approaches capacity is not very realistic. In practice it is observed that as a roadway approaches capacity, *PHF* gets closer to 1. This implies that the flow rate over the peak hour becomes more uniform. This uniformity is the result of, among other factors, motorists adjusting their departure and arrival times to avoid congested periods within the peak hour.

6.5 MULTILANE HIGHWAYS

Multilane highways are similar to freeways in most respects, except for a few key differences:

- Vehicles may enter or leave the roadway at at-grade intersections and driveways (multilane highways do not have full access control).
- Multilane highways may or may not be divided (by a barrier or median separating opposing directions of flow), whereas freeways are always divided.
- Traffic signals may be present.
- Design standards (such as design speeds) are sometimes lower than those for freeways.
- The visual setting and development along multilane highways is usually more distracting to drivers than in the freeway case.

Multilane highways usually have four or six lanes (both directions); have posted speed limits between 40 and 60 mi/h (65 and 100 km/h); and can have physical medians, medians that are two-way left-turn lanes (TWLTLs), or opposing directional volumes that may not be divided by a median at all. Two examples of multilane highways are shown in Fig. 6.3.

The determination of level of service on multilane highways closely mirrors the procedure for freeways. The main differences lie in some of the adjustment factors and their values. The procedure that we will present is valid only for sections of highway that are not significantly influenced by large queue formations and dissipations resulting from traffic signals [this is generally taken as having traffic signals spaced 2.0 mi (3.2 km) apart or more], do not have significant on-street parking, do not have bus stops with high usage, and do not have significant pedestrian activity.

Table 6.11 provides the level-of-service criteria corresponding to traffic density, speed, volume-to-capacity ratio, and the maximum service flow rates for multilane highways. A graphical representation of this table is provided in Fig. 6.4.

Figure 6.3 Examples of multilane highways (divided and undivided).

Table 6.11 LOS Criteria for Multilane Highways

Free-flow speed	Criterion	LOS A	LOS B	LOS C	LOS D	LOS E	Free-flow speed	Criterion	LOS A	LOS B	LOS C	LOS D	LOS E
60 mi/h	Maximum density (pc/mi/ln)	11	18	26	35	40	100 km/h	Maximum density (pc/km/ln)	7	11	16	22	25
	Average speed (mi/h)	60.0	60.0	59.4	56.7	55.0		Average speed (km/h)	100.0	100.0	98.4	91.5	88.0
	Maximum v/c	0.30	0.49	0.70	0.90	1.00		Maximum v/c	0.32	0.50	0.72	0.92	1.00
	Maximum service flow rate (pc/h/ln)	660	1080	1550	1980	2200		Maximum service flow rate (pc/h/ln)	700	1100	1575	2015	2200
55 mi/h	Maximum density (pc/mi/ln)	11	18	26	35	41	90 km/h	Maximum density (pc/km/ln)	7	11	16	22	26
	Average speed (mi/h)	55.0	55.0	54.9	52.9	51.2		Average speed (km/h)	90.0	90.0	89.8	84.7	80.8
	Maximum v/c	0.29	0.47	0.68	0.88	1.00		Maximum v/c	0.30	0.47	0.68	0.89	1.00
	Maximum service flow rate (pc/h/ln)	600	990	1430	1850	2100		Maximum service flow rate (pc/h/ln)	630	990	1435	1860	2100
50 mi/h	Maximum density (pc/mi/ln)	11	18	26	35	43	80 km/h	Maximum density (pc/km/ln)	7	11	16	22	27
	Average speed (mi/h)	50.0	50.0	50.0	48.9	47.5		Average speed (km/h)	80.0	80.0	80.0	77.6	74.1
	Maximum v/c	0.28	0.45	0.65	0.86	1.00		Maximum v/c	0.28	0.44	0.64	0.85	1.00
	Maximum service flow rate (pc/h/ln)	550	900	1300	1710	2000		Maximum service flow rate (pc/h/ln)	560	880	1280	1705	2000
45 mi/h	Maximum density (pc/mi/ln)	11	18	26	35	45	70 km/h	Maximum density (pc/km/ln)	7	11	16	22	28
	Average speed (mi/h)	45.0	45.0	45.0	44.4	42.2		Average speed (km/h)	70.0	70.0	70.0	69.6	67.9
	Maximum v/c	0.26	0.43	0.62	0.82	1.00		Maximum v/c	0.26	0.41	0.59	0.81	1.00
	Maximum service flow rate (pc/h/ln)	490	810	1170	1550	1900		Maximum service flow rate (pc/h/ln)	490	770	1120	1530	1900

Note: The exact mathematical relationship between density and v/c has not always been maintained at LOS boundaries because of the use of rounded values. Density is the primary determinant of LOS. The speed criterion is the speed at maximum density for a given LOS. LOS F is characterized by highly unstable and variable traffic flow. Prediction of accurate flow rate, density, and speed at LOS F is difficult.

Source: Transportation Research Board. *Highway Capacity Manual*. Washington, DC: National Research Council, 2000.

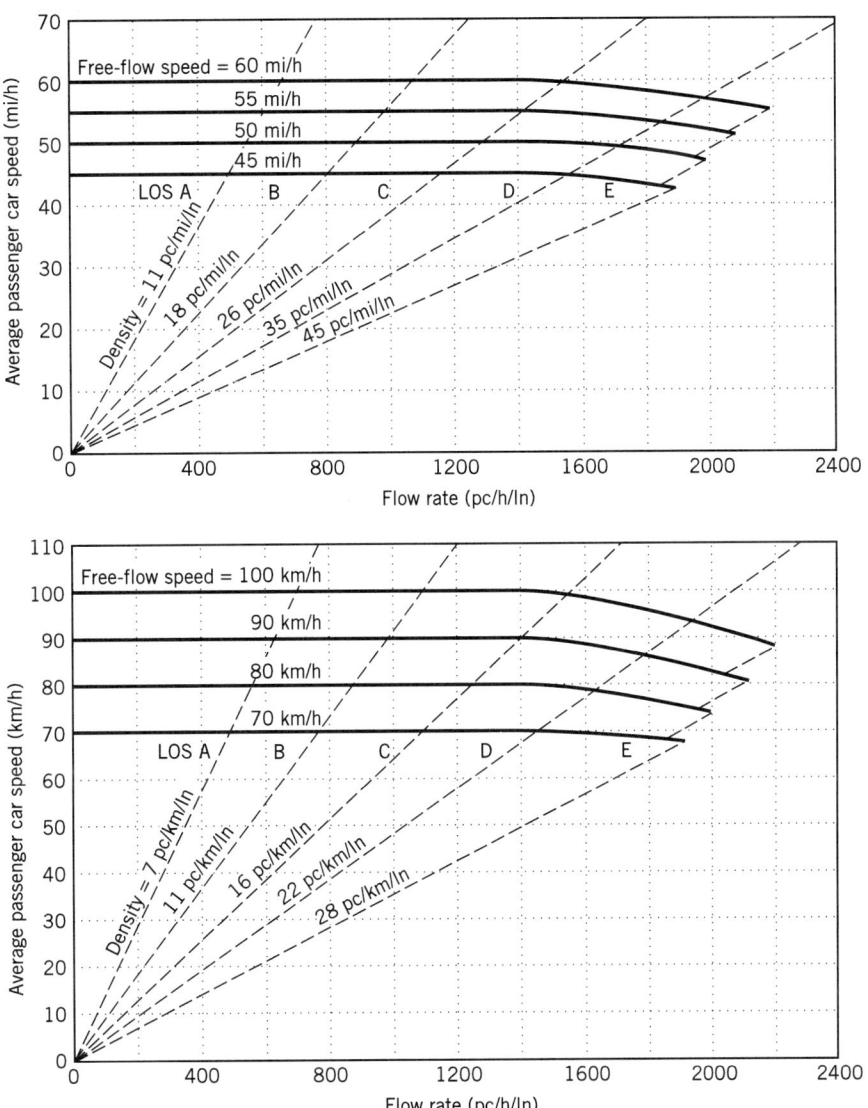

Figure 6.4 Multilane highway speed-flow curves and level-of-service criteria.
Reproduced by permission from Transportation Research Board, *Highway Capacity Manual*, National Research Council, Washington, DC: 2000.

6.5.1 Base Conditions and Capacity

The base conditions for multilane highways are defined as [Transportation Research Board 2000]

- 12-ft (3.6-m) minimum lane widths
- 12-ft (3.6-m) minimum total lateral clearance from roadside objects (right shoulder and median) in the travel direction

- Only passenger cars in the traffic stream
- No direct access points along the roadway
- Divided highway
- Level terrain (no grades greater than 2%)
- Driver population of mostly familiar roadway users
- Free-flow speed of 60 mi/h (100 km/h) or more

As was the case with the freeway level-of-service analysis, adjustments will have to be made when non-base conditions are encountered.

The capacity, c, for multilane highway segments, in pc/h/ln, is given in Table 6.12. From Table 6.11, again note that these capacity values correspond to the maximum service flow rate at LOS E and a v/c of 1.0.

6.5.2 Service Measure

Due to the large degree of similarity between multilane highway and freeway facilities, density is also the service measure (performance measure used for determining level of service) for multilane highways. However, the density threshold for LOS E varies by speed for multilane highways, as can be seen in Table 6.11. The density thresholds for levels of service A–D are the same for multilane highways and freeways.

6.5.3 Determining Free-Flow Speed

FFS for multilane highways is the mean speed of passenger cars operating in flow rates up to 1400 passenger cars per hour per lane (pc/h/ln). If *FFS* is to be estimated rather than measured, the following equation can be used, which accounts for the roadway characteristics of lane width, lateral clearance, presence (or lack) of a median, and access frequency.

$$FFS = BFFS - f_{LW} - f_{LC} - f_M - f_A \tag{6.7}$$

where

FFS = estimated free-flow speed in mi/h (km/h),
$BFFS$ = estimated free-flow speed, in mi/h (km/h), for base conditions,
f_{LW} = adjustment for lane width in mi/h (km/h),
f_{LC} = adjustment for lateral clearance in mi/h (km/h),
f_M = adjustment for median type in mi/h (km/h), and
f_A = adjustment for the number of access points along the roadway in mi/h (km/h).

As you can see, this equation closely resembles Eq. 6.2 in the freeway section. Both include adjustments for lane width and lateral clearance, and the access frequency adjustment is similar to the interchange density adjustment. The main differences are that Eq. 6.7 does not contain an adjustment for number of lanes and does include an adjustment for median type. Research has not identified a relationship between

Table 6.12 Relationship between Free-Flow Speed and Capacity on Multilane Highway Segments

Free-flow speed		Capacity (pc/h/ln)
mi/h	km/h	
60	100	2200
55	90	2100
50	80	2000
45	70	1900

Source: Transportation Research Board. *Highway Capacity Manual.* Washington, DC: National Research Council, 2000.

number of lanes and free-flow speed for multilane highways. The presence of a physical barrier or wide separation between opposing flows (such as a TWLTL) will lead to higher free-flow speeds than if there is no separation or physical barrier between opposing flows. This adjustment is not included for freeways since, by definition, all freeways are divided.

As for *BFFS*, many factors can influence the free-flow speed, with the posted speed limit often being a significant one. For multilane highways, research has found that free-flow speeds, under base conditions, are about 7 mi/h (11 km/h) higher than the speed limit for 40- and 45-mi/h (65- and 70-km/h) posted-speed-limit roadways, and about 5 mi/h (8 km/h) higher for 50-mi/h and higher (80-km/h and higher) posted-speed-limit roadways. The following sections describe the procedures for estimating the adjustment factor values.

Lane Width Adjustment

The same lane width adjustment factor values are used for multilane highways as are used for freeways. Thus, Table 6.3 should be used for multilane highways as well.

Lateral Clearance Adjustment

The adjustment factor for potentially restrictive lateral clearances (f_{LC}) is determined first by computing the total lateral clearance, which is defined as

$$TLC = LC_R + LC_L \tag{6.8}$$

where

TLC = total lateral clearance in ft (m),

LC_R = lateral clearance on the right side of the travel lanes to obstructions (retaining walls, utility poles, signs, trees, etc.), and

LC_L = lateral clearance on the left side of the travel lanes to obstructions.

For undivided highways, there is no adjustment for left-side lateral clearance because this is already taken into account in the f_M term [thus LC_L = 6 ft (1.8 m) in Eq. 6.8]. If an individual lateral clearance (either left or right side) exceeds 6 ft (1.8 m), 6 ft (1.8 m)

is used in Eq. 6.8. Finally, highways with TWLTLs are considered to have LC_L equal to 6 ft (1.8 m). Once Eq. 6.8 is applied, the value for f_{LC} can be determined directly from Table 6.13.

Median Adjustment

Values for the adjustment factor for median type, f_M, are provided in Table 6.14. This table shows that undivided highways have a free-flow speed that is 1.6 mi/h (2.6 km/h) lower than divided highways (which include those with two-way left-turn lanes).

Access Frequency Adjustment

The final adjustment factor in Eq. 6.7 is for the number of access points per mile (kilometer), f_A. Access points are defined to include intersections and driveways (on the right side of the highway in the direction being considered) that significantly influence traffic flow and, as such, do not generally include driveways to individual residences or service driveways at commercial sites. Adjustment values for access point frequency are provided in Table 6.15.

Table 6.13 Adjustment for Lateral Clearance

Total lateral clearance* (ft)	Reduction in free-flow speed (mi/h)		Total lateral clearance* (m)	Reduction in free-flow speed (km/h)	
	Four-lane highways	Six-lane highways		Four-lane highways	Six-lane highways
12	0.0	0.0	3.6	0.0	0.0
10	0.4	0.4	3.0	0.6	0.6
8	0.9	0.9	2.4	1.5	1.5
6	1.3	1.3	1.8	2.1	2.1
4	1.8	1.7	1.2	3.0	2.7
2	3.6	2.8	0.6	5.8	4.5
0	5.4	3.9	0.0	8.7	6.3

*Total lateral clearance is the sum of the lateral clearances of the median [if greater than 6 ft (1.8 m), use 6 ft (1.8 m)] and shoulder [if greater than 6 ft (1.8 m), use 6 ft (1.8 m)]. Therefore, for purposes of analysis, total lateral clearance cannot exceed 12 ft (3.6 m).
Source: Transportation Research Board. *Highway Capacity Manual.* Washington, DC: National Research Council, 2000.

Table 6.14 Adjustment for Median Type

Median type	Reduction in free-flow speed	
	(mi/h)	(km/h)
Undivided highways	1.6	2.6
Divided highways (including TWLTLs)	0.0	0.0

Source: Transportation Research Board. *Highway Capacity Manual.* Washington, DC: National Research Council, 2000.

Table 6.15 Adjustment for Access-Point Frequency

Access points/ mile	Reduction in free-flow speed (mi/h)	Access points/ kilometer	Reduction in free-flow speed (km/h)
0	0.0	0	0.0
10	2.5	6	4.0
20	5.0	12	8.0
30	7.5	18	12.0
≥ 40	10.0	≥ 24	16.0

Source: Transportation Research Board. *Highway Capacity Manual.* Washington, DC: National Research Council, 2000.

6.5.4 Determining Analysis Flow Rate

The analysis flow rate for multilane highways is determined in the same manner as for freeways, using Eq. 6.3 and the remainder of the procedure outlined in Section 6.4.4. There is one minor difference for multilane highways—the guidelines for an extended segment analysis. An extended segment (general terrain type) analysis can be used for multilane highway segments if grades of 3% or less do not extend for more than 1 mi (1.6 km) or any grades greater than 3% do not extend for more than 0.5 mi (0.8 km).

6.5.5 Calculating Density and Determining LOS

The procedure for calculating density and determining LOS for multilane highways is essentially the same as for freeways (see Section 6.4.5). Equation 6.6 is applied to arrive at a density. However, slightly different speed-flow curves and level-of-service criteria are used for multilane highways. Table 6.11 shows the level-of-service criteria for multilane highways, and Fig. 6.4 shows the corresponding speed-flow curves for multilane highways.

The average passenger car speed is found by reading it from the *y*-axis of Fig. 6.4 for the corresponding analysis flow rate (v_p) and free-flow speed. Once the density value is calculated, the level of service can be read from Table 6.11 or Fig. 6.4.

EXAMPLE 6.3

A four-lane undivided highway has 11-ft (3.4-m) lanes, with 4-ft (1.2-m) shoulders on the right side. There are 7 access points per mile (4.35 per km), and the posted speed limit is 50 mi/h (80 km/h). What is the estimated free-flow speed?

SOLUTION

This problem can be solved by direct application of Eq. 6.7 to arrive at an estimated free-flow speed.

$$FFS = BFFS - f_{LW} - f_{LC} - f_M - f_A$$

with

$BFFS = 55$ mi/h (assume FFS = posted speed + 5 mi/h)
$f_{LW} = 1.9$ mi/h (Table 6.3)
$f_{LC} = 0.4$ mi/h (Table 6.13, with $TLC = 4 + 6 = 10$ from Eq. 6.8, with $LC_L = 6$ ft because the highway is undivided)
$f_M = 1.6$ mi/h (Table 6.14)
$f_A = 1.75$ mi/h (Table 6.15, by interpolation)

Substitution gives

$$FFS = 55 - 1.9 - 0.4 - 1.6 - 1.75 = \underline{49.35 \text{ mi/h}}$$

which means that the more restrictive roadway characteristics relative to the base conditions result in a reduction in free-flow speed of 5.65 mi/h.

EXAMPLE 6.4

A six-lane divided highway is on rolling terrain with 2 access points per mile (1.24 per km) and has 10-ft (3.0-m) lanes, with a 5-ft (1.5-m) shoulder on the right side and a 3-ft (0.9-m) shoulder on the left side. The peak-hour factor is 0.80, and the directional peak-hour volume is 3000 vehicles per hour. There are 6% large trucks, 2% buses, and 2% recreational vehicles. A significant percentage of nonfamiliar roadway users are in the traffic stream (the driver population adjustment factor is estimated as 0.95). No speed studies are available, but the posted speed limit is 55 mi/h (88 km/h). Determine the level of service.

SOLUTION

We begin by determining FFS by applying Eq. 6.7:

$$FFS = BFFS - f_{LW} - f_{LC} - f_M - f_A$$

with

$BFFS = 60$ mi/h (assume FFS = posted speed + 5 mi/h)
$f_{LW} = 6.6$ mi/h (Table 6.3)
$f_{LC} = 0.9$ mi/h (Table 6.13, with $TLC = 5 + 3 = 8$ from Eq. 6.8)
$f_M = 0.0$ mi/h (Table 6.14)
$f_A = 0.5$ mi/h (Table 6.15, by interpolation)

Substitution gives

$$FFS = 60.0 - 6.6 - 0.9 - 0.0 - 0.5 = 52.0 \text{ mi/h}$$

Next we determine the analysis flow rate using Eq. 6.3:

$$v_p = \frac{V}{PHF \times N \times f_{HV} \times f_p}$$

with

$$V = 3000 \text{ veh/h (given)}$$
$$PHF = 0.80 \text{ (given)}$$
$$N = 3 \text{ (given)}$$
$$f_p = 0.95 \text{ (given)}$$
$$E_T = 2.5 \text{ (Table 6.7)}$$
$$E_R = 2.0 \text{ (Table 6.7)}$$

From Eq. 6.5,

$$f_{HV} = \frac{1}{1 + 0.08(2.5 - 1) + 0.02(2 - 1)} = 0.877$$

Substitution gives

$$v_p = \frac{3000}{0.80 \times 3 \times 0.877 \times 0.95} = 1500.3 \text{ pc/h/ln}$$

Using Fig. 6.4, construct a speed-flow curve for $FFS = 52$ mi/h (drawing it parallel to the curves for 55 and 50 mi/h) and note that the 1500.3-pc/h/ln flow rate intersects this curve in the LOS D density region. Therefore, this highway is operating at <u>LOS D</u>.

EXAMPLE 6.5

A local manufacturer wishes to open a factory near the segment of highway described in Example 6.4. How many large trucks can be added to the peak-hour directional volume before capacity is reached? (Add only trucks and assume that the *PHF* remains constant.)

SOLUTION

Note that *FFS* will remain unchanged at 52 mi/h. Table 6.11 shows that capacity for $FFS = 55$ mi/h is 2100 pc/h/ln and for $FFS = 50$ mi/h is 2000 pc/h/ln, so a linear interpolation gives us a capacity of 2040 pc/h/ln for $FFS = 52$ mi/h. The current number of large trucks and buses in the peak-hour traffic stream is 240 (0.08×3000) and the current number of recreational vehicles is 60 (0.02×3000). Let us denote the number of new trucks added as V_{nt}, so the combination of Eqs. 6.3 and 6.5 gives

$$v_p = \frac{V + V_{nt}}{(PHF)(N)\left[\frac{1}{1 + \left(\frac{240 + V_{nt}}{V + V_{nt}}\right)(E_T - 1) + \left(\frac{60}{V + V_{nt}}\right)(E_R - 1)}\right](f_p)}$$

with

$$v_p = 2040 \text{ pc/h/ln}$$
$$V = 3000 \text{ veh/h (Example 6.4)}$$
$$PHF = 0.80 \text{ (Example 6.4)}$$
$$N = 3 \text{ (Example 6.4)}$$
$$f_p = 0.95 \text{ (Example 6.4)}$$
$$E_T = 2.5 \text{ (Example 6.4)}$$
$$E_R = 2.0 \text{ (Example 6.4)}$$

$$2040 = \frac{3000 + V_{nt}}{(0.80)(3)\left[\dfrac{1}{1 + \left(\dfrac{240 + V_{nt}}{3000 + V_{nt}}\right)(2.5 - 1) + \left(\dfrac{60}{3000 + V_{nt}}\right)(2 - 1)}\right](0.95)}$$

which gives $V_{nt} = 492.5$ (≈ 492), which is the number of trucks that can be added to the peak-hour volume before capacity is reached.

6.6 TWO-LANE HIGHWAYS

Two-lane highways are defined as roadways with one lane available in each direction. For level-of-service determination, a key distinction between two-lane highways and the freeways and multilane highways previously discussed is that traffic in both directions must now be considered (previously we considered traffic in one direction only). This is because traffic in an opposing direction has a strong influence on level of service. For example, a high opposing traffic volume limits the opportunity to pass slow-moving vehicles (because such a pass requires the passing vehicle to occupy the opposing lane) and thus forces a lower traffic speed—and, as a consequence, a lower level of service. It also follows that any geometric features that restrict passing sight distance (such as sight distance on horizontal and vertical curves) will have an adverse impact on the level of service. Finally, the type of terrain (level, rolling, or mountainous) plays a more critical role in level-of-service calculations, relative to freeways and multilane highways, because of the sometimes limited ability to pass slower-moving vehicles on grades in areas where passing is prohibited due to sight distance restrictions or where opposing traffic does not permit safe passing.

The 2000 edition of the *Highway Capacity Manual* [Transportation Research Board 2000] contains two procedures for operational analysis and level-of-service determination: one for both directions of travel combined, termed a two-way analysis, and another for just one of the two directions of travel, termed a directional analysis. This chapter covers only the two-way analysis procedure. The directional analysis procedure has many aspects in common with the two-way analysis procedure, but coverage of both procedures is beyond the scope of this book.

6.6.1 Base Conditions and Capacity

The base conditions for two-lane highways are defined as [Transportation Research Board 2000]

- 12-ft (3.6-m) minimum lane widths
- 6-ft (1.8-m) minimum shoulder widths
- 0% no-passing zones on the highway segment
- Only passenger cars in the traffic stream
- No direct access points along the roadway
- No impediments to through traffic due to traffic control or turning vehicles
- Level terrain (no grades greater than 2%)
- A 50/50 directional split of traffic (50% traveling one direction and 50% traveling the opposite direction) for two-way analysis

The capacity of extended lengths of two-lane highway under base conditions is 3200 passenger cars per hour (pc/h), total, both directions. The capacity of a single direction of a two-lane highway is 1700 pc/h.

6.6.2 Service Measures

Two service measures have been identified for two-lane highways: (1) percent time spent following and (2) average travel speed. Percent time spent following (*PTSF*) is the average percentage of travel time that vehicles must travel behind slower vehicles due to the lack of passing opportunities (because of geometry and/or opposing traffic). *PTSF* is difficult to measure in the field; thus, it is recommended that the percentage of vehicles traveling with headways less than 3 seconds at a representative location be used as a surrogate measure. *PTSF* is generally representative of a driver's freedom to maneuver in the traffic stream. Average travel speed (*ATS*) is simply the length of the analysis segment divided by the average travel time of all vehicles traversing the segment during the analysis period. *ATS* is an indicator of the mobility on a two-lane highway.

The service measure, and corresponding thresholds, that govern the determination of level of service depends on the functional classification of the two-lane highway. *The Highway Capacity Manual* [Transportation Research Board 2000] has defined two classes of two-lane highway:

- *Class I:* Two-lane highways on which motorists expect to travel at high speeds. Class I highways include inter-city routes, primary arterials connecting major traffic generators, daily commuter routes, and primary links in state or national highway networks.
- *Class II:* Two-lane highways on which motorists do not necessarily expect to travel at high speeds. Scenic or recreational routes, or routes that pass through rugged terrain, are typically assigned to Class II, and these routes generally serve shorter trip lengths than Class I routes.

Note that the level-of-service criteria for two-lane highways are presented later (in Section 6.6.6).

6.6.3 Determining Free-Flow Speed

FFS for two-lane highways is the mean speed of all vehicles operating in flow rates up to 200 pc/h total for both directions. Free-flow speeds on two-lane highways typically range from 45 to 65 mi/h (70 to 105 km/h). If field measurement of *FFS* cannot be made under conditions with a flow rate of 200 pc/h or less, an adjustment can be made with the following equation.

U.S. Customary

$$FFS = S_{FM} + 0.00776 \frac{V_f}{f_{HV}}$$

Metric

$$FFS = S_{FM} + 0.0125 \frac{V_f}{f_{HV}} \quad (6.9)$$

where

FFS = estimated free-flow speed in mi/h (km/h),

S_{FM} = mean speed of traffic measured in the field in mi/h (km/h),

V_f = observed flow rate, in veh/h, for the period when field data were obtained, and

f_{HV} = heavy-vehicle adjustment factor as determined by Eq. 6.5.

If *FFS* is to be estimated rather than measured in the field, the following equation can be used, which accounts for roadway characteristics of lane width, shoulder width, and access frequency.

$$FFS = BFFS - f_{LS} - f_A \quad (6.10)$$

where

FFS = estimated free-flow speed in mi/h (km/h),

$BFFS$ = estimated free-flow speed, in mi/h (km/h), for base conditions,

f_{LS} = adjustment for lane width and shoulder width in mi/h (km/h), and

f_A = adjustment for the number of access points along the roadway in mi/h (km/h).

Specific guidance on choosing a value for *BFFS* is not offered, due to the wide range of speed conditions on two-lane highways and the influence of local and regional factors on driver-desired speeds. Speed data and local knowledge of operating conditions on similar facilities can be used in developing an estimate of *BFFS*. The following discussion describes how to determine the adjustment factor values.

Lane Width and Shoulder Width Adjustment

The adjustment for lane widths and/or shoulder widths that are more restrictive than the base conditions is shown in Table 6.16.

Table 6.16 Adjustment for Lane Width and Shoulder Width

	Reduction in free-flow speed (mi/h) Shoulder width (ft)			
Lane width (ft)	$\geq 0 < 2$	$\geq 2 < 4$	$\geq 4 < 6$	≥ 6
$9 < 10$	6.4	4.8	3.5	2.2
$\geq 10 < 11$	5.3	3.7	2.4	1.1
$\geq 11 < 12$	4.7	3.0	1.7	0.4
≥ 12	4.2	2.6	1.3	0.0
	Reduction in free-flow speed (km/h) Shoulder width (m)			
Lane width (m)	$\geq 0.0 < 0.6$	$\geq 0.6 < 1.2$	$\geq 1.2 < 1.8$	≥ 1.8
$2.7 < 3.0$	10.3	7.7	5.6	3.5
$\geq 3.0 < 3.3$	8.5	5.9	3.8	1.7
$\geq 3.3 < 3.6$	7.5	4.9	2.8	0.7
≥ 3.6	6.8	4.2	2.1	0.0

Source: Transportation Research Board. *Highway Capacity Manual.* Washington, DC: National Research Council, 2000.

Access Frequency Adjustment

The adjustment for access frequency is the same as that for multilane highways, and is shown in Table 6.15.

6.6.4 Determining Analysis Flow Rate

The hourly volume must be adjusted to account for the peak 15-minute flow rate, the terrain, and the presence of heavy vehicles in the traffic stream. The two-way analysis flow rate is calculated with the following equation.

$$v_p = \frac{V}{PHF \times f_G \times f_{HV}} \tag{6.11}$$

where

v_p = 15-min passenger car equivalent flow rate (pc/h),
V = hourly volume (veh/h),
PHF = peak-hour factor,
f_G = grade adjustment factor, and
f_{HV} = heavy-vehicle adjustment factor.

Unlike the analysis flow rate equation (6.3) for freeways and multilane highways, Eq. 6.11 does not contain an adjustment factor for driver population. Although it is reasonable to assume that drivers familiar with the highway will use it more efficiently than recreational or other nonfamiliar users of the facility, studies have yet to identify a significant difference between the two driver populations [Transportation Research Board 2000].

The procedures for determination of the PHF, f_G, and f_{HV} adjustment factor values are described next.

Peak-Hour Factor

PHF for two-lane highways is calculated in a manner consistent with that for freeways and multilane highways. The only distinction is that because the two-lane highway analysis methodology deals with a combined volume of the two (opposing) traffic streams, PHF should be calculated for both directions of traffic flow combined. While a PHF value could be calculated for each individual direction of traffic flow, this could result in an unreasonably high combined analysis volume since the two directions may not peak at the same time.

Grade Adjustment Factor

The grade adjustment factor accounts for the effect of terrain on the traffic flow. For terrain generally classified as level or rolling, Table 6.17 shows values for the grade adjustment factor for average travel speed and percent time spent following.

Heavy-Vehicle Adjustment Factor

Just as for freeways and multilane highways, the heavy-vehicle adjustment factor accounts for the effect on traffic flow due to the presence of trucks, buses, and recreational vehicles in the traffic stream. The passenger car equivalency (PCE) values, however, are different from those for freeway and multilane highway segments. The heavy-vehicle PCE values for level and rolling terrain for both ATS and $PTSF$ are shown in Table 6.18. Two-lane highways in mountainous terrain must be analyzed as specific upgrades and/or downgrades with the directional analysis procedure. For details on the procedure used to evaluate two-lane highways on specific grades [for example, a 5% grade 0.75 mi (1.2 km) long], the reader is referred to the *Highway Capacity Manual* [Transportation Research Board 2000].

Table 6.17 Grade Adjustment Factor for Average Travel Speed (ATS) and Percent Time Spent Following ($PTSF$)

Range of two-way flow rates (pc/h)	Average travel speed		Percent time spent following	
	Level terrain	Rolling terrain	Level terrain	Rolling terrain
0–600	1.00	0.71	1.00	0.77
>600–1200	1.00	0.93	1.00	0.94
>1200	1.00	0.99	1.00	1.00

Source: Transportation Research Board. *Highway Capacity Manual.* Washington, DC: National Research Council, 2000.

Table 6.18 Passenger Car Equivalents for Heavy Vehicles for Average Travel Speed (*ATS*) and Percent Time Spent Following (*PTSF*)

Vehicle type	Range of two-way flow rates (pc/h)	Average travel speed		Percent time spent following	
		Level terrain	Rolling terrain	Level terrain	Rolling terrain
Trucks and buses, E_T	0–600	1.7	2.5	1.1	1.8
	>600–1200	1.2	1.9	1.1	1.5
	>1200	1.1	1.5	1.0	1.0
RVs, E_R	0–600	1.0	1.1	1.0	1.0
	>600–1200	1.0	1.1	1.0	1.0
	>1200	1.0	1.1	1.0	1.0

Source: Transportation Research Board. *Highway Capacity Manual.* Washington, DC: National Research Council, 2000.

It must be noted that the flow rates in Tables 6.17 and 6.18 are in units of passenger cars per hour. However, until Eq. 6.11 is applied, the flow rate is in units of vehicles per hour. This can result in the need to use an iterative approach to arrive at the correct adjustment factors and final analysis flow rate. This will be demonstrated later in Example 6.6.

6.6.5 Calculate Service Measures

If the highway is Class I, both *ATS* and *PTSF* must be calculated. If the highway is Class II, only *PTSF* needs to be calculated. The calculations for these two service measures are described next.

Average Travel Speed

The average travel speed depends on the free-flow speed, the analysis flow rate, and an adjustment factor for the percentage of no-passing zones, and is calculated according to Eq. 6.12.

U.S. Customary

$$ATS = FFS - 0.00776 v_p - f_{np}$$

Metric

$$ATS = FFS - 0.0125 v_p - f_{np} \quad (6.12)$$

where

ATS = estimated average travel speed in mi/h (km/h), for both directions of travel combined,

FFS = free-flow speed in mi/h (km/h), as measured in the field and possibly adjusted by Eq. 6.9 or estimated from Eq. 6.10,

v_p = analysis flow rate in pc/h, as calculated from Eq. 6.11, and

f_{np} = adjustment factor for the percentage of no-passing zones, which is determined from Table 6.19.

Table 6.19 Adjustment for Effect of No-Passing Zones on Average Travel Speed

Two-way flow rate, v_p (pc/h)	Reduction in average travel speed (mi/h)						Reduction in average travel speed (km/h)					
	No-passing zones (%)						No-passing zones (%)					
	0	20	40	60	80	100	0	20	40	60	80	100
0	0.0	0.0	0.0	0.0	0.0	0.0	0.0	0.0	0.0	0.0	0.0	0.0
200	0.0	0.6	1.4	2.4	2.6	3.5	0.0	1.0	2.3	3.8	4.2	5.6
400	0.0	1.7	2.7	3.5	3.9	4.5	0.0	2.7	4.3	5.7	6.3	7.3
600	0.0	1.6	2.4	3.0	3.4	3.9	0.0	2.5	3.8	4.9	5.5	6.2
800	0.0	1.4	1.9	2.4	2.7	3.0	0.0	2.2	3.1	3.9	4.3	4.9
1000	0.0	1.1	1.6	2.9	2.2	2.6	0.0	1.8	2.5	3.2	3.6	4.2
1200	0.0	0.8	1.2	1.6	1.9	2.1	0.0	1.3	2.0	2.6	3.0	3.4
1400	0.0	0.6	0.9	1.2	1.4	1.7	0.0	0.9	1.4	1.9	2.3	2.7
1600	0.0	0.6	0.8	1.1	1.3	1.5	0.0	0.9	1.3	1.7	2.1	2.4
1800	0.0	0.5	0.7	1.0	1.1	1.3	0.0	0.8	1.1	1.6	1.8	2.1
2000	0.0	0.5	0.6	0.9	1.0	1.1	0.0	0.8	1.0	1.4	1.6	1.8
2200	0.0	0.5	0.6	0.9	0.9	1.1	0.0	0.8	1.0	1.4	1.5	1.7
2400	0.0	0.5	0.6	0.8	0.9	1.1	0.0	0.8	1.0	1.3	1.5	1.7
2600	0.0	0.5	0.6	0.8	0.9	1.0	0.0	0.8	1.0	1.3	1.4	1.6
2800	0.0	0.5	0.6	0.7	0.8	0.9	0.0	0.8	1.0	1.2	1.3	1.4
3000	0.0	0.5	0.6	0.7	0.7	0.8	0.0	0.8	0.9	1.1	1.1	1.3
3200	0.0	0.5	0.6	0.6	0.6	0.7	0.0	0.8	0.9	1.0	1.0	1.1

Source: Transportation Research Board. *Highway Capacity Manual*. Washington, DC: National Research Council, 2000.

Percent Time Spent Following

The percent time spent following depends on the analysis flow rate and an adjustment for the combined effect of the percentage of no-passing zones and the directional distribution of traffic, and is calculated according to the two following equations.

$$PTSF = BPTSF + f_{d/np} \qquad (6.13)$$

where

$PTSF$ = percent time spent following for both directions of travel combined,
$BPTSF$ = base percent time spent following for both directions of travel combined, and
$f_{d/np}$ = adjustment factor for the combined effect of the directional distribution of traffic and the percentage of no-passing zones.

The $f_{d/np}$ adjustment value is determined from Table 6.20, and $BPTSF$ is calculated according to Eq. 6.14, which takes into account the analysis flow rate, v_p, as previously defined.

$$BPTSF = 100(1 - e^{-0.000879 v_p}) \qquad (6.14)$$

Table 6.20 Adjustment for Combined Effect of Directional Distribution of Traffic and Percentage of No-Passing Zones on Percent Time Spent Following

Two-way flow rate, v_p (pc/h)	No-passing zones (%)					
	0	20	40	60	80	100
Directional split = 50/50						
≤ 200	0.0	10.1	17.2	20.2	21.0	21.8
400	0.0	12.4	19.0	22.7	23.8	24.8
600	0.0	11.2	16.0	18.7	19.7	20.5
800	0.0	9.0	12.3	14.1	14.5	15.4
1400	0.0	3.6	5.5	6.7	7.3	7.9
2000	0.0	1.8	2.9	3.7	4.1	4.4
2600	0.0	1.1	1.6	2.0	2.3	2.4
3200	0.0	0.7	0.9	1.1	1.2	1.4
Directional split = 60/40						
≤ 200	1.6	11.8	17.2	22.5	23.1	23.7
400	0.5	11.7	16.2	20.7	21.5	22.2
600	0.0	11.5	15.2	18.9	19.8	20.7
800	0.0	7.6	10.3	13.0	13.7	14.4
1400	0.0	3.7	5.4	7.1	7.6	8.1
2000	0.0	2.3	3.4	3.6	4.0	4.3
≥ 2600	0.0	0.9	1.4	1.9	2.1	2.2
Directional split = 70/30						
≤ 200	2.8	13.4	19.1	24.8	25.2	25.5
400	1.1	12.5	17.3	22.0	22.6	23.2
600	0.0	11.6	15.4	19.1	20.0	20.9
800	0.0	7.7	10.5	13.3	14.0	14.6
1400	0.0	3.8	5.6	7.4	7.9	8.3
≥ 2000	0.0	1.4	4.9	3.5	3.9	4.2
Directional split = 80/20						
≤ 200	5.1	17.5	24.3	31.0	31.3	31.6
400	2.5	15.8	21.5	27.1	27.6	28.0
600	0.0	14.0	18.6	23.2	23.9	24.5
800	0.0	9.3	12.7	16.0	16.5	17.0
1400	0.0	4.6	6.7	8.7	9.1	9.5
≥ 2000	0.0	2.4	3.4	4.5	4.7	4.9
Directional split = 90/10						
≤ 200	5.6	21.6	29.4	37.2	37.4	37.6
400	2.4	19.0	25.6	32.2	32.5	32.8
600	0.0	16.3	21.8	27.2	27.6	28.0
800	0.0	10.9	14.8	18.6	19.0	19.4
≥ 1400	0.0	5.5	7.8	10.0	10.4	10.7

Source: Transportation Research Board. *Highway Capacity Manual.* Washington, DC: National Research Council, 2000.

6.6.6 Determine LOS

The first step in the LOS determination is to compare the analysis flow rate, v_p, to the two-way capacity of 3200 pc/h. If v_p exceeds 3200, the LOS is F, and the analysis ends. Additionally, the directional flow rates must be checked against the directional capacity of 1700 pc/h. These flow rates are determined from v_p and the directional traffic split. If either of these exceed the directional capacity, the LOS is F, and the analysis is finished. In this case, *PTSF* is virtually 100%, and speeds are highly variable and difficult to estimate.

If capacity is not exceeded, the calculated *PTSF* and *ATS* values are used with Table 6.21 or Table 6.22 to determine the LOS. For Class I highways, the LOS is determined from Table 6.21. For a particular LOS category to apply, the thresholds for both *PTSF* and *ATS* must be met. For example, for LOS B to apply, *PTSF* must be less than or equal to 50% and *ATS* must be greater than 50 mi/h (80 km/h). If, for a particular two-lane highway, *PTSF* is 45% and *ATS* is 48 mi/h (77 km/h), the LOS would be C. For Class II highways, the LOS is determined from Table 6.22, where *PTSF* is the only determinant.

Table 6.21 LOS Criteria for Class I Two-Lane Highways

LOS	Percent time spent following (PTSF)	Average travel speed (ATS)	
		mi/h	km/h
A	≤ 35	> 55	> 90
B	≤ 50	> 50	> 80
C	≤ 65	> 45	> 70
D	≤ 80	> 40	> 60
E	> 80	≤ 40	≤ 60

Note: LOS F applies whenever the flow rate exceeds the segment capacity.
Source: Transportation Research Board. *Highway Capacity Manual.* Washington, DC: National Research Council, 2000.

Table 6.22 LOS Criteria for Class II Two-Lane Highways

LOS	Percent time spent following (PTSF)
A	≤ 40
B	≤ 55
C	≤ 70
D	≤ 85
E	> 85

Note: LOS F applies whenever the flow rate exceeds the segment capacity.
Source: Transportation Research Board. *Highway Capacity Manual.* Washington, DC: National Research Council, 2000.

EXAMPLE 6.6

One segment of a Class I two-lane highway is on rolling terrain and has an hourly volume of 500 veh/h with $PHF = 0.94$, and the traffic stream contains 5% large trucks, 2% buses, and 6% recreational vehicles. For these conditions determine the analysis flow rate for *ATS* and *PTSF*.

SOLUTION

The first step is to calculate the flow rate, in veh/h, that will be used to determine the grade adjustment and PCE values. This is done by dividing the hourly volume by the *PHF*.

$$\frac{V}{PHF} = \frac{500}{0.94} = 532$$

The values for *ATS* will be selected first.

For rolling terrain,

$f_G = 0.71$ (Table 6.17)
$E_T = 2.5$ (Table 6.18)
$E_R = 1.1$ (Table 6.18)

Substituting the PCE values into Eq. 6.5 gives

$$f_{HV} = \frac{1}{1 + 0.07(2.5 - 1) + 0.06(1.1 - 1)} = 0.90$$

Substituting the f_{HV} and f_G values into Eq. 6.11 gives

$$v_p = \frac{500}{0.94 \times 0.71 \times 0.90} = 832 \text{ pc/h}$$

This flow rate is higher than the first flow rate category (832 > 600) for which the adjustment factor values were chosen. Thus, the calculation needs to be repeated using the second flow rate category (600–1200) for Tables 6.17 and 6.18.

Again, for rolling terrain,

$f_G = 0.93$ (Table 6.17)
$E_T = 1.9$ (Table 6.18)
$E_R = 1.1$ (Table 6.18)

Substituting the PCE values into Eq. 6.5 gives

$$f_{HV} = \frac{1}{1 + 0.07(1.9 - 1) + 0.06(1.1 - 1)} = 0.935$$

Substituting the f_{HV} and f_G values into Eq. 6.11 gives

$$v_p = \frac{500}{0.94 \times 0.93 \times 0.935} = \underline{\underline{612 \text{ pc/h}}}$$

Since this flow rate is in the range for which the values from Tables 6.17 and 6.18 were selected (600 < 612 < 1200), this will be the analysis flow rate.

Repeating this process for *PTSF* results in the following final values:

$f_G = 0.94$ (Table 6.17)
$E_T = 1.5$ (Table 6.18)
$E_R = 1.0$ (Table 6.18)
$f_{HV} = 0.966$ (Eq. 6.5)
$v_p = \underline{\underline{586 \text{ pc/h}}}$ (Eq. 6.11)

EXAMPLE 6.7

The two-lane highway segment in Example 6.6 has the following additional characteristics: 11-ft (3.4-m) lanes, 2-ft (0.6-m) shoulders, access frequency of 10 per mile (6.2 per km), 50% no-passing zones, base *FFS* of 55 mi/h (88.5 km/h), and a directional traffic split of 60/40. Using the analysis flow rates for *ATS* and *PTSF* from Example 6.6, determine the level of service for this two-lane highway segment.

SOLUTION

We begin by checking whether the highway segment is over capacity.

The analysis flow rates of 612 and 586 for *ATS* and *PTSF*, respectively, are both well below the two-way capacity of 3200 pc/h. Because both two-way flow rates are also below the directional capacity of 1700 pc/h, it is clear just by inspection that the directional capacity is also not exceeded.

Since the facility is not over capacity, we can proceed with the LOS determination. We first estimate the free-flow speed using Eq. 6.10.

$$FFS = BFFS - f_{LS} - f_A$$

with

$BFFS = 55$ mi/h (given)
$f_{LS} = 3.0$ mi/h (Table 6.16)
$f_A = 2.5$ mi/h (Table 6.15)

Substituting these values into Eq. 6.10 gives,

$$FFS = 55 - 3.0 - 2.5 = 49.5 \text{ mi/h}$$

The average travel speed will be calculated first, using Eq. 6.12.

$$ATS = FFS - 0.00776v_p - f_{np}$$

with

$FFS = 49.5$ mi/h (from previous calculation)
$v_p = 612$ pc/h (Example 6.6)
$f_{np} = 2.7$ mi/h (Table 6.19, by linear interpolation for v_p and percent no-passing zones)

Substituting these values into Eq. 6.12 gives

$$ATS = 49.5 - 0.00776(612) - 2.7 = 42.05 \text{ mi/h}$$

The percent time spent following is calculated next, using Eq. 6.13.

$$PTSF = BPTSF + f_{d/np}$$

BPTSF is calculated from Eq. 6.14.

$$BPTSF = 100(1 - e^{-0.000879v_p})$$

Substituting the v_p value of 586 pc/h determined from Example 6.6 gives

$$BPTSF = 100(1 - e^{-0.000879(586)}) = 40.3\%$$

$f_{d/np}$ is found to be 17.0% from Table 6.20, again by linear interpolation of flow rate and percent no-passing zones. Note that a three-way linear interpolation is possible with this table if the directional split does not equal one of the five predefined categories. Substituting these values into Eq. 6.13 gives

$$PTSF = 40.3 + 17.0 = 57.3\%$$

We now determine the LOS for the calculated ATS and PTSF values from Table 6.21, because it is a Class I highway. From this table, the LOS is D. Although PTSF falls within the LOS C category, ATS falls within the LOS D category; thus, ATS governs the level of service for this two-lane highway under these roadway and traffic conditions.

6.7 DESIGN TRAFFIC VOLUMES

In the preceding sections of this chapter, consideration was given to the determination of level of service, given some hourly volume. However, a procedure for selecting an appropriate hourly volume is needed to compute the level of service and to determine the number of lanes that need to be provided in a new roadway design to achieve some specified level of service. The selection of an appropriate hourly volume is complicated by two issues. First, there is considerable variability in traffic volume by time of day, day of week, time of year, and type of roadway. Figure 6.5 shows such variations in traffic volumes by hour of day and day of week for typical intra-city and inter-city routes. Figure 6.6 gives variations by time of year by comparing monthly percentages

212 Chapter 6 Highway Capacity and Level-of-Service Analysis

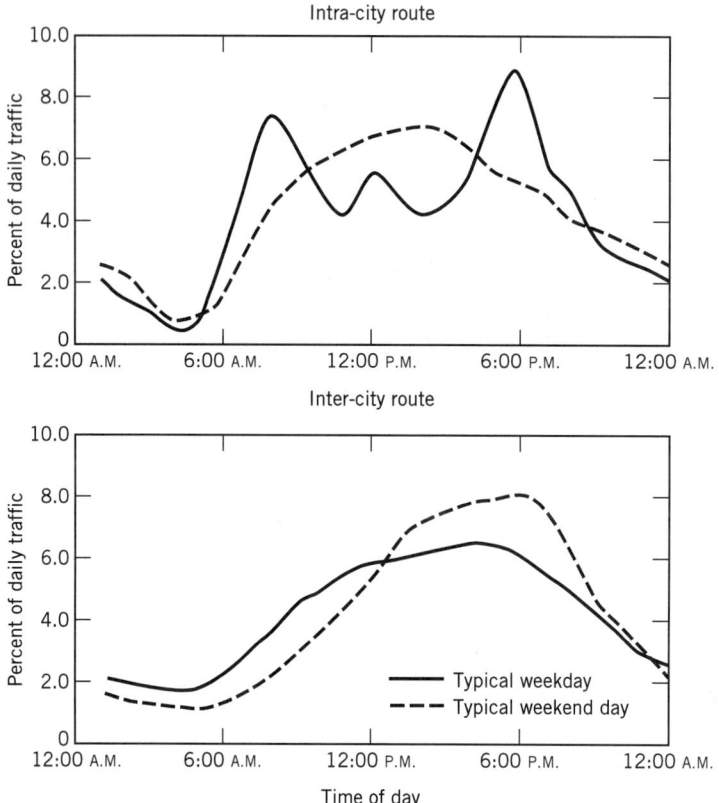

Figure 6.5 Examples of hourly and daily traffic variations for intra-city and inter-city routes.

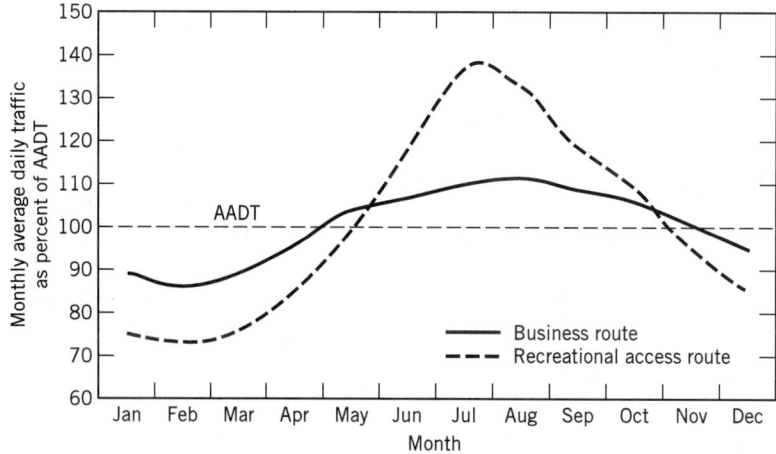

Figure 6.6 Example of monthly traffic volume variations for business and recreational access routes.

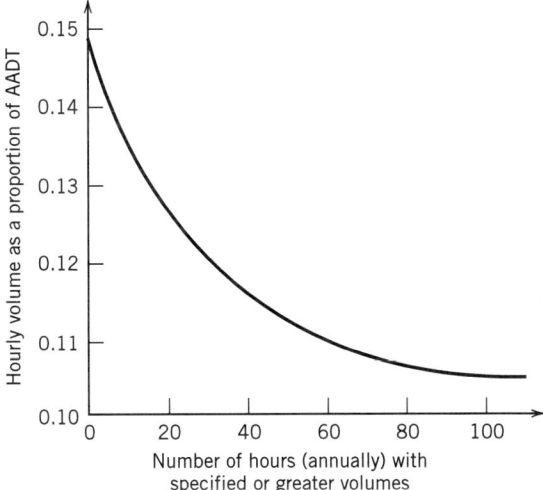

Figure 6.7 Highest 100 hourly volumes over a one-year period for a typical roadway.

of the annual average daily traffic, AADT (in units of vehicles per day and computed as the total yearly traffic volume divided by the number of days in the year). The second concern is an outgrowth of the first: Given the temporal variability in traffic flow, what hourly volume should be used for design and/or analysis? To answer this question, consider the example diagram shown in Fig. 6.7. This figure plots hourly volume (as a percentage of AADT) against the cumulative number of hours that exceed this volume, per year. For example, the highest traffic flow in the year, on this sample roadway, would have an hourly volume of $0.148 \times$ AADT (a volume that is exceeded by zero other hours). Sixty hours in the year would have a volume that exceeds $0.11 \times$ AADT. In determining the number of lanes that should be provided on a new or redesigned roadway, it is obvious that using the worst single hour in a year (the hour with the highest traffic flow, which would be $0.148 \times$ AADT from Fig. 6.7) would be a wasteful use of resources because additional lanes would be provided for a relative rare occurrence. In contrast, if the 100th highest volume is used for design, the design level of service will be exceeded 100 times a year, which will result in considerable driver delay. Clearly, some compromise between the expense of providing additional capacity (such as additional lanes) and the expense of incurring additional driver delay must be made.

A common practice in the United States is to use a design hour-volume (DHV) that is between the 10th and 50th highest volume hours of the year, depending on the type and location of the roadway (urban freeway, rural/suburban multilane highway, etc.), local traffic data, and engineering judgment. Perhaps the most common hourly volume used for roadway design is the 30th highest of the year. In practice, the

K-factor is used to convert annual average daily traffic (AADT) to the 30th highest hourly volume. K is defined as

$$K = \frac{\text{DHV}}{\text{AADT}} \tag{6.15}$$

where

K = factor used to convert annual average daily traffic to a specified annual hourly volume,

DHV = design hour-volume (typically, the 30th highest annual hourly volume), and

AADT = roadway's annual average daily traffic in veh/day.

For example, Fig. 6.7 shows that the K-value corresponding to the 30th highest hourly volume is 0.12. More generally, K_i can be defined as the K-factor corresponding to the ith highest annual hourly volume. Again, for example, the 20th highest annual hourly volume would have a K-value, K_{20}, of 0.126, from Fig. 6.7. If K is not subscripted, the 30th highest annual hourly volume is assumed ($K = K_{30}$).

Finally, in the design and analysis of some highway types (such as freeways and multilane highways), the concern lies with directional traffic flows. Thus a factor is needed to reflect the proportion of peak-hour traffic volume traveling in the peak direction. This factor is denoted D and is used to arrive at the directional design-hour volume (DDHV) by application of

$$\text{DDHV} = K \times D \times \text{AADT} \tag{6.16}$$

where

DDHV = directional design-hour volume,

D = directional distribution factor to reflect the proportion of peak-hour traffic volume traveling in the peak direction, and

Other terms are as defined previously.

EXAMPLE 6.8

A freeway is to be designed as a passenger-car–only facility for an AADT of 35,000 vehicles per day. It is estimated that the freeway will have a free-flow speed of 70 mi/h (112.6 km/h). The design will be for commuters, and the peak-hour factor is estimated to be 0.85 with 65% of the peak-hour traffic traveling in the peak direction. Assuming that Fig. 6.7 applies, determine the number of lanes required to provide at least LOS C using the highest annual hourly volume and the 30th highest annual hourly volume.

SOLUTION

By inspection of Fig. 6.7, the highest annual hourly volume has $K_1 = 0.148$. Application of Eq. 6.16 gives

$$\text{DDHV} = K_1 \times D \times \text{AADT}$$
$$= 0.148 \times 0.65 \times 35{,}000 = 3367 \text{ veh/h}$$

The next step is to determine the maximum service flow rate that can be accommodated at LOS C for $FFS = 70$ mi/h. From Table 6.1, we see that this value is 1770 pc/h/ln. Thus, we must provide enough lanes such that the per-lane traffic flow is less than or equal to this value.

We can use Eq. 6.3 to find v_p, based on an assumed number of lanes. Comparing the calculated value of v_p to the maximum service flow rate of 1770 will determine whether we have an adequate number of lanes.

Assuming a four-lane freeway, Eq. 6.3 gives

$$v_p = \frac{3367}{0.85 \times 2 \times 1.0 \times 1.0} = 1980.6 \text{ pc/h/ln}$$

with

$V = 3367$ (DDHV from above)
$f_{HV} = 1.0$ (no heavy vehicles)
$f_p = 1.0$ (commuters)

This value is higher than 1770, so we need to provide more lanes. The calculation is repeated, this time with an assumed six-lane freeway:

$$v_p = \frac{3367}{0.85 \times 3 \times 1.0 \times 1.0} = 1320.4 \text{ pc/h/ln}$$

Since this value is less than 1770, a six-lane freeway (three lanes each direction) is necessary to provide LOS C operation for the design traffic flow rate.

For the 30th highest hourly annual volume, Fig. 6.7 gives $K_{30} = K = 0.12$, which when used in Eq. 6.16 gives

$$\text{DDHV} = K \times D \times \text{AADT}$$
$$= 0.12 \times 0.65 \times 35{,}000 = 2730 \text{ veh/h}$$

Again applying Eq. 6.3, with an assumed four-lane freeway, yields

$$v_p = \frac{2730}{0.85 \times 2 \times 1.0 \times 1.0} = 1605.9 \text{ pc/h/ln}$$

This value is less than 1770, so a four-lane freeway (two lanes each direction) is adequate for this design traffic flow rate.

This example demonstrates the impact that the chosen design traffic flow rate has on roadway design. Only a four-lane freeway is necessary to provide LOS C for the 30th highest annual hourly volume, as opposed to a six-lane freeway needed to satisfy the level-of-service requirement for the highest annual hourly volume.

NOMENCLATURE FOR CHAPTER 6

AADT	annual average daily traffic	f_{LS}	free-flow speed adjustment factor for lane and shoulder width(s) (two-lane highways)
ATS	average travel speed (two-lane highways)		
BFFS	estimated free-flow speed for base conditions	f_{LW}	free-flow speed adjustment factor for lane width (freeways and multilane highways)
c	roadway capacity		
D	factor for directional distribution of traffic	f_M	free-flow speed adjustment factor for median type (multilane highways)
DHV	design-hour volume		
DDHV	directional design-hour volume	f_{np}	adjustment factor for the percentage of no-passing zones (two-lane highways)
E_R	passenger car equivalents for recreational vehicles	f_p	driver population adjustment factor (freeways and multilane highways)
E_T	passenger car equivalents for large trucks and buses	K_i	factor used to convert AADT to ith highest annual hourly volume
FFS	measured or estimated free-flow speed	LC_L	left-side lateral clearance (multilane highways)
f_A	free-flow speed adjustment factor for access point frequency (multilane and two-lane highways)	LC_R	right-side lateral clearance (multilane highways)
$f_{d/np}$	adjustment factor for the combined effect of the directional distribution of traffic and the percentage of no-passing zones (two-lane highways)	N	number of lanes in one direction
		PHF	peak-hour factor
		PTSF	percent time spent following (two-lane highways)
f_G	grade adjustment factor (two-lane highways)	S_{FM}	mean speed of traffic measured in the field (two-lane highways)
f_{HV}	heavy-vehicle adjustment factor		
f_{ID}	free-flow speed adjustment factor for interchange density (freeways)	TLC	total lateral clearance (multilane highways)
		v_p	analysis flow rate
f_{LC}	free-flow speed adjustment factor for lateral clearance (freeways and multilane highways)	V	hourly volume
		V_{15}	highest 15-minute flow
		v/c	volume-to-capacity ratio

REFERENCES

Transportation Research Board. *Highway Capacity Manual.* Washington, DC: National Research Council, 2000.

PROBLEMS

6.1 A six-lane rural freeway has regular weekday users and currently operates at maximum LOS C conditions. The base free-flow speed is 65 mi/h, lanes are 11 ft wide, the right-side shoulder is 4 ft wide, and the interchange density is 0.25 per mile. The highway is on rolling terrain with 10% large trucks and buses (no recreational vehicles), and the peak-hour factor is 0.90. Determine the hourly volume for these conditions.

6.2 Consider the freeway in Problem 6.1. At one point along this freeway there is a 4% upgrade with a directional hourly traffic volume of 5435 vehicles. If all other conditions are as described in Problem 6.1, how long can this grade be without the freeway LOS dropping to F?

6.3 A four-lane urban freeway is located on rolling terrain and has 12-ft lanes, no lateral obstructions within 6 ft of the pavement edges, and an interchange every 2 miles. The traffic stream consists of cars, buses, and large trucks (no recreational vehicles). A weekday directional peak-hour volume of 1800 vehicles (familiar users) is observed, with 700 arriving in the most congested 15-min period. If a level of service no worse than C is desired, determine the maximum number of large trucks and buses that can be present in the peak-hour traffic stream.

6.4 Consider the freeway and traffic conditions described in Problem 6.3. If 180 of the 1800 vehicles observed in the peak hour were large trucks and buses, what would the level of service of this freeway be on a 5-mi, 6% downgrade?

6.5 A six-lane freeway in a scenic area has a measured free-flow speed of 55 mi/h. The peak-hour factor is 0.80, and there are 8% large trucks and buses and 6% recreational vehicles in the traffic stream. One upgrade is 5% and 0.5 mi long. An analyst has determined that the freeway is operating at capacity on this upgrade during the peak hour. If the peak-hour traffic volume is 3900 vehicles, what value for the driver population factor was used?

6.6 A four-lane freeway is on mountainous terrain with 11-ft lanes, a 5-ft right-side shoulder, interchange spacing of one every 10 miles, and a 60-mi/h base free-flow speed. During the peak hour there are 12% large trucks and buses and 6% recreational vehicles. PHF is 0.88 and the driver population adjustment is determined to be 0.90. The freeway currently operates at capacity during the peak hour, in the direction in question. If an additional 11-ft lane is added (each direction), and all other factors stay the same, what will be the new level of service?

6.7 A segment of urban four-lane freeway has a 3% upgrade that is 1500 ft long followed by a 1000-ft 4% upgrade. It has 12-ft lanes and 3-ft shoulders. The base free-flow speed is 65 mi/h, and the directional hourly traffic flow is 2000 vehicles with 5% large trucks and buses (no recreational vehicles). There are no interchanges in the vicinity of this freeway segment. If the peak-hour factor is 0.90 and all of the drivers are regular users, what is the level of service of this compound-grade freeway segment?

6.8 Consider Example 6.2, in which it was determined that 1717 vehicles could be added to the peak hour before capacity is reached. Assuming rolling terrain as in Example 6.1, how many passenger cars could be added to the original traffic mix before peak-hour capacity is reached? (Assume only passenger cars are added and that the number of large trucks and buses originally in the traffic stream remains constant.)

6.9 An eight-lane urban freeway is on rolling terrain and has 11-ft lanes with a 4-ft right-side shoulder. The interchange density is 1.25 per mile. The base free-flow speed is 70 mi/h. The directional peak-hour traffic volume is 5400 vehicles with 6% large trucks and 5% buses (no recreational vehicles). The traffic stream consists of regular users and the peak-hour factor is 0.95. It has been decided that large trucks will be banned from the freeway during the peak hour. What will the freeway's density and level of service be before and after the ban? (Assume that the trucks are removed and all other traffic attributes are unchanged.)

6.10 A 5% upgrade on a six-lane freeway is 1.25 mi long. On this segment of freeway, the directional peak-hour volume is 3800 vehicles with 2% large trucks and 4% buses (no recreational vehicles), the peak-hour factor is 0.90, and all drivers are regular users. The base free-flow speed is 65 mi/h, the lanes are 12 ft wide, there are no lateral obstructions within 10 ft of the roadway, and the interchange density is 1.5 per mile. A bus strike will eliminate all bus traffic, but it is estimated that for each bus removed from the roadway, six additional passenger cars will be added as travelers seek other means of travel. What is the density, volume-to-capacity ratio, and level of service of the upgrade segment before and after the bus strike?

6.11 A multilane highway (two lanes in each direction) is on level terrain. The free-flow speed has been measured at 45 mi/h. The peak-hour directional traffic flow is 1300 vehicles with 6% large trucks and buses and 2% recreational vehicles ($f_p = 0.95$). If the peak-hour factor is 0.85, determine the highway's level of service.

6.12 Consider the multilane highway in Problem 6.11. If the proportion of vehicle types and peak-hour factor remain constant, how many vehicles can be added to the directional traffic flow before capacity is reached?

6.13 A six-lane multilane highway has a peak-hour factor of 0.90, 11-ft lanes with a 4-ft right-side shoulder, and a two-way left-turn lane in the median. The directional peak-hour traffic flow is 4000 vehicles with 8% large trucks and buses and 2% recreational vehicles. The driver population factor has been estimated at 0.95. What will the level of service of this highway be on a 4% upgrade that is 1.5 miles long if the speed limit is 55 mi/h and there are 15 access points per mile?

6.14 A divided multilane highway in a recreational area ($f_p = 0.90$) has four lanes and is on rolling terrain. The highway has 10-ft lanes with a 6-ft right-side shoulder and a 3-ft left-side shoulder. The posted speed is 50 mi/h. Formerly there were 4 access points per mile, but recent development has increased the number of access points to 12 per mile. Before the development, the peak-hour factor was 0.95 and the directional hourly volume was 2300 vehicles with 10% large trucks and buses and 3% recreational vehicles. After the development, the peak-hour directional flow is 2700 vehicles with the same vehicle percentages and peak-hour factor. What is the level of service before and after the development?

6.15 A multilane highway has four lanes and a measured free-flow speed of 55 mi/h. One upgrade is 5% and is 0.62 mi long. Currently trucks are not permitted on the highway, but there are 2% buses (no recreational vehicles) in the directional peak-hour volume of 1900 vehicles (the peak-hour factor is 0.80). Local authorities are considering allowing trucks on this upgrade. If this is done, they estimate that 150 large trucks will use the highway during the peak hour. What would be the level of service before and after the trucks are allowed (assuming the driver population adjustment to be 1.0 before and 0.97 after)?

6.16 A four-lane undivided multilane highway has 11-ft lanes, 4-ft shoulders on both sides, and 10 access points per mile. It is determined that the roadway currently operates at capacity with $PHF = 0.80$ and a driver population adjustment of 0.9. If the highway is on level terrain with 8% trucks and buses (no recreational vehicles) and the speed limit is 55 mi/h, what is the directional hourly volume?

6.17 A new four-lane divided multilane highway is being planned with 12-ft lanes, 6-ft shoulders on both sides, and a 50-mi/h speed limit. One 3% downgrade is 4.5 mi long, and there will be 4 access points per mile. The peak-hour directional volume along this grade is estimated to consist of 1800 passenger cars, 140 large trucks, 40 buses, and 10 recreational vehicles. If the peak-hour factor is estimated to be 0.85, what level of service will this segment of highway operate under?

6.18 A six-lane divided multilane highway has a measured free-flow speed of 50 mi/h. It is on mountainous terrain with a traffic stream consisting of 6% large trucks and buses and 2% recreational vehicles. The driver population adjustment is 0.92. One direction of the highway currently operates at maximum LOS C conditions, and it is known that the highway has $PHF = 0.90$. How many vehicles can be added to this highway before capacity is reached, assuming the proportion of vehicle types remains the same but the peak-hour factor increases to 0.95?

6.19 A four-lane undivided multilane highway has 11-ft lanes and 5-ft shoulders. At one point along the highway there is a 4% upgrade that is 0.62 mi long. There are 15 access points along this grade. The peak-hour traffic volume has 2500 passenger cars and 200 trucks and buses (no recreational vehicles), and 720 of these vehicles arrive in the most congested 15-min period. This traffic stream is primarily commuters. The measured free-flow speed is 55 mi/h. To improve the level of service, the local transportation agency is considering reducing the number of access points by blocking some driveways and rerouting their traffic. How many of the 15 access points must be blocked to achieve LOS C?

6.20 A Class I two-lane highway is on level terrain, has a measured free-flow speed of 65 mi/h, and has 50% no-passing zones. The peak-hour traffic volume is 180 vehicles, with $PHF = 0.90$. There are 15% trucks and buses (no RVs) and the directional split is 70/30. Determine the level of service.

6.21 A Class I two-lane highway carries a peak-hour volume of 540 vehicles and has a peak-hour factor of

0.87. The traffic stream has a 60/40 directional split with 5% large trucks, 10% recreational vehicles, no buses, and 85% passenger cars. The highway is on rolling terrain and has 80% no-passing zones. The free-flow speed was measured at 57 mi/h, but this was during a flow rate of 275 pc/h. Determine the level of service.

6.22 A Class I two-lane highway is on level terrain with passing permitted throughout. The highway has 11-ft lanes with 4-ft shoulders. There are 16 access points per mile. The base *FFS* is 60 mi/h. The highway is oriented north and south, and during the peak hour, 440 vehicles are going northbound and 360 vehicles are going southbound. If the *PHF* is 0.87 and there are 4% large trucks, 3% buses, and 1% recreational vehicles, what is the level of service?

6.23 A two-lane highway is currently operating at its two-way capacity in rolling terrain. The traffic stream consists of cars and trucks only. A recent traffic count revealed 720 vehicles (total, both directions) arriving in the most congested 15-min interval. What is the percentage of trucks in the traffic stream based on the *ATS* service measure?

6.24 A Class II two-lane highway needs to be redesigned for an area with rolling terrain. The peak-hour traffic volume is 500 vehicles, with a directional split of 60/40 and a *PHF* of 0.85. The traffic stream includes 8% large trucks, 2% buses, and 5% recreational vehicles. What is the maximum percentage of no-passing zones that can be built into the design with LOS B maintained?

6.25 A four-lane freeway segment consists of passenger cars only, a driving population of regular users, a peak-hour directional distribution of 0.70, a peak-hour factor of 0.80, and a measured free-flow speed of 70 mi/h. Assuming Fig. 6.7 applies, if the AADT is 30,000 veh/day, determine the level of service for the 10th, 50th, and 100th highest annual hourly volumes.

6.26 A four-lane urban freeway operates at capacity during the peak hour. It has 11-ft lanes and 4-ft shoulders. The freeway has only regular users and there are 8% large trucks and buses (no recreational vehicles), and the freeway is on rolling terrain with a peak-hour factor of 0.85. There is one interchange every 2 miles. It is known that 12% of the AADT occurs in the peak hour and that the directional factor is 0.6. What is the freeway's AADT?

6.27 A six-lane highway has regular weekday users and currently operates at maximum LOS C conditions. The measured free-flow speed is 55 mi/h. The highway is on rolling terrain with 10% large trucks and buses and 5% recreational vehicles, and the peak-hour factor is 0.94. If 20% of all directional traffic occurs during the peak hour, determine the total directional traffic volume.

Chapter 7

Traffic Control and Analysis at Signalized Intersections

7.1 INTRODUCTION

Due to conflicting traffic movements, roadway intersections are a source of great concern to traffic engineers. Intersections can be a major source of crashes and vehicle delays (as vehicles yield to avoid conflicts with other vehicles). Most roadway intersections are not signalized due to low traffic volumes and adequate sight distances. However, at some point, traffic volumes and crash frequency/severity (and other factors) reach a level that warrants the installation of a traffic signal.

The installation and operation of a traffic signal to control conflicting traffic and pedestrian movements at an intersection has advantages and disadvantages. The advantages include a potential reduction of some types of crashes (particularly angle crashes), provision for pedestrians to cross the street, provision for side-street vehicles to enter the traffic stream, provision for the progressive flow of traffic in a signal-system corridor, possible improvements in capacity, and possible reductions in delay. However, signals are by no means the perfect solution for delay or crash problems at an intersection. A poorly timed signal or one that is not justified can have a negative impact on the operation of the intersection by increasing vehicle delay, increasing the rate of vehicle crashes (particularly rear-end crashes), causing a disruption to traffic progression (adversely impacting the through movement of traffic), and encouraging the use of routes not intended for through traffic (such as routes through residential neighborhoods). Traffic signals are also costly to install, with even some basic signal installations costing in excess of $100,000. Therefore, the decision to install a signal must be weighed and studied carefully. To assist transportation engineers in this process, the Federal Highway Administration of the U.S. Department of Transportation publishes the *Manual on Uniform Traffic Control Devices* (MUTCD) [U.S. Federal Highway Administration 2000], which contains a section on warrants for the installation of a traffic signal. There are a total of eight warrants, which include consideration of vehicle volumes, pedestrian volumes, school crossings, signal coordination, and crash experience. The reader is referred to the MUTCD for details on these warrants.

Unlike uninterrupted flow, in which vehicle movement is affected only by other vehicles and the roadway environment, the introduction of a traffic control device such as a signal exerts a significant influence on the flow of vehicles. Thus, the analysis of traffic flow at signalized intersections can become very complex. This chapter will make several simplifying assumptions to keep the material at an accessible level.

The chapter begins by providing an overview of the physical elements of intersection configuration and traffic signal control. A basic understanding of these principles provides the foundation for designing intersection geometry and traffic movement sequence plans. This is followed by a presentation of concepts, definitions, and analytical techniques that are used in the design and analysis of signal timing plans at signalized intersections.

7.2 INTERSECTION AND SIGNAL CONTROL CHARACTERISTICS

An intersection is defined as an at-grade crossing of two or more roadways. For analysis, the roadways entering the intersection are segmented into approaches, which are defined by lane groups (groups of one or more lanes). These lane groups are usually based on the allowed movements (left, through, right) within each lane and the sequencing of allowed movements by the traffic signal. The establishment of lane groups will be discussed in more detail in Section 7.5.2.

To illustrate these concepts, note that approach 1 of the intersection depicted in Fig. 7.1 consists of a lane for the exclusive use of left turns, a lane for the exclusive use of right turns, and two lanes for the exclusive use of through movements. Approach 3 is similar to approach 1 but does not include an exclusive right-turn lane; instead, the right turns share the outside lane with the through movements. Because the lanes for the exclusive use of left and right turns are short in length, they are usually referred to as bays and are intended to hold a limited number of queued vehicles. Queuing analysis can be used to determine the length of bay necessary to prevent queued turning vehicles from overflowing the bay and blocking the through lanes (known as spillover) and/or the length necessary to prevent queued through vehicles from blocking the entrance of the turn bay (known as spillback). Approach 2 consists of a shared through/right-turn lane and an exclusive left-turn lane (not a bay in this case because it extends upstream). Approach 4 is similar to approach 2, but the inside lane is a shared through/left-turn lane.

From a driver's perspective, a traffic signal is just a collection of light-emitting devices [usually incandescent bulbs or light-emitting diodes (LEDs)] and lenses that are housed in casings of various configurations (referred to as signal heads) whose purpose is to display red, yellow, and green full circles and/or arrows. Figure 7.2 shows typical configurations of signal heads in the United States. These signal heads are usually mounted to mast arms or wire spanned across the intersection.

The following terminology is commonly used in the design of traffic signal controls.

Indication. The illumination of one or more signal lenses (greens, yellows, reds) indicating an allowed or prohibited traffic movement.

Interval. A period of time during which all signal indications (greens, yellows, reds) remain the same for all approaches.

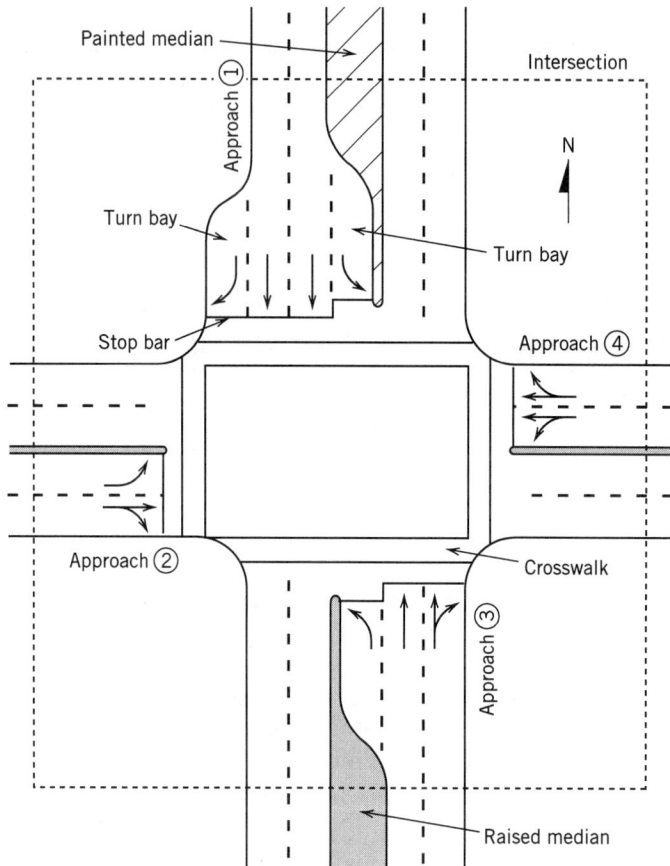

Figure 7.1 Typical signalized intersection elements.

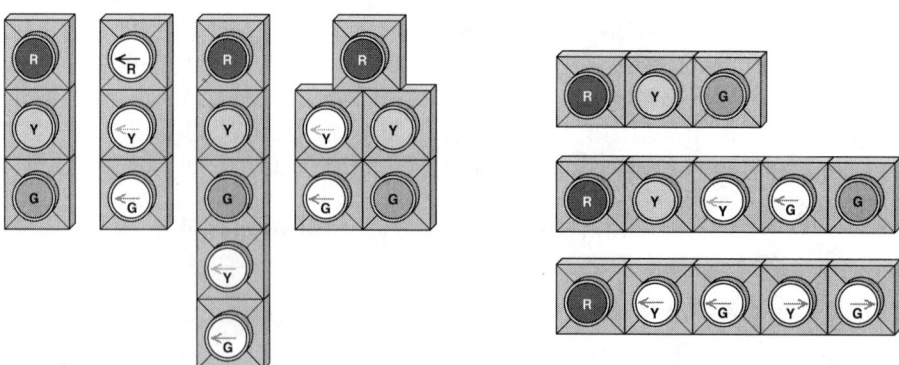

Figure 7.2 Typical signal head configurations in the United States (G = green; Y = yellow; R = red).

Cycle. One complete sequence (for all approaches) of signal indications (greens, yellows, reds).

Cycle length. The total time for the signal to complete one cycle (given the symbol C and usually expressed in seconds).

Green time. The amount of time within a cycle for which a movement or combination of movements receives a green indication (the illumination of a signal lens). This is expressed in seconds and given the symbol G.

Yellow time. The amount of time within a cycle for which a movement or combination of movements receives a yellow indication. This is expressed in seconds and given the symbol Y. This time is referred to as the change interval, as it alerts drivers that the signal indication is about to change from green to red.

Red time. The amount of time within a cycle for which a movement or combination of movements receives a red indication. This is expressed in seconds and given the symbol R.

All-red time. The time within a cycle in which all approaches have a red indication (expressed in seconds and given the symbol AR). This time is referred to as the clearance interval, because it allows vehicles that might have entered at the end of the yellow interval to clear the intersection before the green phase starts for the next conflicting movement(s). This type of interval is becoming increasingly common for safety reasons because the rate of vehicles entering at the end of the yellow and beginning of the red indication has steadily increased in recent years.

Phase. The sum of the displayed green, yellow, and red times for a movement or combination of movements that receive the right of way simultaneously during the cycle. The sum of the phase lengths (in seconds) is the cycle length.

The term "movement" is used frequently in the preceding definitions. In addition to a directional descriptor, such as left, through, or right, a distinction is made by categorizing movements as either protected or permitted.

Protected movement. A movement that has the right-of-way and does not need to yield to conflicting movements, such as opposing vehicle traffic or pedestrians. Through movements, which are always protected, are given a green full circle indication (or in some geometric configurations, a green arrow pointing up). Left- or right-turn movements that are protected are given a green arrow indication (pointing either left or right).

Permitted movement. A movement that must yield to opposing traffic flow or a conflicting pedestrian movement. This turn is made during gaps (time headways) in opposing traffic and conflicting pedestrian movements. Left- or right-turn movements with a green full circle indication are permitted movements. Left-turning vehicles in this situation must wait for gaps in the opposing through and right-turning traffic before making their turns. Right-turning vehicles must yield to pedestrians in the adjacent crosswalk before making their turns.

To understand how these control characteristics are implemented, it is useful to analyze the physical implementation of these concepts. The display of the various signal indications (green, yellow, red, protected, permitted) at an intersection is handled by a signal controller (which is typically located in a cabinet next to the intersection). Modern signal controllers are sophisticated pieces of electronic equipment. These controllers, when combined with a method of vehicle detection, offer great flexibility in controlling phase duration and sequence. Traffic signal controllers are designed to operate in one or more of the following modes: pretimed, semi-actuated, or fully actuated.

Pretimed. A signal whose timing (cycle length, green time, etc.) is fixed over specified time periods and does not change in response to changes in traffic flow at the intersection. No vehicle detection is necessary with this mode of operation.

Semi-actuated. A signal whose timing (cycle length, green time, etc.) is affected when vehicles are detected (by video, pavement-embedded inductance loop detectors, etc.) on some, but not all, approaches. This mode of operation is usually found where a low-volume road intersects a high-volume road, often referred to as the minor and major streets, respectively. In such cases, green time is allocated to the major street until vehicles are detected on the minor street; then the green indication is briefly allocated to the minor street and then returned to the major street.

Fully actuated. A signal whose timing (cycle length, green time, etc.) is completely influenced by the traffic volumes, when detected, on all of the approaches. Fully actuated signals are most commonly used at intersections of two major streets and where substantial variations exist in all approach traffic volumes over the course of a day.

Most modern signal controllers are designed to operate in what is termed a dual-ring configuration. This configuration allows maximum flexibility for controlling phase duration and sequencing, which is necessary for operating in a fully actuated mode. The dual-ring concept can be best explained with the use of a graphical illustration (see Fig. 7.3a).

In this figure, the three-letter notation in each numbered box refers to direction and movement type. For example, WBL means westbound left and NBT means northbound through. At an intersection with four approaches and separate/exclusive left-turn phases for each approach, a total of eight separate movements are possible. The dual-ring terminology comes from four movements (1–4) being represented on the top ring, and four movements (5–8) being represented on the bottom ring. The logic behind this configuration is straightforward. Any movement in ring 1 can occur simultaneously with any movement in ring 2 as long as both are on the same side of the barrier (a term used figuratively to separate conflicting traffic movements). For example, the first phase at an intersection can consist of opposing WB and EB left-turn movements (1 and 5). However, if no vehicles are detected in the EB left-turn lane, this movement can be skipped and the first phase would consist of movement 1 (WBL) and movement 6 (WBT). If no left turns are detected for both the WB and EB directions, then the first phase will consist of movements 2 (EBT) and 6 (WBT). The same applies to movements on the right side of the barrier (3, 4, 7, and 8).

Furthermore, if the volume for one of the movements during a phase subsides sooner than the volumes for the other movement(s), the green time for this movement

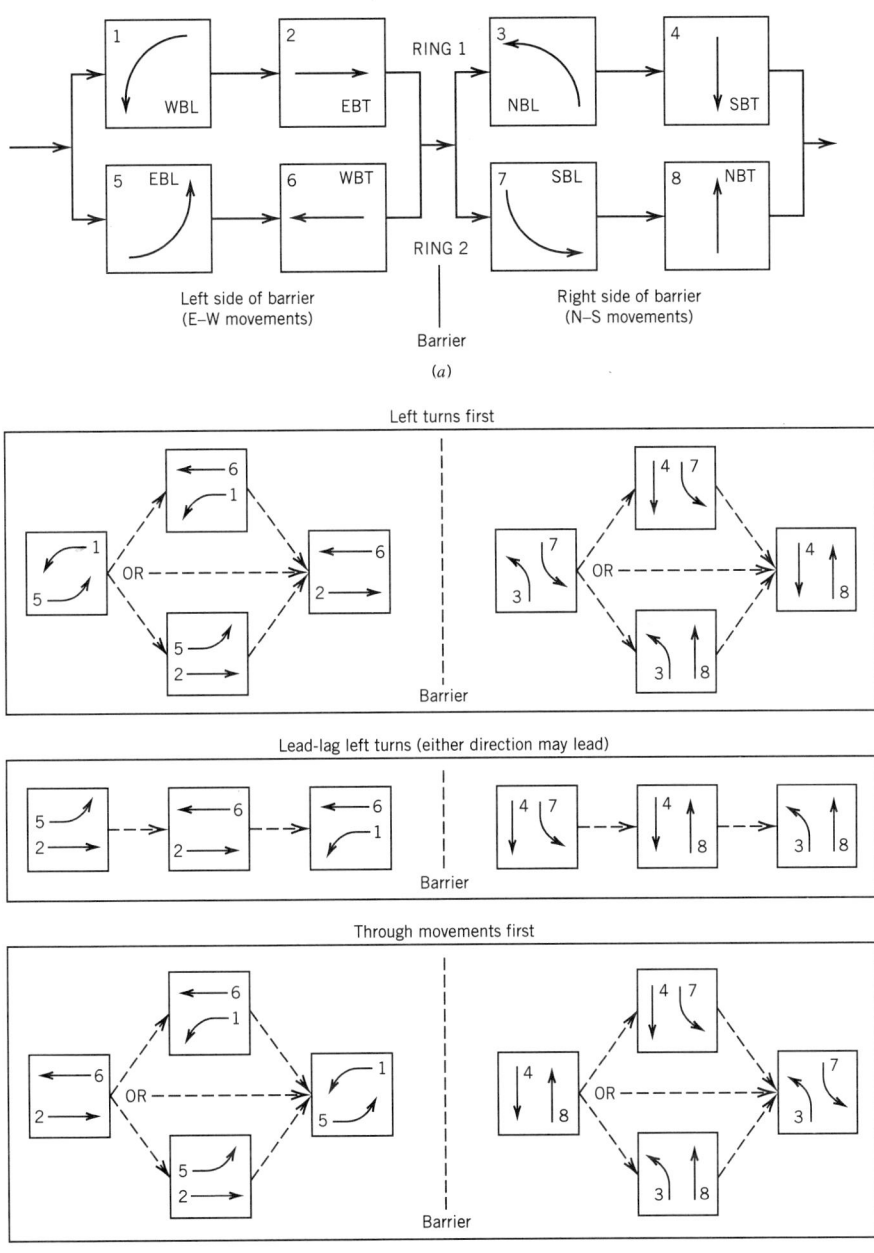

Figure 7.3 Dual-ring signal control. (*a*) Movement-based representation of dual-ring logic. (*b*) Phase-based representation of dual-ring logic.
Part (*a*) reproduced by permission from Transportation Research Board, Highway Capacity Manual, National Research Council, Washington, DC, 2000.

can be terminated and another movement can be initiated, according to the previously described logic. For example, suppose phase 1 consists of movements 1 and 5, but the volume for movement 1 is considerably larger than the volume for movement 5. If both movements 1 and 5 received enough green time to satisfy the vehicle demand for movement 1, this would result in wasted green time for movement 5. With the dual-ring configuration, movement 5 can be terminated before movement 1, and movement 6 can be initiated while movement 1 continues. This results in a more efficient allocation of green time and reduced delay. The phase sequence and phase durations can therefore vary from one cycle to the next at a fully actuated intersection, especially with highly variable approach volumes. Consequently, the cycle length can vary from one cycle to the next. Figure 7.3b illustrates the typical dual-ring phase sequence options as discussed above. Note that phases can be skipped entirely if no traffic demand is present, and other phase options are possible depending on intersection geometry/lane movement assignment.

It must be pointed out that although no two phases are required to either start or terminate at the same time in a dual-ring configuration, all movements on the left side of the barrier must be terminated before any movement on the right side of the barrier can be initiated, and vice versa. This is a safety feature; no movement on the left side of the barrier can be allowed to move simultaneously with any movement on the right side of the barrier, or else conflicting traffic streams will intersect.

7.3 ANALYSIS OF TRAFFIC AT SIGNALIZED INTERSECTIONS

This section will utilize and build upon the elements of traffic flow theory introduced in Chapter 5 to enable the basic analysis of traffic flow at signalized intersections.

7.3.1 Concepts and Definitions

Before presenting the analytical principles and techniques used for signalized intersections, it is important to introduce several key concepts and definitions used in the analysis of traffic at signalized intersections.

Saturation Flow Rate

The saturation flow rate is the maximum hourly volume that can pass through an intersection, from a given lane or group of lanes, if that lane (or lanes) were allocated constant green over the course of an hour. Saturation flow rate is given by

$$s = \frac{3600}{h} \qquad (7.1)$$

where

s = saturation flow rate in veh/h,
h = saturation headway in s/veh, and
3600 = number of seconds per hour.

Note the similarity between Eq. 7.1 and Eq. 5.4. The difference is that h in Eq. 7.1 is a constant minimum headway value maintained for saturated conditions, as opposed to an average headway value as used in Eq. 5.4 (Eq. 7.1 also directly yields units of veh/h for s because of the numerator). The use of the term "saturation" is an important qualifier in this definition, as it implies the presence of constant vehicle demand in measuring the headway. If the measure of interest is simply the traffic flow through the intersection for some period of time, then the appropriate equation would be 5.1 or 5.4.

Research has found that a typical maximum saturation flow rate of 1900 passenger cars per hour per lane (pc/h/ln) is possible at signalized intersections, and this is referred to as the base saturation flow rate. This corresponds to a saturation headway of about 1.9 seconds. Just as for the analysis of uninterrupted flow, a number of roadway and traffic factors can affect the maximum flow rate through an intersection. These factors include lane widths; grades; curbside parking maneuvers; the distribution of traffic among multiple approach lanes; the level of roadside development; bus stops; and the influence of pedestrians, bicycles, and heavy vehicles (as they occupy more roadway space and have poorer acceleration/deceleration capabilities). Additionally, lanes that allow left or right turns usually have lower saturation flow rates because drivers reduce speed to make a turning maneuver (especially heavy vehicles, with their increased turning radii). Furthermore, if a turning movement is permitted rather than protected, its saturation flow rate will be reduced as a result of the turning vehicles yielding to conflicting through and right-turning vehicles (for left turns only), bicycles, and/or pedestrians. All of these factors are accommodated by applying adjustments to the base saturation flow rate. The end result is usually a value less than 1900 pc/h/ln for each approach lane at an intersection, and is referred to as the adjusted saturation flow rate. Additionally, the units are converted to vehicles per hour per lane (veh/h/ln) due to adjustment of the heavy-vehicle volume with passenger car equivalents (in a manner similar to the procedures of Chapter 6). The process for making adjustments to the base saturation flow rate for the preceding factors to arrive at an adjusted saturation flow rate is extensive and beyond the scope of this book (see the *Highway Capacity Manual* [Transportation Research Board 2000]). Of course, saturation flow rates can be measured directly in the field, in which case no further adjustments are necessary. For the rest of this chapter, it should be assumed that the provided saturation flow rates have been adjusted for the given conditions; the term "adjusted" has been dropped just for notational convenience.

Lost Time

Due to the traffic signal's function of continuously alternating the right-of-way between conflicting movements, traffic streams are continuously started and stopped. Every time this happens, a portion of the cycle length is not being completely utilized, which translates to lost time (time that is not effectively serving any movement of traffic). Total lost time is a combination of start-up and clearance lost times. Start-up lost times occur because when a signal indication turns from red to green, drivers in the queue do not instantly start moving at the saturation flow rate; there is an initial lag due to drivers reacting to the change of signal indication. This start-up delay results in a portion of the green time for that movement not being completely utilized. This start-up lost time has a typical value of around 2 seconds. This concept is illustrated in Fig. 7.4.

In this figure, note that the headway for the first several vehicles is larger than the saturation headway. The saturation headway is typically reached after the fourth vehicle in the queue. The summation of the amount of headway time greater than the saturation headway for each of the first four vehicles yields the total start-up lost time for the movement.

The stopping of a traffic movement also results in lost time. When the signal indication turns from green to yellow, the latter portion of time during the yellow interval is generally not utilized by traffic. Additionally, if there is an all-red interval, this time period is generally not utilized by traffic. These periods of time during the change and clearance intervals that are not effectively used by traffic are referred to as clearance lost time. Typically, the last second of the yellow interval and the entire all-red interval are included in the estimate of clearance lost time. However, for intersections with significant red-light running, the clearance lost time may be negligible.

Start-up and clearance lost times are summed to arrive at a total lost time for the phase, given as

$$t_L = t_{sl} + t_{cl} \tag{7.2}$$

where

t_L = total lost time for a movement during a cycle in seconds,
t_{sl} = start-up lost time in seconds, and
t_{cl} = clearance lost time in seconds.

This amount of time remains fixed, regardless of phase or cycle length. Thus, for shorter cycle lengths, the lost time will comprise a larger percentage of the cycle length and will result in a larger amount of lost time over the course of a day compared with longer cycle lengths. However, longer cycle lengths usually have more phases than shorter cycle lengths, which may result in similar proportions of lost time.

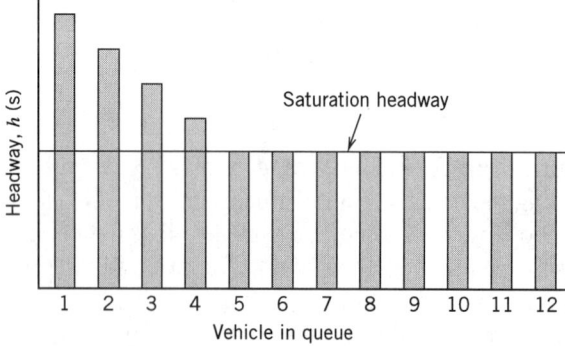

Figure 7.4 Concept of saturation headway and lost time.

Effective Green and Red Times

For analysis purposes, the time during a cycle that is effectively (or not effectively) utilized by traffic must be used rather than the time for which green, yellow, and red signal indications are actually displayed, because they are most likely different. This results in two measures of interest: the effective green time and the effective red time. The effective green time is the time during which a traffic movement is effectively utilizing the intersection. The effective green time for a given movement or phase is calculated as

$$g = G + Y + AR - t_L \qquad (7.3)$$

where

g = effective green time for a traffic movement in seconds,
G = displayed green time for a traffic movement in seconds,
Y = displayed yellow time for a traffic movement in seconds,
AR = displayed all-red time in seconds, and
t_L = total lost time for a movement during a cycle in seconds.

The effective red time is the time during which a traffic movement is not effectively utilizing the intersection. The effective red time for a given movement or phase is calculated as

$$r = R + t_L \qquad (7.4)$$

where

r = effective red time for a traffic movement in seconds,
R = displayed red time for a traffic movement in seconds, and
t_L = total lost time for a movement during a cycle in seconds.

Alternatively, the effective red time can be calculated as follows, assuming the cycle length and effective green time have already been determined:

$$r = C - g \qquad (7.5)$$

where

C = cycle length in seconds, and
Other terms are as defined previously.

Likewise, the effective green time can be calculated by subtracting the effective red time from the cycle length.

Capacity

Because movements on an intersection approach do not receive a constant green indication (as assumed in the definition for saturation flow rate), another measure must be defined that accounts for the hourly volume that can be accommodated on an intersection approach given that the approach will receive less than 100% green time. This measure is capacity and is given by

$$c = s \times g/C \qquad (7.6)$$

where

c = capacity (the maximum hourly volume that can pass through an intersection from a lane or group of lanes under prevailing roadway, traffic, and control conditions) in veh/h,

s = saturation flow rate in veh/h, and

g/C = ratio of effective green time to cycle length.

7.3.2 Signalized Intersection Analysis with D/D/1 Queuing

The assumption of $D/D/1$ queuing (as discussed in Chapter 5) provides a strong intuitive appeal that helps in understanding the analytical fundamentals underlying traffic analysis at signalized intersections. To begin applying $D/D/1$ queuing to signalized intersections, we consider the case where the approach capacity exceeds the approach arrivals. Under these conditions the assumption of $D/D/1$ queuing will result in a queuing system as shown in Fig. 7.5.

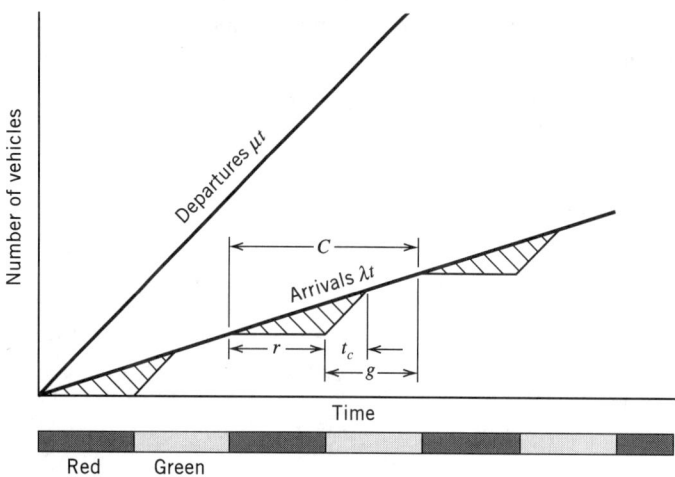

Figure 7.5 $D/D/1$ signalized intersection queuing with approach capacity (μg) exceeding arrivals (λC) for all cycles.

In this figure,

λ = arrival rate, typically in veh/s,

μ = departure rate, typically in veh/s,

t = total transpired time in seconds,

t_c = time from the start of the effective green until queue clearance in seconds,

r = effective red time in seconds,

g = effective green time in seconds, and

C = cycle length in seconds.

7.3 Analysis of Traffic at Signalized Intersections

The "arrivals λt" line gives the total number of vehicle arrivals at time t, and the "departures μt" line gives the slope of vehicle departures (number of vehicles that depart) during the effective greens. Note that the per-cycle approach arrivals will be λC and the corresponding approach capacity (maximum departures) per cycle will be μg. Figure 7.5 is predicated on the assumption that μg exceeds λC for all cycles (no queues exist at the beginning or end of a cycle).

Given the properties of $D/D/1$ queues presented in Chapter 5, a number of general equations can be derived by simple inspection of Fig. 7.5:

1. The time to queue clearance after the start of the effective green, t_c [note that $\lambda(r + t_c) = \mu t_c$ and traffic intensity $\rho = \lambda/\mu$ (as in Chapter 5)],

$$t_c = \frac{\rho r}{(1-\rho)} \tag{7.7}$$

2. The proportion of the cycle with a queue, P_q,

$$P_q = \frac{r + t_c}{C} \tag{7.8}$$

3. The proportion of vehicles stopped, P_s,

$$P_s = \frac{\lambda(r + t_c)}{\lambda(r + g)} = \frac{r + t_c}{C} = P_q$$

also, $P_s = \frac{\lambda(r + t_c)}{\lambda(r + g)} = \frac{\mu t_c}{\lambda C} = \frac{t_c}{\rho C} \tag{7.9}$

4. The maximum number of vehicles in the queue, Q_{max},

$$Q_{max} = \lambda r \tag{7.10}$$

5. The total vehicle delay per cycle, D_t,

$$D_t = \frac{\lambda r^2}{2(1-\rho)} \tag{7.11}$$

6. The average delay per vehicle, d_{avg},

$$d_{avg} = \frac{\lambda r^2}{2(1-\rho)} \times \frac{1}{\lambda C}$$
$$= \frac{r^2}{2C(1-\rho)} \tag{7.12}$$

7. The maximum delay of any vehicle, assuming a FIFO queuing discipline, d_{max},

$$d_{max} = r \tag{7.13}$$

EXAMPLE 7.1

An approach at a pretimed signalized intersection has a saturation flow rate of 2400 veh/h and is allocated 24 seconds of effective green in an 80-second signal cycle. If the flow at the approach is 500 veh/h, provide an analysis of the intersection assuming D/D/1 queuing.

SOLUTION

Putting arrival and departure rates into common units of vehicles per second,

$$\lambda = \frac{500 \text{ veh/h}}{3600 \text{ s/h}} = 0.139 \text{ veh/s}$$

$$\mu = \frac{2400 \text{ veh/h}}{3600 \text{ s/h}} = 0.667 \text{ veh/s}$$

This gives a traffic intensity of

$$\rho = \frac{0.139 \text{ veh/s}}{0.667 \text{ veh/s}} = 0.208$$

Checking to make certain that capacity exceeds arrivals, note that the capacity (μg) is 16 veh/cycle (0.667×24), which is greater than (permitting fractions of vehicles for the sake of clarity) the 11.12 arrivals ($\lambda C = 0.139 \times 80$). Therefore, Eqs. 7.7 to 7.13 are valid. By definition,

$$r = C - g$$
$$= 80 - 24 = 56 \text{ s}$$

This leads to the following values:

1. Time to queue clearance after the start of the effective green (Eq. 7.7),

$$t_c = \frac{0.208(56)}{(1 - 0.208)}$$
$$= 14.71 \text{ s}$$

2. Proportion of the cycle with a queue (Eq. 7.8),

$$P_q = \frac{56 + 14.71}{80}$$
$$= 0.884$$

3. Proportion of vehicles stopped (Eq. 7.9),

$$P_s = \frac{14.71}{0.208(80)}$$
$$= 0.884$$

7.3 Analysis of Traffic at Signalized Intersections

4. Maximum number of vehicles in the queue (Eq. 7.10),

$$Q_{max} = 0.139(56)$$
$$= \underline{\underline{7.78}}$$

5. Total vehicle delay per cycle (Eq. 7.11),

$$D_t = \frac{0.139(56)^2}{2(1-0.208)}$$
$$= \underline{\underline{275.19 \text{ veh-s}}}$$

6. Average delay per vehicle (Eq. 7.12),

$$d_{avg} = \frac{56^2}{2(80)(1-0.208)}$$
$$= \underline{\underline{24.75 \text{ s}}}$$

7. Maximum delay of any vehicle (Eq. 7.13),

$$d_{max} = r$$
$$= \underline{\underline{56 \text{ s}}}$$

Recall that Eqs. 7.7 through 7.13 are valid only when the approach capacity exceeds approach arrivals. For the case when approach arrivals exceed capacity for some signal cycles, $D/D/1$ queuing can again be used, as illustrated in the following example.

EXAMPLE 7.2

An approach to a pretimed signalized intersection has a saturation flow rate of 1700 veh/h. The signal's cycle length is 60 seconds and the approach's effective red is 40 seconds. During three consecutive cycles 15, 8, and 4 vehicles arrive. Determine the total vehicle delay over the three cycles assuming $D/D/1$ queuing.

SOLUTION

For all cycles, the departure rate is

$$\mu = \frac{1700 \text{ veh/h}}{3600 \text{ s/h}} = 0.472 \text{ veh/s}$$

During the first cycle, the number of vehicles that will depart from the signal is (permitting fractions for the sake of clarity)

$$\mu g = 0.472(20)$$
$$= 9.44 \text{ veh}$$

Therefore, 5.56 vehicles (15 − 9.44) will not be able to pass through the intersection on the first cycle even though they arrive during the first cycle. At the end of the second cycle, 23 vehicles (15 + 8) will have arrived, but only 18.88 ($2\mu g$) will have departed, leaving 4.12 vehicles waiting at the beginning of the third cycle. At the end of the third cycle, a total of 27 vehicles will have arrived and as many as 28.32 ($3\mu g$) could have departed, so the queue that began to form during the first cycle will dissipate at some time during the third cycle. This process is shown graphically in Fig. 7.6.

From this figure, the total vehicle delay of the first cycle is (the area between arrival and departure curves) is

$$D_1 = \tfrac{1}{2}(60)(15) - \tfrac{1}{2}(20)(9.44)$$
$$= 355.6 \text{ veh-s}$$

Similarly, the delay in the second cycle is

$$D_2 = \tfrac{1}{2}(60)(15+23) - (40)(9.44) - \tfrac{1}{2}(20)(9.44+18.88)$$
$$= 479.2 \text{ veh-s}$$

To determine the delay in the third cycle, it is necessary to first know exactly when, in this cycle, the queue dissipates. The time to queue clearance after the start of the effective green, t_c, is (with λ_3 being the arrival rate during the third cycle and n_3 being the number of vehicles in the queue at the start of the third cycle)

$$n_3 + \lambda_3(r + t_c) = \mu t_c$$

where

$$\lambda_3 = \frac{4 \text{ veh}}{60 \text{ s}} = 0.067 \text{ veh/s}$$

Therefore,

$$(23 - 18.88) + 0.067(40 + t_c) = 0.472 t_c$$

which gives $t_c = 16.8$ seconds. Thus the queue will clear 56.8 seconds (40 + 16.8) after the start of the third cycle, at which time a total of 26.8 vehicles (0.067 × 56.8 + 15 + 8) will have arrived at, and departed from, the intersection. The vehicle delay for the third cycle is

$$D_3 = \tfrac{1}{2}(56.8)(23+26.8) - (40)(18.88) - \tfrac{1}{2}(16.8)(18.88+26.8)$$
$$= 275.4 \text{ veh-s}$$

giving the total delay over all three cycles as

$$D_t = D_1 + D_2 + D_3$$
$$= 355.6 + 479.2 + 275.4$$
$$= \underline{1110.2 \text{ veh-s}}$$

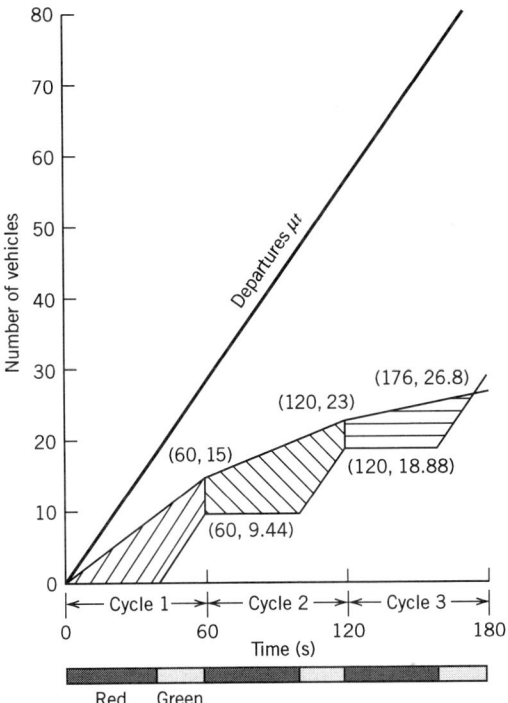

Figure 7.6 *D/D*/1 queuing diagram for Example 7.2.

It is also possible to handle the case where intersection arrivals and/or departures are deterministic but time-varying in a fashion similar to that shown in Chapter 5. An example of time-varying arrivals is presented next.

EXAMPLE 7.3

The saturation flow rate of an approach to a pretimed signal is 6000 veh/h. The signal has a 60-second cycle with 20 seconds of effective red allocated to the approach. At the beginning of an effective red (with no vehicles remaining in the queue from a previous cycle) vehicles start arriving at a rate $\lambda(t) = 0.4 + 0.01t + 0.00057t^2$ [where $\lambda(t)$ is in vehicles per second and t is the number of seconds from the beginning of the cycle]. Thirty seconds into the cycle the arrival rate remains constant at its 30-second level and stays at that rate until the end of the cycle. What is the total vehicle delay over the cycle (in vehicle-seconds) assuming *D/D*/1 queuing?

SOLUTION

Vehicle arrivals for the first 30 seconds (again allowing fractions of vehicles for the sake of clarity) are

$$\int_0^{30} 0.4 + 0.01t + 0.00057t^2 = 0.4t + 0.005t^2 + 0.00019t^3 \Big|_0^{30}$$
$$= 21.63 \text{ veh}$$

The arrival rate after 30 seconds is 1.213 veh/s [0.4 + 0.01(30) + 0.00057(30)²] and is no longer time-varying. During the effective green, the departure rate is 1.667 veh/s (6000/3600). To determine when the queue will clear, let t' be the time after 30 seconds (the time after which the arrival rate is no longer time-varying), so

$$21.63 + 1.213t' = 1.667(t' + 10)$$

which gives $t' = 10.93$ seconds. Thus the queue clears at 40.93 seconds after the beginning of the cycle. For delay, the area under the arrival curve is

$$\int_0^{30} 0.4t + 0.005t^2 + 0.00019t^3 + \tfrac{1}{2}[21.63 + 1.213(10.93) + 21.63](10.93)$$

$$= 0.2t^2 + 0.00167t^3 + 0.0000475t^4 \Big|_0^{30} + 308.87$$
$$= 263.48 + 308.87$$
$$= 572.35 \text{ veh-s}$$

The area under the departure curve is

$$\tfrac{1}{2}(34.90 \times 20.94) = 365.42 \text{ veh-s}$$

So total vehicle delay over the cycle is <u>206.93 veh-s</u> (572.35 − 365.42).

7.3.3 Signalized Intersection Analysis for Level of Service

The level-of-service concept was introduced in Chapter 6 for uninterrupted-flow facilities. This same concept also applies to interrupted-flow facilities. Although a variety of performance measures can be calculated for signalized intersections, as demonstrated in the previous section, only one measure has been chosen as the service measure. This measure is control delay, and it applies to both signalized and unsignalized intersections. Control delay (also referred to as signal delay for the case of a signalized intersection) represents the total delay experienced by the driver as a result of the control, which includes delay due to deceleration time, queue move-up time, stop time, and acceleration time.

The main limitation with the delay equation (Eq. 7.12) presented in the previous section is that its derivation is based on the assumption of uniform traffic arrivals. In practice, this is an overly strict assumption and is likely to yield considerable error compared with delays measured in the field. As discussed in Chapter 5, nonuniform arrivals are very likely in most traffic flow situations, and are frequently the case for signalized arterials. What is needed for a more complete analysis of delay at signalized intersections is a formula that accounts for uniform and random arrivals, as well as signal control influences on this arrival pattern. Such a formula has been developed for the *Highway Capacity Manual* [Transportation Research Board 2000] and is

$$d = d_1 \times PF + d_2 + d_3 \tag{7.14}$$

where

d = average signal delay per vehicle in seconds,
d_1 = average delay per vehicle due to uniform arrivals in seconds,
PF = progression adjustment factor,
d_2 = average delay per vehicle due to random arrivals in seconds, and
d_3 = average delay per vehicle due to initial queue at start of analysis time period, in seconds.

Note that this equation is applied to each established lane group in the intersection. Individual lane group delays must be aggregated to arrive at an overall intersection delay, which will be discussed later, in Section 7.5.9. As can be seen, Eq. 7.14 is a composite of several different component equations, with each component equation accounting for a specific contribution to overall delay. The first is the delay due to uniform arrivals (uniform delay), given as

$$d_1 = \frac{0.5 C \left(1 - \frac{g}{C}\right)^2}{1 - \left[\min(1, X)\frac{g}{C}\right]} \tag{7.15}$$

where

d_1 = average delay per vehicle due to uniform arrivals in seconds,
C = cycle length in seconds,
g = effective green time for lane group in seconds, and
X = v/c ratio for lane group.

Note that X, the volume-to-capacity ratio, can be put in queuing notation terms as $\lambda C/\mu g$, where all terms are as previously defined. Thus, it can be shown that Eq. 7.15 is equivalent to Eq. 7.12, that is, $d_1 = d_{avg}$ (based on the D/D/1 queuing analysis of Section 7.3.2).

The progression adjustment factor, *PF*, accounts for the effect of signal progression quality on delay. Progression quality between adjacent signals is a function of the level of coordination of the phase timings between the two signals, which is influenced by several factors, such as cycle length, vehicle travel speeds, and distance between the intersections. Good signal coordination will lead to a high percentage of vehicles arriving during the green portion of the phase for an approach, thus significantly reducing delay for that approach. On the other hand, poor signal coordination will lead to a high percentage of vehicle arrivals on the red and increased delay. It is uniform delay that is most affected by progression quality, and thus the adjustment is applied only to d_1.

More in-depth discussion about incorporating the effects of adjacent signals into the analysis of a signalized intersection is beyond the scope of this chapter (for more information, see the *Highway Capacity Manual* [Transportation Research Board 2000]). For the purposes of this chapter, it is assumed that the signalized intersection undergoing analysis is operating in an isolated mode, meaning it is free from the effects of adjacent signals. For analysis purposes, intersections operating in an isolated mode are considered to be progression neutral; that is, the arriving traffic flow has neither good nor bad progression quality, and thus *PF* is set equal to 1.0.

Although the assumption of uniform arrivals is often not very realistic, Eq. 7.15 does provide reasonably accurate results for intersections with low to moderate flow rates, generally up to v/c ratios of about 0.5. However, as the traffic intensity increases from moderate to a level nearing the capacity of the intersection, the probability of having cycle failures, where not all queued vehicles get through during a particular cycle, increases substantially. These cycle failures are random occurrences for the most part, but must be accounted for in the estimation of overall delay to achieve reasonably accurate results under higher flow conditions. Whereas the formula for delay assuming uniform (deterministic) arrivals is based on a purely theoretical derivation, the need to account for stochastic vehicle arrival effects adds considerable complexity to the analysis of delay at signalized intersections. Numerous researchers over the last several decades have proposed delay formulas and refinements to meet this need, based on combinations of analytical, empirical, and simulation-based methods. One particular formulation that has become refined over the past decade and has been adopted by the *Highway Capacity Manual* [Transportation Research Board 2000] is

$$d_2 = 900T \left[(X-1) + \sqrt{(X-1)^2 + \frac{8kIX}{cT}} \right] \qquad (7.16)$$

where

d_2 = average delay per vehicle due to random arrivals, in seconds,
T = duration of analysis period in h,
X = v/c ratio for lane group,
k = delay adjustment factor that is dependent on signal controller mode,
I = upstream filtering/metering adjustment factor, and
c = lane group capacity, in veh/h.

Assuming the analysis flow rate is based on the peak 15-minute traffic flow within the analysis hour, T is equal to 0.25.

The signal-controller–mode delay adjustment factor, k, accounts for whether the intersection is operating in an actuated or pretimed mode. A value of 0.5 is used for intersections under pretimed control. Given that actuated intersection control usually results in more efficient handling of traffic volumes, k can take on values less than 0.5 to account for this efficiency and resultant reduced delay. For actuated control, k depends upon the v/c ratio and the unit extension (an actuated controller setting related to the minimum headway for consecutive vehicle arrivals required to extend the green time). For the purposes of the example and end-of-chapter problems, pretimed signal control is assumed.

The upstream filtering/metering factor is used to adjust for the effect that an upstream signal has on the randomness of the arrival pattern at a downstream intersection. An upstream signal will typically have the effect of reducing the variance of the number of arrivals at the downstream intersection. I is defined as

$$I = \frac{\text{variance of the number of arrivals per cycle}}{\text{mean number of arrivals per cycle}} \quad (7.17)$$

Recall from Chapter 5 that for a Poisson distribution (which is used to represent random arrivals), the variance is equal to the mean. Thus, if the downstream arrival pattern of vehicles conforms to the Poisson distribution, this factor will be equal to 1.0. This is considered to be the case for signalized intersections operating in an isolated mode (intersections sufficiently distant from adjacent signalized intersections). For nonisolated signalized intersections, the I value will be less than 1.0 due to the reduced variance of vehicle arrivals. The I value is dependent on the v/c ratio of the upstream movements that contribute to the downstream intersection volume. Again, for the purposes of this chapter, it will be assumed that a signal is operating in isolated mode, and thus the I value will be equal to 1.0.

The last component of the overall delay equation, d_3, is intended to account for the delay caused by an initial queue of vehicles at the beginning of the analysis time period. It is assumed for the relevant example and end-of-chapter problems that there is no initial queue; thus, this term will simply be equal to zero. For more information about this delay component, refer to the *Highway Capacity Manual* [Transportation Research Board 2000].

EXAMPLE 7.4

Compute the average approach delay per cycle using Eq. 7.14, given the conditions described in Example 7.1. Assume the traffic flow accounts for the peak 15-min period and that there is no initial queue at the start of the analysis period.

SOLUTION

In this example, the uniform delay is computed as (using Eq. 7.15)

$$d_1 = \frac{0.5C\left(1 - \frac{g}{C}\right)^2}{1 - \left[\min(1, X)\frac{g}{C}\right]}$$

with

$C = 80$ s
$g = 24$ s

The v/c ratio, X, is calculated as

$$\frac{v}{c} = \frac{v}{s \times g/C}$$

with

v = 500 veh/h
s = 2400 veh/h

giving

$$\frac{v}{c} = \frac{500}{2400 \times 24/80} = \frac{500}{720} = 0.694$$

Alternatively, X can be calculated as

$$X = \frac{\lambda C}{\mu g}$$

$$= \frac{0.139(80)}{0.667(24)}$$

$$= 0.695$$

This value checks with the previous result, with the small difference due to rounding in the latter calculation. Uniform delay is calculated as

$$d_1 = \frac{0.5(80)\left(1 - \frac{24}{80}\right)^2}{1 - \left[0.694 \times \frac{24}{80}\right]} = 24.75 \text{ s}$$

This value matches the delay value computed in Example 7.1, using the D/D/1 queuing analysis approach. Random delay will be computed as (using Eq. 7.16)

$$d_2 = 900T\left[(X-1) + \sqrt{(X-1)^2 + \frac{8kIX}{cT}}\right]$$

with

T = 0.25 (15 min, from problem statement)
X = 0.694 (from above)
k = 0.5 (pretimed control)
I = 1.0 (isolated mode)
c = 720 veh/h (from above)

$$d_2 = 900(0.25)\left[(0.694 - 1) + \sqrt{(0.694 - 1)^2 + \frac{8(0.5)(1.0)0.694}{(720)0.25}}\right]$$

$$= 5.45 \text{ s}$$

Now the total signal delay is computed using Eq. 7.14:
$$d = d_1 \times PF + d_2 + d_3$$

with

$d_1 = 24.75$ s
$PF = 1.0$ (progression neutral)
$d_2 = 5.45$ s
$d_3 = 0$ s (given)

$$d = 24.75 \times 1.0 + 5.45 + 0 = \underline{30.20 \text{ s}}$$

7.4 OPTIMAL TRAFFIC SIGNAL TIMING

Allocating effective green times to competing approaches in an optimal fashion has been a goal of traffic engineers since traffic signals were first used. Unfortunately, the problem of optimal timing is complicated by a number of factors. For example, although the distribution of traffic can be approximated by some arrival assumptions (deterministic or Poisson), to properly optimize signal timing the arrival pattern must be known with certainty. In recent years, vehicle sensors at intersections have allowed traffic signal systems to respond to variations in traffic flow, and this has gone a long way toward satisfying the need for traffic arrival information. However, a more fundamental question pervades signal optimization: On what basis should traffic signal timings be optimized? Should vehicle delay be minimized, should the number of vehicles stopped be minimized, should green time available to the major street be maximized, or should some other factor serve as the optimization criterion? Because different optimization criteria almost always provide different results, many signal timing strategies seek to balance the values of more than one factor (such as vehicle stops and vehicle delays). Needless to say, the optimization-measurement problem makes this aspect of traffic analysis a fruitful area for future research.

To demonstrate one possible signal optimization strategy (putting aside the optimization issues raised above), assume that the sole objective of signal timing is to minimize vehicle delay and that traffic can be represented by a simple $D/D/1$ queue. Such a strategy is illustrated by the following example.

EXAMPLE 7.5

A pretimed signal controls a four-way intersection with no turning permitted and zero lost time. The eastbound (EB) and westbound (WB) traffic volumes are 700 and 800 veh/h, respectively, and the two movements share the same effective green and effective red portions of the cycle. The northbound (NB) and southbound (SB) directions also share cycle times, with volumes of 400 and 250 veh/h, respectively. If the saturation flow of all approaches is 1800 veh/h, the cycle length is 60 seconds, and $D/D/1$ queuing applies, determine the effective red and effective green times that must be allocated to each directional combination (north-south, east-west) to minimize total vehicle delay, and compute the total delay per cycle.

SOLUTION

Putting arrival and departure rates into common units of vehicles per second,

$$\lambda_{EB} = \frac{700 \text{ veh/h}}{3600 \text{ s/h}} = 0.194 \text{ veh/s}$$

$$\lambda_{WB} = \frac{800 \text{ veh/h}}{3600 \text{ s/h}} = 0.222 \text{ veh/s}$$

$$\lambda_{NB} = \frac{400 \text{ veh/h}}{3600 \text{ s/h}} = 0.111 \text{ veh/s}$$

$$\lambda_{SB} = \frac{250 \text{ veh/h}}{3600 \text{ s/h}} = 0.069 \text{ veh/s}$$

$$\mu = \frac{1800 \text{ veh/h}}{3600 \text{ s/h}} = 0.5 \text{ veh/s}$$

Since the departure rate is the same for all approaches, the traffic intensities are $\rho_{EB} = 0.388$, $\rho_{WB} = 0.444$, $\rho_{NB} = 0.222$, and $\rho_{SB} = 0.138$. If it is assumed that approach capacity exceeds approach arrivals for all approaches (an assumption that will be tested later), the total vehicle delay at the intersection is (using Eq. 7.11)

$$D_t = \frac{\lambda_{EB} r_{EB}^2}{2(1-\rho_{EB})} + \frac{\lambda_{WB} r_{WB}^2}{2(1-\rho_{WB})} + \frac{\lambda_{NB} r_{NB}^2}{2(1-\rho_{NB})} + \frac{\lambda_{SB} r_{SB}^2}{2(1-\rho_{SB})}$$

or, substituting the computed values of the λ's and ρ's,

$$D_t = 0.1585 r_{EB}^2 + 0.1996 r_{WB}^2 + 0.07115 r_{NB}^2 + 0.04 r_{SB}^2$$

The problem states that east and west effective reds are equal, and north and south effective reds are equal. So let r_{EW} be the effective red of eastbound and westbound directions ($r_{EW} = r_{EB} = r_{WB}$) and let r_{NS} be the effective red of northbound and southbound directions ($r_{NS} = r_{NB} = r_{SB}$). By definition, with a 60-second signal cycle, $r_{NS} = 60 - r_{EW}$. Substituting this into the total delay expression gives

$$D_t = 0.1585 r_{EW}^2 + 0.1996 r_{EW}^2 + 0.07115(60 - r_{EW})^2 + 0.04(60 - r_{EW})^2$$

$$= 0.46925 r_{EW}^2 - 13.338 r_{EW} + 400.14$$

At minimum total delay $dD_t/dr_{EW} = 0$. Differentiating yields

$$\frac{dD_t}{dr_{EW}} = 0.9357 r_{EW} - 13.338 = 0$$

which gives $r_{EW} = \underline{14.2 \text{ s}}$ ($g_{EW} = \underline{45.8 \text{ s}}$) and $r_{NS} = \underline{45.8 \text{ s}}$ ($g_{NS} = \underline{14.2 \text{ s}}$). For total delay,

$$D_t = 0.1585(14.2)^2 + 0.1996(14.2)^2 + 0.07115(45.8)^2 + 0.04(45.8)^2$$

$$= \underline{\underline{305.36 \text{ veh-s}}}$$

Finally, a check of the earlier assumption that approach capacity exceeds approach arrivals for all approaches must be undertaken. In the 60-second cycle, eastbound and westbound vehicle arrivals are 11.64 (0.194 × 60) and 13.32 (0.222 × 60), respectively. With 45.8 seconds of effective green, the capacity for both approaches is 22.9 (0.5 × 45.8) vehicles, so the assumption is satisfied. The northbound and southbound vehicle arrivals are 6.67 (0.111 × 60) and 4.14 (0.069 × 60), respectively, and with 14.2 seconds of effective green, the capacity for both approaches is 7.1 (0.5 × 14.2) vehicles. Again, the assumption is satisfied and the method used in this example is valid.

7.5 DEVELOPMENT OF A TRAFFIC SIGNAL PHASING AND TIMING PLAN

Assuming the decision to install a traffic signal at an intersection has been made, an appropriate phasing and timing plan must be developed. The development of a traffic signal phasing and timing plan can be complex, particularly if the intersection has multiple-lane approaches and requires protected turning movements (a turn arrow). However, the timing plan analysis can be simplified by dealing with each approach separately. This section provides the basic process and fundamentals needed to develop a phasing and timing plan for an isolated, fixed-time (pretimed) traffic signal. As timing plans become more complex, they simply build on these fundamental principles. The reader is encouraged to review the material in other references to see how actuated and progressive timing plans are developed [Kell and Fullerton 1991; Transportation Research Board 2000]. This section describes the basic process that results in the development of a signal phasing and timing plan.

7.5.1 Select Signal Phasing

Recall that a cycle is the sum of individual phases. The most basic traffic signal cycle is made up of two phases, as shown in Fig. 7.7. In this case, phase 1 accommodates the movement of the northbound and southbound vehicles, and phase 2 accommodates the movement of the eastbound and westbound vehicles. These phases will alternate during the continuous operation of the signal. This phasing scheme, however, could prove to be very inefficient if one or more of the approaches includes a high left-turn volume. Given that each approach consists of one lane, vehicles will be delayed behind a left-turning vehicle waiting for a gap in the opposing traffic stream. If the high volume of left turns is present on both the northbound and southbound approaches, for example, each of these approaches could be given a separate phase. This would be more efficient because left-turning vehicles on these two approaches would not have to wait for gaps in the opposing traffic stream, thus greatly reducing delays for all vehicles. This would result in a three-phase operation, as shown in Fig. 7.7. When movements on opposing approaches are given separate phases, as in this case, it is referred to as split phasing.

It is important to remember, however, that there is lost time (start-up and clearance) associated with each phase. Thus, with each phase added to a cycle, the lost

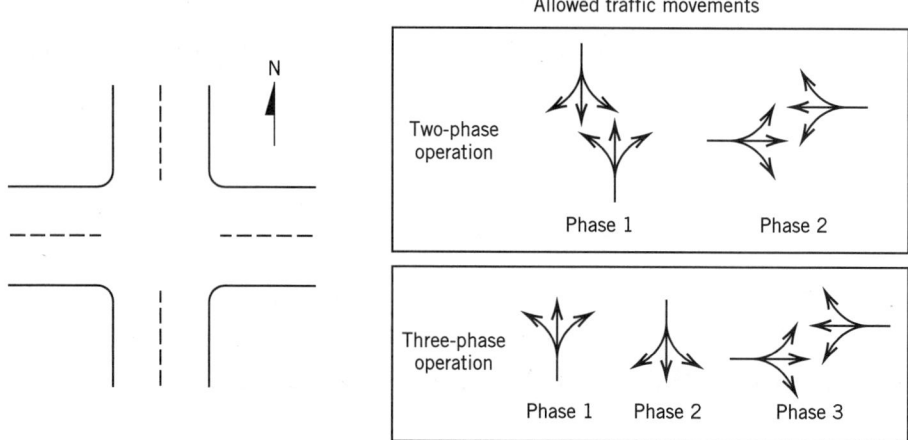

Figure 7.7 Illustration of two-phase and three-phase signal operation.

time increases. Although the lost time may be only 3 to 5 seconds per phase, the accumulated lost time throughout the day can be significant. Given this, a point of diminishing returns is reached with the addition of phases as the efficiencies gained by separating traffic movements eventually become outweighed by the inefficiencies of increased lost time. Thus, a primary concern in signal timing is to keep the number of phases to a minimum. Because protected-turn phases add to lost time, they should be used only when warranted. Because of opposing motor vehicle traffic, left-turn movements typically require a protected-turn phase much more often than right turns. There are no nationally established guidelines on when protected left-turn phasing should be used, so local policies and practices should be consulted before a decision is made about whether to provide a protected left-turn phase. In general, decisions on whether to provide a protected left-turn phase are based on one or more of the following factors:

- Volume (just left turn or combination of left turn and opposing volume)
- Delay
- Queuing (spillover)
- Traffic progression
- Opposing traffic speeds
- Geometry (number of left-turn lanes, crossing distance, sight distance)
- Crash experience (which may also be related to any of the above factors)

More specific guidance on this issue can be found in several references, including the *Highway Capacity Manual* [Transportation Research Board 2000], the *Traffic Control Devices Handbook* [ITE 2001], and the *Manual of Traffic Signal Design* [Kell and Fullerton 1991].

7.5 Development of a Traffic Signal Phasing and Timing Plan **245**

One of the more common guidelines is the use of the cross product of left-turn volume and opposing through and right-turn volumes. The *Highway Capacity Manual* offers the following criteria for this guideline: The use of a protected left-turn phase should be considered when the product of left-turning vehicles and opposing traffic volume exceeds 50,000 during the peak hour for one opposing lane, 90,000 for two opposing lanes, or 110,000 for three or more opposing lanes.

Some common signal phasing configurations and sequences are shown in Fig. 7.8. In this figure, note that the dashed lines represent permitted movements and the solid lines represent protected movements.

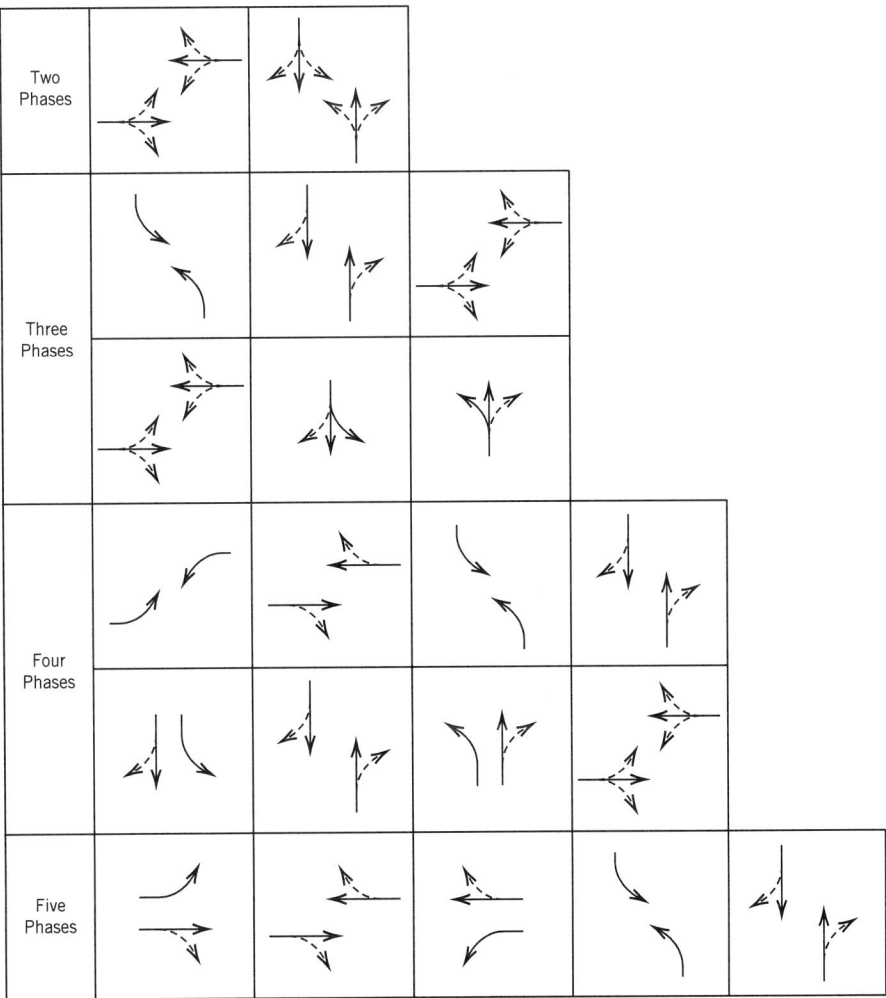

Figure 7.8 Typical phasing configurations and sequencing.

When the left turns precede the through and right-turn movements in the phasing sequence for an approach, they are referred to as leading left turns. When the left turns follow the through and right-turn movements, they are referred to as lagging left turns. Although not shown in this figure, it is also possible for a movement to be protected for a period of time and then permitted for a period of time, or vice versa. This is most commonly seen with left-turn movements, and is referred to as protected plus permitted or permitted plus protected, depending on the sequence.

EXAMPLE 7.6

Refer to the intersection shown in Fig. 7.9. Use the cross product guideline to determine if protected left-turn phases should be provided for any of the approaches.

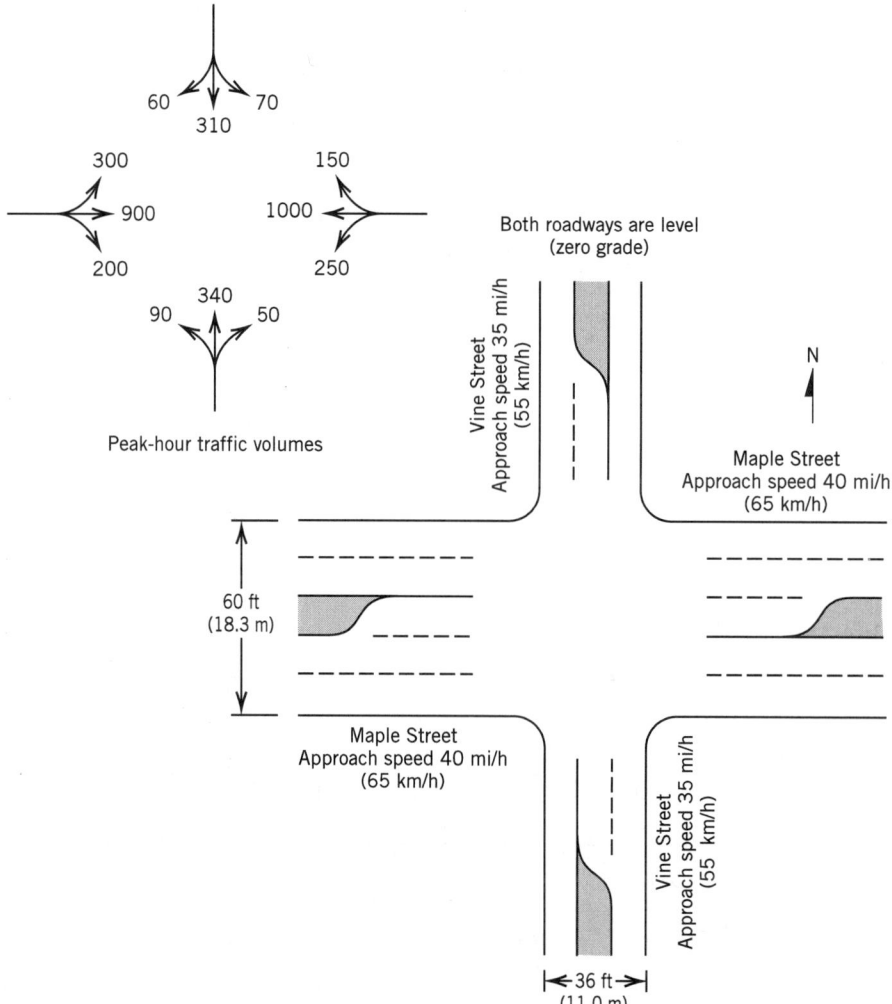

Figure 7.9 Intersection geometry and peak traffic volumes for example problems.

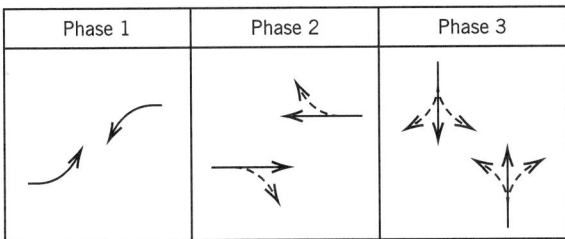

Figure 7.10 Recommended signal phasing plan for the intersection in Example 7.6.

SOLUTION

There are 250 westbound vehicles that turn left during the peak hour. The product of the westbound left-turning vehicles and the opposing eastbound traffic (right-turn and straight-through vehicles) is 275,000 [250 × (900 + 200)]. There are 300 eastbound vehicles that turn left during the peak hour. The product of the eastbound left-turning vehicles and the opposing westbound traffic (right-turn and straight-through vehicles) is 345,000 [300 × (1000 + 150)].

Because the cross product for each of these approaches is greater than 90,000 (the requirement for two opposing lanes), a protected left-turn phase is suggested for the WB and EB left turn movements. The NB and SB approaches do not require a protected left-turn phase using this criterion because the cross products for these approaches are less than 50,000 (for one opposing lane). Therefore, a three-phase traffic-signal plan is recommended, as shown in Fig. 7.10.

7.5.2 Establish Analysis Lane Groups

Each intersection approach is initially treated separately, and the results are later aggregated. Thus, each approach must be subdivided into logical groupings of traffic movements for analysis purposes. Based on the lane and traffic movement distribution on an approach, lane groups can be readily determined. The following general guidelines are offered for establishing lane groups [Transportation Research Board 2000].

- Movements made simultaneously from the same lane must be treated as a lane group.
- If an exclusive turn lane (or lanes) is present, it is usually treated as a separate lane group.
- If an approach includes an exclusive left-turn and/or right-turn lane, the remaining lanes are usually considered as a single lane group.

- If a multiple-lane approach includes a lane (or lanes) with shared movements, it must first be determined whether it really serves multiple movements or whether it is a de facto lane for one of the movements. For example, an approach that includes a shared left-turn and through movement lane may be operating primarily as a left-turn lane if another through lane is present. Likewise, it may be operating primarily as a through lane if an exclusive left-turn lane is also present.

Figure 7.11 shows some typical lane groupings for analysis purposes. Note that when multiple lanes are combined into a lane group, the subsequent analysis calculations for this lane group should treat these lanes as a single unit.

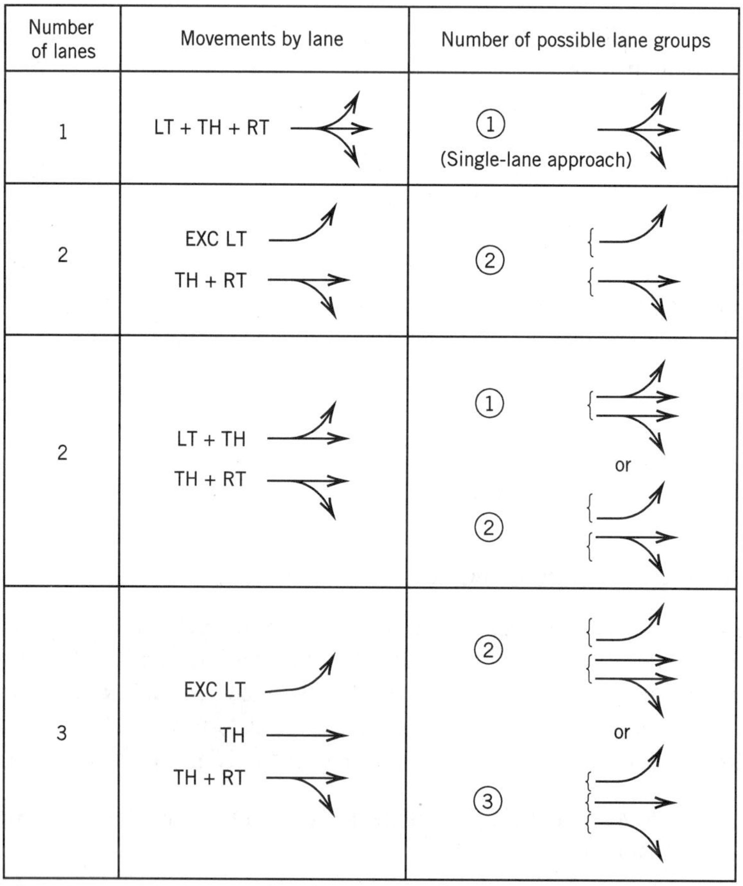

Number of lanes	Movements by lane	Number of possible lane groups
1	LT + TH + RT	① (Single-lane approach)
2	EXC LT / TH + RT	②
2	LT + TH / TH + RT	① or ②
3	EXC LT / TH / TH + RT	② or ③

Figure 7.11 Typical lane groupings for analysis (LT = left turn; TH = through; RT = right turn; EXC = exclusive).
Reproduced by permission from Transportation Research Board, *Highway Capacity Manual*, National Research Council, Washington, DC, 2000.

7.5 Development of a Traffic Signal Phasing and Timing Plan

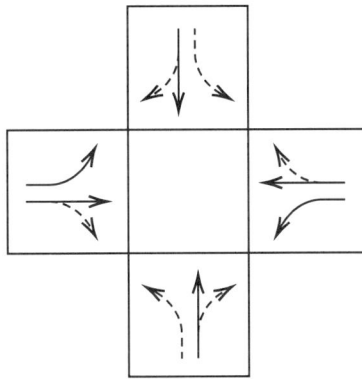

Figure 7.12 Analysis lane groups for the three-phase design at the intersection of Maple Street and Vine Street.

EXAMPLE 7.7

Determine the lane groups to use for analysis of the Maple and Vine streets intersection.

SOLUTION

The EB and WB left-turn movements will each be a lane group because they have a separate lane and move in a separate phase from the through/right-turn movements. Likewise, the EB and WB through/right-turn movements proceed together in a separate phase, and will therefore be separate lane groups. Although the right turns use only the outside lane, this movement's impact on the saturation flow rate for the two lanes combined will be determined. The NB and SB left turns will also each be a separate lane group. Even though they move during the same phase as the adjacent through and right-turn movements, these left turns are permitted and will have very different operating characteristics from the through and right-turn movements. Because the through and right-turn movements use the same lane, they will be an individual lane group for both the NB and SB approaches. The recommended lane groups for analysis for each of the approaches are shown in Fig. 7.12.

7.5.3 Calculate Analysis Flow Rates and Adjusted Saturation Flow Rates

Just as for the analysis of uninterrupted flow, the hourly traffic volume arriving on each intersection approach must be converted to an analysis flow rate that accounts for the peak 15-minute flow within that hour (typically the peak hour). This is accomplished by calculating the peak-hour factor (*PHF*) and dividing this into the hourly volume (as shown in Chapter 6), which yields the analysis flow rate.

One note about adjusting for the *PHF*. With the multiple traffic streams entering an intersection, a separate *PHF* can be calculated for each approach's traffic stream.

However, adjusting each approach volume by its specific *PHF* can yield unrealistically high combined analysis volumes, because the different approach volumes usually do not peak during the same 15-minute period. Applying a single *PHF* determined for the intersection as a whole will result in more reasonable analysis volumes.

The adjustment of the saturation flow rate was discussed in Section 7.3.1. Again, it is assumed that the approach volumes and saturation flow rates provided in this chapter have already been adjusted.

7.5.4 Determine Critical Lane Groups and Total Cycle Lost Time

For any combination of lane group movements during a particular phase, one of these lane groups will control the necessary green time for that phase. This lane group is referred to as the critical lane group. When the traffic movements of each lane group occur during only one phase of the signal cycle, the determination of the critical lane group for each phase is straightforward. In this case, the critical lane group for each phase is simply the lane group with the highest ratio of vehicle arrival rate to vehicle departure rate (λ/μ). This quantity is referred to as the flow ratio (which was called the traffic intensity, ρ, in Chapter 5) and is designated v/s (arrival flow rate divided by saturation flow rate). If the allocation of green time for each phase is based on the flow ratio of the critical lane group, then the noncritical lane group movements will be accommodated as well.

As previously discussed, with dual-ring controllers, a wide variety of phasing sequences are possible. The situation where a movement starts in one signal phase and continues in the next signal phase is referred to as an overlapping phase (see Fig. 7.8). For the case of overlapping phases in a signal cycle, the identification of the critical lane groups is more complex, as the lane group with the highest flow ratio in each phase is not necessarily the critical lane group for that phase. The remainder of this chapter will focus only on nonoverlapping phases. The reader is referred to the *Highway Capacity Manual* [Transportation Research Board 2000] for details on determining critical lane groups for overlapping phases.

In addition, the sum of the flow ratios for the critical lane groups can be used to calculate a suitable cycle length, which will be discussed in the next section. This is given by

$$Y_c = \sum_{i=1}^{n} \left(\frac{v}{s}\right)_{ci} \qquad (7.18)$$

where

Y_c = sum of flow ratios for critical lane groups,
$(v/s)_{ci}$ = flow ratio for critical lane group i, and
n = number of critical lane groups.

The total lost time for the cycle will also be used in the calculation of cycle length. In determining the total lost time for the cycle, the general rule is to apply the lost time for a critical lane group when its movements are initiated (the start of its green interval). The total cycle lost time is given as

7.5 Development of a Traffic Signal Phasing and Timing Plan 251

$$L = \sum_{i=1}^{n} (t_L)_{ci} \tag{7.19}$$

where

L = total lost time for cycle in seconds,
$(t_L)_{ci}$ = total lost time for critical lane group i in seconds, and
n = number of critical lane groups.

EXAMPLE 7.8

Calculate the sum of the flow ratios for the critical lane groups for the three-phase timing plan determined in Example 7.6 given the saturation flow rates in Table 7.1.

SOLUTION

Note that the saturation flow rates are relatively low for the SB and NB left (L) turns because they are permitted only, and the opposing through and right-turn (T/R) vehicles limit the number of usable gaps for these vehicles. The saturation flow rates for the WB and EB through and right turn (T/R) movements account for both through lanes.

The flow ratios will now be calculated, with the critical lane group for each phase indicated with a check mark in Table 7.2.

Table 7.1 Saturation Flow Rates for Three-Phase Design at Intersection of Maple Street and Vine Street

Phase 1	Phase 2	Phase 3
EB L: 1750 veh/h	EB T/R: 3400 veh/h	SB L: 450 veh/h
		NB L: 475 veh/h
WB L: 1750 veh/h	WB T/R: 3400 veh/h	SB T/R: 1800 veh/h
		NB T/R: 1800 veh/h

Table 7.2 Flow Ratios and Critical Lane Groups for Three-Phase Design at Intersection of Maple Street and Vine Street

Phase 1	Phase 2	Phase 3
EB L: $\dfrac{300}{1750} = 0.171$ √	EB T/R: $\dfrac{1100}{3400} = 0.324$	SB L: $\dfrac{70}{450} = 0.156$
		NB L: $\dfrac{90}{475} = 0.189$
WB L: $\dfrac{250}{1750} = 0.143$	WB T/R: $\dfrac{1150}{3400} = 0.338$ √	SB T/R: $\dfrac{370}{1800} = 0.206$
		NB T/R: $\dfrac{390}{1800} = 0.217$ √

As indicated in the table, the critical lane group for phases 1, 2, and 3, respectively, are the EB left turn, the WB through and right turn, and the NB through and right turn.

The sum of the flow ratios for the critical lane groups for this phasing plan will be needed for the next section. Since this phasing plan does not include any overlapping phases, this value is simply the sum of the highest lane group v/s ratios for the three phases, as follows:

$$Y_c = \sum_{i=1}^{n}\left(\frac{v}{s}\right)_{ci}$$

$$= 0.171 + 0.338 + 0.217 = \underline{0.726}$$

Assuming 2 seconds of start-up lost time and 2 seconds of clearance lost time (1 second of yellow time plus 1 second of all-red time), for each critical lane group, gives a lost time of 4 s/phase. The total lost time for the cycle is then 12 seconds (3 phases × 4 s/phase).

EXAMPLE 7.9

Suppose it is necessary to run the NB and SB movements in a split-phase configuration (with phase 3 for SB movements and a new phase 4 for NB movements). Calculate the sum of the flow ratios for the critical lane groups and total cycle lost time for this situation, assuming the EB and WB movement phasing remains the same.

Table 7.3 summarizes the calculation of the flow ratios and the identification of the critical lane groups.

SOLUTION

The sum of the flow ratios for the critical lane groups for this phasing plan is

$$\sum_{i=1}^{n}\left(\frac{v}{s}\right)_{ci} = 0.171 + 0.338 + 0.206 + 0.217 = \underline{0.932}$$

The total lost time for the cycle is $\underline{16 \text{ seconds}}$ (4 phases × 4 s/phase).

Table 7.3 Flow Ratios and Critical Lane Groups for Four-Phase Design (Split Phase for N-S Movements) at Intersection of Maple Street and Vine Street

Phase 1	Phase 2	Phase 3	Phase 4
EB L: $\frac{300}{1750} = 0.171$ ✓	EB T/R: $\frac{1100}{3400} = 0.324$	SB L: $\frac{70}{1750} = 0.040$	NB L: $\frac{90}{1750} = 0.051$
WB L: $\frac{250}{1750} = 0.143$	WB T/R: $\frac{1150}{3400} = 0.338$ ✓	SB T/R: $\frac{370}{1800} = 0.206$ ✓	NB T/R: $\frac{390}{1800} = 0.217$ ✓

7.5.5 Calculate Cycle Length

The cycle length is simply the summation of the individual phase lengths. In practice, cycle lengths are generally kept as short as possible, typically between 40 and 60 seconds. However, complex intersections with five or more phases can have cycle lengths of 120 seconds or more. The minimum cycle length necessary for the lane group volumes and phasing plan of an intersection is given by

$$C_{min} = \frac{L \times X_c}{X_c - \sum_{i=1}^{n}\left(\frac{v}{s}\right)_{ci}} \quad (7.20)$$

where

C_{min} = minimum necessary cycle length in seconds (typically rounded up to the nearest 5-second increment in practice),
L = total lost time for cycle in seconds,
X_c = critical v/c ratio for the intersection,
$(v/s)_{ci}$ = flow ratio for critical lane group i, and
n = number of critical lane groups.

In this equation, the total lost time for the cycle and the sum of the flow ratios for the critical lane groups are predetermined. However, a critical intersection volume/capacity ratio, X_c, must be chosen for the desired degree of utilization. In other words, if it is desired for the intersection to operate at its full capacity, a value of 1.0 is used for X_c. A value of 1.0 is not generally recommended, however, due to the randomness of vehicle arrivals, which can result in occasional cycle failures. Note that this equation gives the minimum cycle length necessary for the intersection to operate at a specified degree of capacity utilization. This cycle length does not necessarily minimize the average vehicle delay experienced by motorists at the intersection.

A practical equation for the calculation of the cycle length that seeks to minimize vehicle delay was developed by Webster [1958]. Webster's optimum cycle length formula is

$$C_{opt} = \frac{1.5 \times L + 5}{1.0 - \sum_{i=1}^{n}\left(\frac{v}{s}\right)_{ci}} \quad (7.21)$$

where

C_{opt} = cycle length to minimize delay in seconds, and
Other terms are as defined previously.

The cycle length determined from this calculation is only approximate. Webster noted that values between $0.75 C_{opt}$ and $1.5 C_{opt}$ will likely give similar values of delay. Calculating an accurate optimal cycle length (and phase length) can be a very computationally intensive exercise for all but the most simple signalized intersections, especially if coordination between multiple signals is involved.

It should be noted that regardless of the minimum or optimal cycle length calculated, practical maximum cycle lengths must generally be observed. Public acceptance or tolerance of large cycle lengths will vary by location (urban vs. rural), but as a rule, cycle lengths in excess of 3 minutes (180 seconds) should be used only in exceptional circumstances.

EXAMPLE 7.10

Calculate the minimum and optimal cycle lengths for the intersection of Maple and Vine streets, using the information provided in the preceding examples, for both the three-phase and four-phase design.

SOLUTION

For the three-phase design (Example 7.8), the sum of the flow ratios for the critical lane groups and the total cycle lost time were determined to be 0.726 and 12 seconds, respectively. For the minimum cycle length, a somewhat conservative value of 0.9 will be used for the critical intersection v/c ratio to minimize the potential of cycle failures due to occasionally high arrival volumes. Using these values in Eq. 7.20 gives

$$C_{min} = \frac{12 \times 0.9}{0.9 - 0.726} = 62.1 \rightarrow \underline{\underline{65 \text{ s}}} \text{ (rounding up to nearest 5 seconds)}$$

Using Eq. 7.21 for the optimal cycle length gives

$$C_{opt} = \frac{1.5 \times 12 + 5}{1.0 - 0.726} = 83.9 \rightarrow \underline{\underline{85 \text{ s}}} \text{ (rounding up to nearest 5 seconds)}$$

For the four-phase design (Example 7.9), the sum of the critical flow ratios and the total cycle lost time were determined to be 0.932 and 16 seconds, respectively. The first issue with this design is that a higher X_c will need to be used because the sum of flow ratios for critical lane groups is higher than the 0.90 used for the three-phase design (otherwise the denominator of Eq. 7.20 will be negative). To minimize the cycle length, the maximum value of 1.0 will be used for X_c in Eq. 7.20, as follows:

$$C_{min} = \frac{16 \times 1.0}{1.0 - 0.932} = \underline{\underline{235.3 \text{ s}}}$$

The second issue is that despite the use of an X_c value of 1.0 (the intersection operating at capacity) to minimize the cycle length, an unreasonably high cycle length is still required for this design. Thus, this design is not nearly as desirable as the three-phase design.

Generally a split-phase design is recommended only under one or more of the following conditions:

- The left turns are the dominant movement.
- The left turns share a lane with the through movement.

- There is a large difference in the total approach volumes.
- There are unusual opposing approach geometrics.

It should also be noted that serving pedestrians in an efficient manner on split-phase approaches can be difficult.

7.5.6 Allocate Green Time

After a cycle length has been calculated, the next step in the traffic signal timing process is to determine how much green time should be allocated to each phase. The cycle length is the sum of all effective green times plus the total lost time. Thus, after subtracting the total lost time from the cycle length, the remaining time can be distributed as green time among the phases of the cycle.

There are several strategies for allocating the green time to the various phases. One of the most popular and simplest is to distribute the green time such that the v/c ratios are equalized for the critical lane groups, as by the following equation:

$$g_i = \left(\frac{v}{s}\right)_{ci} \left(\frac{C}{X_i}\right) \qquad (7.22)$$

where

g_i = effective green time for phase i,
$(v/s)_{ci}$ = flow ratio for critical lane group i,
C = cycle length in seconds, and
X_i = v/c ratio for lane group i.

EXAMPLE 7.11

Determine the green-time allocations for the 65-second cycle length found in Example 7.10, using the method of v/c ratio equalization.

SOLUTION

Because the calculated cycle length was rounded up a few seconds, the critical intersection v/c ratio for this rounded cycle length will be calculated for use in the green-time allocation calculations. Equation 7.20 can be rearranged to solve for X_c as follows:

$$X_c = \frac{\sum_{i=1}^{n} \left(\frac{v}{s}\right)_{ci} \times C}{C - L}$$

Using this equation with

$(v/s)_{ci}$ = 0.726 (Example 7.8)
C = 65 s (Example 7.10)
L = 12 s (Example 7.8)

gives

$$X_c = \frac{0.726 \times 65}{65 - 12} = 0.890$$

Therefore, the cycle length of 65 seconds and X_c of 0.890 are used to calculate the effective green times for the three phases, as follows:

$$g_1 = \left(\frac{v}{s}\right)_{c1}\left(\frac{C}{X_1}\right)$$

$$= 0.171 \times \frac{65}{0.890} = \underline{\underline{12.5 \text{ s}}} \quad \text{(EB and WB left-turn movements)}$$

$$g_2 = \left(\frac{v}{s}\right)_{c2}\left(\frac{C}{X_2}\right)$$

$$= 0.338 \times \frac{65}{0.890} = \underline{\underline{24.7 \text{ s}}} \quad \text{(EB and WB through and right-turn movements)}$$

$$g_3 = \left(\frac{v}{s}\right)_{c3}\left(\frac{C}{X_3}\right)$$

$$= 0.217 \times \frac{65}{0.890} = \underline{\underline{15.8 \text{ s}}} \quad \text{(NB and SB left, through, and right-turn movements)}$$

The cycle length is checked by summing these effective green times and the lost time, giving

$$C = g_1 + g_2 + g_3 + L$$
$$= 12.5 + 24.7 + 15.8 + 12 = 65.0$$

Therefore, all calculations are correct.

7.5.7 Calculate Change and Clearance Intervals

Recall that the change interval corresponds to the yellow time and the clearance interval corresponds to the all-red time. If an all-red interval does not exist, then the yellow time is considered as both the change and clearance intervals. The change interval alerts drivers that the green interval is about to end and that they should come to stop before entering the intersection, or continue through the intersection if they are too close to come to a safe stop. The clearance interval allows for those vehicles that might have entered the intersection at the end of the yellow to clear the intersection before conflicting traffic movements are given a green signal indication. In the past, the yellow indication was intended to also allow for clearance time. Today, however, there is routine red-indication abuse and frequent running of red indications after the yellow time. As a result, the all-red indication is often implemented.

7.5 Development of a Traffic Signal Phasing and Timing Plan

Typically, the yellow time is in the range of 3 to 5 seconds. Warning times that are shorter than 3 seconds and longer than 5 seconds are not practical because long warning times encourage motorists to continue to enter the intersection whereas short times can place the driver in a dilemma zone. A dilemma zone is created for the driver if a safe stop before the intersection cannot be accomplished, and continuing through the intersection at a constant speed (without accelerating) will result in the vehicle entering the intersection during a red indication. If a dilemma zone exists, drivers always make the wrong decision, whether they decide to stop or to continue through the intersection. Figure 7.13 illustrates the dilemma zone. Referring to this figure, suppose a vehicle traveling at a constant speed requires distance x_s to stop. If the vehicle is closer to the intersection than distance d_d, then it can enter before the all-red indication. If the vehicle is in the shaded area ($x_s - d_d$ from the intersection) when the yellow light is displayed, the driver is in the dilemma zone and can neither stop in time nor continue through the intersection at a constant speed without passing through a red indication.

Formulas and policies for calculating yellow (Y) and all-red (AR) times vary by agency, but one set of commonly accepted formulas is provided in the *Traffic Engineering Handbook* [ITE 1999] and are as follows:

$$Y = t_r + \frac{V}{2a + 2gG} \qquad (7.23)$$

where

Y = yellow time (usually rounded up to the nearest 0.5 second),
t_r = driver perception/reaction time, usually taken as 1.0 second,
V = speed of approaching traffic in ft/s (m/s),
a = deceleration rate for the vehicle, usually taken as 10.0 ft/s² (3.05 m/s²),
g = acceleration due to gravity [32.2 ft/s² (9.807 m/s²)], and
G = percent grade divided by 100.

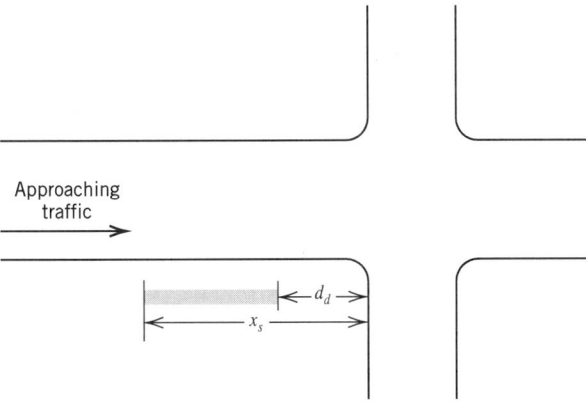

Figure 7.13 The dilemma zone for traffic approaching a signalized intersection.

and

$$AR = \frac{w+l}{V} \quad (7.24)$$

where

   AR = all-red time (usually rounded up to the nearest 0.5 second),
   w = width of the cross street in ft (m),
   l = length of the vehicle, usually taken as 20 ft (6 m), and
   V = speed of approaching traffic in ft/s (m/s).

To avoid a dilemma zone and the possibility of a vehicle being in the intersection when a conflicting movement receives a green-signal indication, the total of the change and clearance intervals (yellow plus all-red times) should always be equal to or greater than the sum of Eqs. 7.23 and 7.24.

EXAMPLE 7.12

Determine the yellow and all-red times for vehicles traveling on Vine and Maple streets as shown in Fig 7.9.

SOLUTION

For the Vine Street phasing (applying Eqs. 7.23 and 7.24),

$$Y = 1.0 + \frac{(35 \times 5280/3600)}{2(10)}$$

$$= 3.6 \quad \rightarrow \quad \underline{4.0 \text{ s}} \text{ (rounding up to the nearest 0.5 s)}$$

$$AR = \frac{60 + 20}{35 \times 5280/3600}$$

$$= 1.6 \quad \rightarrow \quad \underline{2.0 \text{ s}} \text{ (rounding up to the nearest 0.5 s)}$$

For the Maple Street phasing (applying Eqs. 7.23 and 7.24),

$$Y = 1.0 + \frac{(40 \times 5280/3600)}{2(10)}$$

$$= 3.9 \quad \rightarrow \quad \underline{4.0 \text{ s}} \text{ (rounding up to the nearest 0.5 s)}$$

$$AR = \frac{36 + 20}{40 \times 5280/3600}$$

$$= \underline{1.0 \text{ s}}$$

Note that separate calculations are usually required for exclusive left-turn phases, as vehicle approach speeds are often lower than for through vehicles and intersection crossing distances may be longer (due to the width of the opposing direction and the circular travel path).

7.5.8 Check Pedestrian Crossing Time

In urban areas and other locations where pedestrians are present, the signal-timing plan should be checked for its ability to provide adequate pedestrian crossing time. At locations where streets are wide and green times are short, it is possible that pedestrians can be caught in the middle of the intersection when the phase changes. To avoid this problem, the minimum green time required for pedestrian crossing should be checked against the apportioned green time for the phase. If there is not enough green time for a pedestrian to safely cross the street, the apportioned green time should be increased to meet the pedestrian needs. If pedestrian pushbuttons are provided at an intersection (for actuated control), the green time can be increased to meet pedestrian crossing needs only when the pushbuttons are activated.

The minimum pedestrian green time is given by

$$G_p = 3.2 + \frac{L}{S_p} + (0.27 N_{ped}) \quad \text{for } W_E \leq 10 \text{ ft (3.05 m)} \quad (7.25)$$

$$G_p = 3.2 + \frac{L}{S_p} + \left(2.7 \frac{N_{ped}}{W_E}\right) \quad \text{for } W_E > 10 \text{ ft (3.05 m)} \quad (7.26)$$

where

G_p = minimum pedestrian green time in seconds,
3.2 = pedestrian start-up time in seconds,
L = crosswalk length in ft (m),
S_p = walking speed of pedestrians, usually taken as 4.0 ft/s (1.2 m/s),
N_{ped} = number of pedestrians crossing during an interval, and
W_E = effective crosswalk width in ft (m).

The generally recommended walking speed of 4.0 ft/s (1.2 m/s) represents a slower-than-average speed. However, at intersections where a significant number of slower pedestrians (elderly, vision impaired, etc.) are served, the use of a slower walking speed may be warranted.

EXAMPLE 7.13

Determine the minimum amount of pedestrian green time required for the intersection of Vine and Maple streets. Assume a maximum of 15 pedestrians crossing either street during any one phase and a crosswalk width of 8 ft (2.44 m).

SOLUTION

A pedestrian who crosses Maple Street will cross while Vine Street has a green interval. The minimum pedestrian green time needed on Vine Street is (using Eq. 7.25, as the effective crosswalk width is less than or equal to 10 ft)

$$G_p = 3.2 + \frac{36}{4.0} + (0.27 \times 15) = \underline{16.25 \text{ s}}$$

In Example 7.11, Vine Street was assigned 15.8 seconds of effective green time [13.8 seconds of displayed green time (from Eq. 7.3)]. This amount of time is insufficient for pedestrians crossing Maple Street. Therefore, the green time for this phase will have to be increased to accommodate crossing pedestrians, and the overall signal timing plan adjusted accordingly (although we will continue to use the previously computed green time in subsequent examples). The minimum pedestrian green time needed on Maple Street (for the through/right-turn phase, when pedestrian movement would be permitted) is

$$G_p = 3.2 + \frac{60}{4.0} + (0.27 \times 15) = \underline{22.25 \text{ s}}$$

In Example 7.11, Maple Street was assigned 24.7 seconds of effective green time (23.7 seconds of displayed green time) for this phase, so this green time is adequate for pedestrians crossing Vine Street.

7.6 LEVEL-OF-SERVICE DETERMINATION

Before the implementation of any developed signal phasing and timing plan, the level of service should be determined to assess whether the intersection will operate at an acceptable level under this plan. As previously mentioned, the service measure (the performance measure by which level of service is assessed) for signalized intersections is delay. In previous examples we calculated the delay for a specific lane group, but now we must do this for all lane groups and then aggregate the delay values to arrive at an overall intersection delay measure and corresponding level of service.

The first step is to aggregate the delays of all lane groups for an approach, and then repeat the procedure for each approach of the intersection. This will result in approach-specific delays and levels of service. The aggregated lane group delay for each approach is given by

$$d_A = \frac{\sum_i d_i v_i}{\sum_i v_i} \tag{7.27}$$

7.6 Level-of-Service Determination

where

d_A = average delay per vehicle for approach A in seconds,
d_i = average delay per vehicle for lane group i (on approach A) in seconds, and
v_i = analysis flow rate for lane group i in veh/h.

Once all the approach delays have been calculated, they can be aggregated to arrive at the overall intersection delay. The aggregated approach delay for the intersection is given by

$$d_I = \frac{\sum_A d_A v_A}{\sum_A v_A} \qquad (7.28)$$

where

d_I = average delay per vehicle for the intersection in seconds,
d_A = average delay per vehicle for approach A in seconds, and
v_A = analysis flow rate for approach A in veh/h.

The delay level-of-service criteria for signalized intersections are specified in the *Highway Capacity Manual* [Transportation Research Board 2000] and are given in Table 7.4. These delay criteria can be used to determine the level of service for a lane group, an approach, and the intersection.

EXAMPLE 7.14

Determine the level of service for the eastbound approach of Maple Street assuming no initial queue at the start of the analysis period.

SOLUTION

Lane groups for this intersection were established in Example 7.7. Two lane groups were established for the eastbound (EB) approach, one for the left-turn movement and the other for the combined through and right-turn movements. The delay will be calculated for the left-turn lane group first, followed by the through/right-turn lane group.

Table 7.4 Level-of-Service Criteria for Signalized Intersections

LOS	Control delay per vehicle (s/veh)
A	≤ 10
B	> 10–20
C	> 20–35
D	> 35–55
E	> 55–80
F	> 80

Source: Transportation Research Board. *Highway Capacity Manual.* Washington, DC: National Research Council, 2000.

For the left-turn lane group, the uniform delay is computed using Eq. 7.15 with

$$C = 65 \text{ s}$$
$$g = 12.5 \text{ s}$$
$$X = \frac{v}{c} = \frac{v}{s \times g/C} = \frac{300}{1750 \times 12.5/65} = \frac{300}{337} = 0.891$$

giving

$$d_1 = \frac{0.5(65)\left(1 - \frac{12.5}{65}\right)^2}{1 - \left[0.891 \times \frac{12.5}{65}\right]} = 25.6 \text{ s}$$

Random delay is computed using Eq. 7.16 with

$T = 0.25$ (15 min)
$X = 0.891$ (from above)
$k = 0.5$ (pretimed control)
$I = 1.0$ (isolated mode)
$c = 337$ veh/h (from above)

giving

$$d_2 = 900(0.25)\left[(0.891 - 1) + \sqrt{(0.891 - 1)^2 + \frac{8(0.5)(1.0)0.891}{(337)0.25}}\right]$$
$$= 27.8 \text{ s}$$

Now, the total signal delay is computed using Eq. 7.14 with

$d_1 = 25.6$ s
$PF = 1.0$ (progression neutral)
$d_2 = 27.8$ s
$d_3 = 0$ s (given)

$$d = 25.6 \times 1.0 + 27.8 + 0 = 53.4 \text{ s}$$

From Table 7.4, this lane group delay corresponds to a level of service of D.

For the through/right-turn lane group, the uniform delay is computed using Eq. 7.15 with

$$C = 65 \text{ s}$$
$$g = 24.7 \text{ s}$$
$$X = \frac{v}{c} = \frac{v}{s \times g/C} = \frac{1100}{3400 \times 24.7/65} = \frac{1100}{1292} = 0.851$$

giving

$$d_1 = \frac{0.5(65)\left(1 - \frac{24.7}{65}\right)^2}{1 - \left[0.851 \times \frac{24.7}{65}\right]} = 18.5 \text{ s}$$

Random delay is computed using Eq. 7.16 with

$T = 0.25$ (15 min)
$X = 0.851$ (from above)
$k = 0.5$ (pretimed control)
$I = 1.0$ (isolated mode)
$c = 1292$ veh/h (from above)

giving

$$d_2 = 900(0.25)\left[(0.851 - 1) + \sqrt{(0.851 - 1)^2 + \frac{8(0.5)(1.0)0.851}{(1292)0.25}}\right]$$

$$= 7.2 \text{ s}$$

Now, the average signal delay for this lane group is computed using Eq. 7.14 with

$d_1 = 18.5$ s
$PF = 1.0$ (progression neutral)
$d_2 = 7.2$ s
$d_3 = 0$ s (given)

$$d = 18.5 \times 1.0 + 7.2 + 0 = 25.7 \text{ s}$$

From Table 7.4, this lane group delay corresponds to a level of service of C.

Now, to compute the volume-weighted aggregate delay for the approach, we use Eq. 7.27:

$$d_A = \frac{\sum d_i v_i}{\sum v_i}$$

with

d_{LT} = delay for left-turn lane group
$d_{T/R}$ = delay for through/right-turn lane group
v_{LT} = analysis flow rate for left-turn lane group
$v_{T/R}$ = analysis flow rate for through/right-turn lane group

giving

$$d_{EB} = \frac{v_{LT} \times d_{LT} + v_{T/R} \times d_{T/R}}{v_{LT} + v_{T/R}}$$

$$= \frac{300 \times 53.4 + 1100 \times 25.7}{300 + 1100}$$

$$= \frac{44{,}290}{1400}$$

$$= 31.6 \text{ s}$$

From Table 7.4, this approach delay corresponds to a level of service of C.

EXAMPLE 7.15

Determine the level of service for the intersection of Maple and Vine streets.

SOLUTION

The delay for each of the other three approaches (WB, NB, SB) can be determined by the exact same process used in Example 7.14 for the EB approach. Due to the length of the calculations involved for the remaining three approaches, the results are summarized in Table 7.5.

In this table, note that the v/c ratios for the critical lane groups match (rounding differences aside) the calculated critical intersection v/c ratio, X_c, as they should, because green time was allocated based on the strategy of equalizing v/c ratios for the critical lane groups in each phase (using Eq. 7.22).

The overall intersection delay calculation will be shown for the sake of clarity. Using Eq. 7.28, the intersection delay is given by

$$d_I = \frac{31.6 \times 1400 + 30.2 \times 1400 + 49.7 \times 480 + 42.3 \times 440}{1400 + 1400 + 480 + 440}$$

$$= \frac{128{,}988}{3{,}720}$$

$$= 34.7 \text{ s}$$

It is worth pointing out that although all but two lane groups (EB T/R and WB T/R) have a level of service of D or higher, the much higher volumes for those two approaches relative to the others keeps the level of service at C (albeit barely) due to the volume weighting in the delay aggregation.

Table 7.5 Summary of Delay and Level-of-Service Calculations for the Intersection of Maple and Vine Streets

Approach	EB		WB		NB		SB	
Lane group	LT	T/R	LT	T/R	LT	T/R	LT	T/R
Analysis flow rate (v)	300	1100	250	1150	90	390	70	370
Saturation flow rate (s)	1750	3400	1750	3400	475	1800	450	1800
Flow ratio (v/s)	0.171	0.324	0.143	0.338	0.189	0.217	0.156	0.206
Critical lane group ($\sqrt{\ }$)	√			√		√		
Y_c				0.726				
Lost time/phase				4				
Total lost time				12				
Cycle length (C_{min})				65				
X_c				0.891				
Eff. green time (g)	12.5	24.7	12.5	24.7	15.8	15.8	15.8	15.8
g/C	0.192	0.380	0.192	0.380	0.243	0.243	0.243	0.243
Lane group capacity (c)	337	1292	337	1292	115	438	109	438
v/c (X)	0.891	0.851	0.743	0.890	0.779	0.891	0.640	0.846
d_1	25.6	18.5	24.7	18.9	23.0	23.8	22.1	23.4
PF				1.0				
k				0.5				
l				1.0				
T				0.25				
d_2	27.8	7.2	13.8	9.5	39.4	23.0	25.3	17.9
Lane group delay (d)	53.4	25.7	38.5	28.3	62.4	46.7	47.3	41.4
Lane group LOS	D	C	D	C	E	D	D	D
Approach delay	31.6		30.2		49.7		42.3	
Approach LOS	C		C		D		D	
Intersection delay				34.7				
Intersection LOS				C				

NOMENCLATURE FOR CHAPTER 7

a	deceleration rate for vehicle at an intersection
AR	all-red time
C	cycle length
c	capacity
d_{avg}	average vehicle delay per cycle ($D/D/1$ queuing)
d_1	average vehicle delay per cycle assuming uniform arrivals
d_2	average vehicle delay per cycle assuming random arrivals
d_3	average vehicle delay per cycle due to initial queue at start of analysis time period
d_d	distance from the intersection for which the dilemma zone is avoided
d_{max}	maximum delay of any vehicle ($D/D/1$ queuing)
D	deterministic arrivals or departures
D_t	total vehicle delay ($D/D/1$ queuing)
g	effective green time or acceleration due to gravity
G	displayed green time or grade of roadway
G_p	pedestrian green time
I	upstream filtering/metering adjustment factor
k	delay adjustment factor dependent on signal controller mode
l	vehicle length
L	total cycle lost time or crosswalk length
n	number of phases or number of vehicles or number of critical lane groups
N_{ped}	number of crossing pedestrians per phase
P_q	proportion of the signal cycle with a queue ($D/D/1$ queuing)
P_s	proportion of stopped vehicles ($D/D/1$ queuing)
PF	progression adjustment factor
Q	number of vehicles in the queue
Q_{max}	maximum number of vehicles in the queue ($D/D/1$ queuing)
r	effective red time
R	displayed red time
s	saturation flow rate
S_p	pedestrian walking speed
t	time
t_c	time after the start of effective green until queue clearance ($D/D/1$ queuing)
t_L	total lost time for a movement during a cycle
t_r	driver perception/reaction time
V	travel speed of vehicle
w	width of street
W_E	effective crosswalk width
X_i	volume-to-capacity ratio for lane group i
X_c	critical volume-to-capacity ratio for the intersection
x_s	distance required to stop
Y	displayed yellow time
Y_c	sum of flow ratios for critical lane groups
λ	arrival rate
μ	departure rate
ρ	traffic intensity

REFERENCES

Institute of Transportation Engineers (ITE). *Traffic Engineering Handbook*, 5th ed. Washington, DC, 1999.

Institute of Transportation Engineers (ITE). *Traffic Control Devices*, 2nd ed. Washington, DC, 2001.

Kell, James H., and Iris J. Fullerton. *Manual of Traffic Signal Design*, 2nd ed, Washington, DC: Institute of Transportation Engineers, 1991.

Transportation Research Board. *Traffic Flow Theory: A Monograph*. Special Report 165. Washington, DC: National Research Council, 1975.

Transportation Research Board. *Highway Capacity Manual*, Washington, DC: National Research Council, 2000.

U.S. Federal Highway Administration. *Manual on Uniform Traffic Control Devices for Streets and Highways*. Washington, DC: U.S. Government Printing Office, 2000.

Webster, F. V. *Traffic Signal Settings*, Road Research Technical Paper No. 39. London: Great Britain Road Research Laboratory, 1958.

PROBLEMS

7.1 An intersection approach has a saturation flow rate of 1500 veh/h, and vehicles arrive at the approach at the rate of 800 veh/h. The approach is controlled by a pretimed signal with a cycle length of 60 seconds and $D/D/1$ queuing holds. Local standards dictate that signals should be set such that all approach queues dissipate 10 seconds before the end of the effective green portion of the cycle. Assuming that approach capacity exceeds arrivals, determine the maximum length of effective red that will satisfy the local standards.

7.2 An approach to a pretimed signal has 30 seconds of effective red, and $D/D/1$ queuing holds. The total delay at the approach is 83.33 veh-s/cycle and the saturation flow rate is 1000 veh/h. If the capacity of the approach equals the number of arrivals per cycle, determine the approach flow rate and cycle length.

7.3 An approach to a pretimed signal has 25 seconds of effective green in a 60-second cycle. The approach volume is 500 veh/h and the saturation flow rate is 1400 veh/h. Calculate the average vehicle delay assuming $D/D/1$ queuing.

7.4 An observer notes that an approach to a pretimed signal has a maximum of eight vehicles in a queue in a given cycle. If the saturation flow rate is 1440 veh/h and the effective red time is 40 seconds, how much time will it take this queue to clear after the start of the effective green (assuming that approach capacity exceeds arrivals and $D/D/1$ queuing applies)?

7.5 An approach to a pretimed signal with a 60-second cycle has 8.9 vehicles in the queue at the beginning of the effective green. Four of the 8.9 vehicles in the queue are left over from the previous cycle (at the end of the previous cycle's effective green). The saturation flow rate of the approach is 1500 veh/h, total delay for the cycle is 5.78 vehicle-minutes, and at the end of the effective green there are 2 vehicles left in the queue. Determine the arrival rate assuming that it is unchanged over the duration of the observation period (from the beginning to the end of this 5.78–vehicle-minute delay cycle). (Assume $D/D/1$ queuing.)

7.6 At the beginning of an effective red, vehicles are arriving at an approach at the rate of 500 veh/h and 16 vehicles are left in the queue from the previous cycle (at the end of the previous cycle's effective green). However, due to the end of a major sporting event, the arrival rate is continuously increasing at a constant rate of 200 veh/h/min (after 1 minute the arrival rate will be 700 veh/h, after 2 minutes 900 veh/h, etc.). The saturation flow rate of the approach is 1800 veh/h, the cycle length is 60 seconds, and the effective green time is 40 seconds. Determine the total vehicle delay until complete queue clearance. (Assume $D/D/1$ queuing.)

7.7 The saturation flow rate for an intersection approach is 3600 veh/h. At the beginning of a cycle (effective red) no vehicles are queued. The signal is timed so that when the queue (from the continuously arriving vehicles) is 13 vehicles long, the effective green begins. If the queue dissipates 8 seconds before the end of the cycle and the cycle length is 60 seconds, what is the arrival rate assuming $D/D/1$ queuing?

7.8 The saturation flow rate for a pretimed signalized intersection approach is 1800 veh/h. The cycle length is 80 seconds. It is known that the arrival rate during the effective green is twice the arrival rate during the effective red. During one cycle, there are 2 vehicles in the queue at the beginning of the cycle (the beginning of the effective red) and there are 7.9 vehicles in the queue at the end of the effective red (the beginning of the effective green). If the queue clears exactly at the end of the effective green, and $D/D/1$ queuing applies, determine the total vehicle delay in the cycle (in veh-s).

7.9 An approach to a signalized intersection has a saturation flow rate of 1800 veh/h. At the beginning of an effective red, there are six vehicles in the queue and vehicles arrive at 900 veh/h. The signal has a 60-second cycle with 25 seconds of effective red. What is the total vehicle delay after one cycle (assume $D/D/1$ queuing)?

7.10 An approach to a signalized intersection has a saturation flow rate of 2640 veh/h. For one cycle, the approach has 3 vehicles in queue at the beginning of an effective red, and vehicles arrive at 1064 veh/h. The signal for the approach is timed such that the effective green starts 8 seconds after the approach's vehicle queue reaches 10 vehicles, and lasts 15 seconds. What is the total delay for this signal cycle?

7.11 An approach has a saturation flow rate of 1800 veh/h. During one 80-second cycle, there are 4 vehicles queued at the beginning of the cycle (the start of the effective red) and 2 vehicles queued at the end of the cycle (the end of the effective green). At the beginning of the effective green there are 10 vehicles in the queue. The arrival rate is constant and the process is $D/D/1$. If the effective red is known to be less than 40 seconds, what is the total vehicle delay for this signal cycle?

7.12 Vehicles arrive at an approach to a pretimed signalized intersection. The arrival rate over the cycle is given by the function $\lambda(t) = 0.22 + 0.012t$ [$\lambda(t)$ is in veh/s and t is in seconds]. There are no vehicles in the queue when the cycle (effective red) begins. The cycle length is 60 seconds and the saturation flow rate is 3600 veh/h. Determine the effective green and red times that will allow the queue to clear exactly at the end of the cycle (the end of the effective green), and determine the total vehicle delay over the cycle (assuming $D/D/1$ queuing).

7.13 At the start of the effective red at an intersection approach to a pretimed signal, vehicles begin to arrive at a rate of 800 veh/h for the first 40 seconds and 500 veh/h from then on. The approach has a saturation flow rate of 1200 veh/h and an effective green of 20 seconds, and the cycle length is 40 seconds. What is the total vehicle delay two full cycles after the 800-veh/h arrival rate begins? (Assume $D/D/1$ queuing.)

7.14 An approach to a pretimed signal has 25 seconds of effective green, a saturation flow rate of 1300 veh/h, and a volume-to-capacity ratio less than 1. If the cycle length is 60 seconds and the overall delay formula (Eq. 7.14) estimates an average delay that is 34 s higher than that estimated by using just the uniform delay formula, determine the vehicle arrival rate. (Assume the signal is isolated and $d_3 = 0$.)

7.15 Recent computations at an approach to a pretimed-signalized intersection indicate that the average delay per vehicle is 16.6 seconds (from Eq. 7.14). The volume-to-capacity ratio ($\lambda C/\mu g$) is 0.8, the saturation flow rate is 1600 veh/h, and the effective green time is 50 seconds. If the uniform delay (assuming $D/D/1$ queuing) is 11.25 seconds per vehicle, determine the arrival flow (in veh/h) and the cycle length. (Assume the signal is isolated and $d_3 = 0$.)

7.16 An intersection has a three-phase signal with the movements allowed in each phase, and corresponding analysis and saturation flow rates shown in Table 7.6. Calculate the sum of the flow ratios for the critical lane groups.

7.17 An intersection has a four-phase signal with the movements allowed in each phase, and corresponding analysis and saturation flow rates shown in Table 7.7. Calculate the sum of the flow ratios for the critical lane groups.

7.18 The minimum cycle length for an intersection is determined to be 95 seconds. The critical lane group flow ratios were calculated as 0.235, 0.250, 0.170, and 0.125, for phases 1–4, respectively. What X_c was used in the determination of this cycle length, assuming a lost time of 5 seconds per phase?

7.19 A pretimed four-phase signal has critical lane group flow rates for the first three phases of 200, 187, and 210 veh/h (saturation flow rates are 1800 veh/h/ln for all phases). The lost time is known to be 4 seconds for each phase. If the cycle length is 60 seconds, what is the estimated effective green time of the fourth phase?

Table 7.6 Data for Problem 7.16

Phase	1	2	3
Allowed movements	NB L, SB L	NB T/R, SB T/R	EB L, WB L EB T/R, WB T/R
Analysis flow rate	330, 365 veh/h	1125, 1075 veh/h	110, 80 veh/h 250, 285 veh/h
Saturation flow rate	1700, 1750 veh/h	3400, 3300 veh/h	650, 600 veh/h 1750, 1800 veh/h

Table 7.7 Data for Problem 7.17

Phase	1	2	3	4
Allowed movements	EB L, WB L	EB T/R, WB T/R	SB L, SB T/R	NB L, NB T/R
Analysis flow rate	245, 230 veh/h	975, 1030 veh/h	255, 235 veh/h	225, 215 veh/h
Saturation flow rate	1750, 1725 veh/h	3350, 3400 veh/h	1725, 1750 veh/h	1700, 1750 veh/h

7.20 A four-phase traffic signal has critical lane group flow ratios of 0.225, 0.175, 0.200, and 0.150. If the lost time per phase is 5 seconds, and a critical intersection v/c of 0.85 is desired, calculate the minimum cycle length and the phase effective green times such that the lane group v/c ratios are equalized.

7.21 For Problem 7.16, calculate the minimum cycle length and the effective green time for each phase (balancing v/c for the critical lane groups). Assume lost time is 4 seconds per phase and a critical intersection v/c of 0.90 is desired.

7.22 For Problem 7.16, calculate the northbound approach delay and level of service.

7.23 For Problem 7.16, calculate the southbound approach delay and level of service.

7.24 For Problem 7.16, calculate the westbound approach delay and level of service.

7.25 For Problem 7.16, calculate the eastbound approach delay and level of service.

7.26 For Problem 7.16, calculate the overall intersection delay and level of service.

7.27 For Problem 7.16, calculate the optimal cycle length (Webster's formulation) and the corresponding effective green times (based on lane group v/c equalization).

7.28 For Problem 7.17, calculate the minimum cycle length and the effective green time for each phase (balancing v/c for the critical movements). Assume lost time is 4 seconds per phase and a critical intersection v/c of 0.95 is desired.

7.29 For Problem 7.17, calculate the northbound approach delay and level of service.

7.30 For Problem 7.17, calculate the southbound approach delay and level of service.

7.31 For Problem 7.17, calculate the westbound approach delay and level of service.

7.32 For Problem 7.17, calculate the eastbound approach delay and level of service.

7.33 For Problem 7.17, calculate the overall intersection delay and level of service.

7.34 For Problem 7.17, calculate the optimal cycle length (Webster's formulation) and the corresponding effective green times (based on lane group v/c equalization).

7.35 A new shopping center opens near the intersection of Vine and Maple streets (the intersection shown in Fig. 7.9). The net effect is to increase the approaching traffic volumes by 10%. Calculate the new level of service for the westbound approach, assuming all else remains the same.

7.36 For Problem 7.35, calculate the new level of service for the northbound approach.

7.37 Consider the intersection of Vine and Maple streets as shown in Fig. 7.9. Suppose Vine Street's northbound and southbound approaches are both on an 8% upgrade, and the assumed vehicle approach speed is 30 mi/h. What should the yellow and all-red times be?

7.38 Consider Example 7.8. Two additional 12-ft through lanes are added to Vine Street (the street in the intersection shown in Fig. 7.9), one lane in each direction. If the peak-hour traffic volumes are unchanged, but the Vine Street left-turn saturation flow rates increase by 100 veh/h because of the added through lanes, what would the revised effective green time, yellow time, and all-red time be for each phase?

7.39 Calculate the overall intersection level of service for Problem 7.38.

7.40 Consider Problem 7.38. Calculate the new minimum pedestrian green time, assuming the effective crosswalk width is 6 ft and the maximum number of crossing pedestrians in any phase is 20.

7.41 Consider Problem 7.38. How much traffic volume can be added to the southbound approach (assuming the same turning movement percentage) before LOS D is reached for the approach?

7.42 Consider Problem 7.38. How much traffic volume must be diverted from the eastbound approach (assuming the same turning movement percentage) to achieve LOS B for the approach.

Chapter 8

Travel Demand and Traffic Forecasting

8.1 INTRODUCTION

The modification of a highway network either by new road construction or operating improvements on existing roads (use or retiming of traffic signals) must be predicated on some expectation or forecast of traffic volumes. For new road construction, forecasts of traffic are needed to determine an appropriate pavement (number of equivalent axle loads, as discussed in Chapter 4) and geometric design (number of lanes, shoulder widths, etc.) that will provide an acceptable level of service. For operational improvements, traffic forecasts are needed to estimate the effectiveness of alternative improvement options.

In forecasting vehicle traffic, two interrelated elements must be considered: the overall regional traffic growth/decline and possible traffic diversion. The long-term trend of traffic growth/decline is clearly an important concern, because projects such as highway construction and operational improvements must be undertaken with some idea of what future traffic conditions will be. In the design of these projects, engineers must seek to provide a sufficient highway level of service and an acceptable pavement ride quality for future traffic volumes. One would expect that factors affecting long-term regional traffic growth/decline trends are primarily economic and, to a historically lesser extent, social in nature. The economics of the region in which a highway action is being undertaken determine the amount of traffic-generating activities (work, social/recreational, and shopping) and the spatial distribution of residential, industrial, and commercial areas. The social aspects of the population determine attitudes and behavioral tendencies with regard to traffic-generating decisions. For example, some regional populations may have social characteristics that make them more likely than other regional populations to make fewer trips, to carpool, to vanpool, or to take public transportation (buses or subways), all of which significantly impact the amount of highway traffic.

In addition to overall regional traffic growth/decline, there is the more microscopic, short-term phenomenon of traffic diversion. As new roads are constructed, as operational improvements are made, and/or as roads gradually become more congested, traffic will shift as drivers change routes or trip departure times in an effort to avoid congestion and improve the level of service that they experience. Thus the highway network must be viewed as a system and, consequently, with the realization that a capacity or level-of-service change in any one segment of the highway network will have an impact on traffic flows in many of the surrounding highway segments.

From this discussion, it is obvious that travel demand and traffic forecasting is a formidable problem, because it requires accurate regional economic forecasts as well as accurate forecasts of highway users' social and behavioral responses regarding trip-oriented decisions, in order to predict growth/decline trends and traffic diversion. Virtually everyone is aware of how inaccurate economic forecasts can be, which is testament to the complexity and uncertainty associated with such forecasts. Similarly, one can readily imagine the difficulty associated with forecasting individuals' trip-oriented decisions.

In spite of the enormous obstacles to accurate forecasting of traffic, over the years highway engineers and analysts have persisted in the development and refinement of a wide variety of travel demand and traffic forecasting techniques. An overriding consideration has always been the ease with which a technique can be implemented in terms of input data requirements and the ability of users to comprehend the underlying methodological approach. The field has evolved such that traffic analysts can legitimately argue that recent developments in travel demand and traffic forecasting are largely beyond the reach of practice-oriented implementation. In many respects, this is an expected development because, due to the complexity of the problem, there will always be a tendency for theoretical work to exceed the limits of practical implementability. Unfortunately, the methodological gap between theory and practice has resulted in the use of a wide variety of travel demand and traffic forecasting techniques, and selecting among them is a function of the technical expertise of forecasting agency personnel as well as time and financial factors.

In the past, textbooks have attempted to cover the full range of travel demand and traffic forecasting techniques from the readily implementable, simplistic approaches to the more theoretically refined methods. In so doing, such textbooks have often sacrificed depth of coverage, and as a consequence, travel demand and traffic forecasting frequently had the appearance of being confusing and disjointed. This chapter attempts to convey the basic principles underlying travel demand and traffic forecasting as opposed to reviewing the many techniques available to forecast traffic. This is achieved by focusing on an approach that is fairly advanced technically, and one that effectively and efficiently conveys the fundamental concepts of travel demand and traffic forecasting. For more information on travel demand and traffic forecasting techniques that are more implementable or more theoretically advanced than the concepts provided in this chapter, the reader is referred to other sources [Meyer and Miller 2001; Sheffi 1985; Washington et al. 2003].

8.2 TRAVELER DECISIONS

Forecasts of highway traffic should, at least in theory, be predicated on some understanding of traveler decisions, because the various decisions that travelers make regarding trips will ultimately determine the quantity, spatial distribution (by route), and temporal distribution of vehicles on a highway network. Within this context, travelers can be viewed as making four distinct but interrelated decisions regarding trips: temporal decisions, destination decisions, modal decisions, and spatial or route decisions. The temporal decision includes the decision to travel and, more important, when to travel. The destination decision is concerned with the selection of a specific destination (shopping center, recreational facility, etc.), and the modal decision relates to how the trip is to be made (by automobile, bus, walking, or bicycling). Finally, spatial decisions focus on which route is to be taken from the traveler's origin (the traveler's initial location) to the desired destination. Being able to understand, let alone predict, such decisions is a monumental task. The remaining sections of this chapter seek to define the dimensions of this decision-prediction task and, through examples and illustrations, to demonstrate methods of forecasting traveler decisions and, ultimately, traffic volumes.

8.3 SCOPE OF THE TRAVEL DEMAND AND TRAFFIC FORECASTING PROBLEM

Because travel demand and traffic forecasting is predicated on the accurate forecasting of traveler decisions, two factors must be addressed in the development of an effective travel demand and traffic forecasting methodology: the complexity of the traveler decision-making process and system equilibration.

To begin the development of a fuller understanding of the complexity of traveler decisions, consider the schematic presented in Fig. 8.1. This figure indicates that traveler socioeconomics and activity patterns constitute a major driving force in the decision-making process. Socioeconomics, including factors such as household income, number of household members, and traveler age, affect the types of activities that the traveler is likely to be involved in (work, yoga classes, shopping, children's day care, dancing lessons, community meetings, etc.), which in turn are primary factors in many travel decisions. Socioeconomics can also have a direct effect on travel-related decisions by, for example, limiting modal availability (e.g., travelers in low-income households may be forced to take a bus due to the nonavailability of a household automobile).

If we look more directly at the decision to travel, mode/destination choice, and highway route choice, Fig. 8.1 indicates that both long-term and short-term factors affect these decisions. For the decision to travel, as well as mode/destination choice, the long-term factors of modal availability, residential and commercial distributions, and modal infrastructure play a significant role. These factors are considered long term because they change relatively slowly over time. For example, the development and/or relocation of residential neighborhoods and commercial centers is a process that may take years. Changes in modal infrastructure (construction/relocation of highways, subways, commuter rail systems) and modal availability (changes in automobile ownership, bus routing/scheduling) are also factors that evolve over relatively long periods of time. In contrast, a short-term factor, such as modal traffic, is one that can vary within a short period of time as discussed in Chapter 6.

8.3 Scope of the Travel Demand and Traffic Forecasting Problem

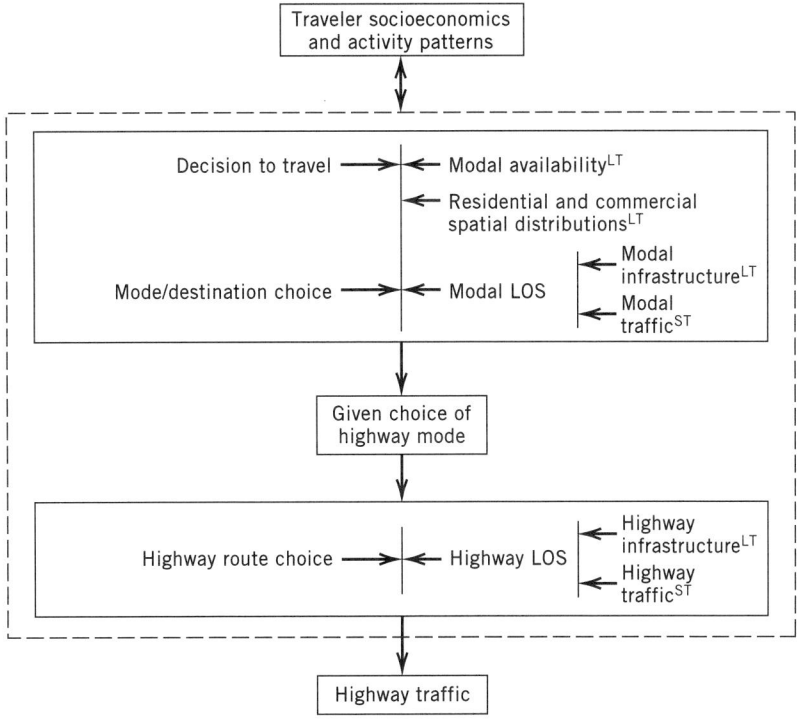

LT: Long-term factors.
ST: Short-term factors.

Figure 8.1 Overview of the process by which highway traffic is determined.

Moving down the illustration presented in Fig. 8.1, we see that the choice of a traveler's highway route is also determined by both long-term (highway infrastructure) and short-term (highway traffic) factors. The outcome of the combination of these traveler decisions is, of course, highway traffic, the prediction of which is the objective of travel demand and traffic forecasting.

Aside from the complexities involved in the traveler decision-making process, the issue of system equilibration (mentioned at the beginning of this section) must also be considered. Note that Fig. 8.1 indicates not only that long- and short-term factors affect traveler decisions and choices, but also that these decisions and choices in turn affect the long- and short-term factors. Such a reciprocal relationship is most apparent in considering the relationship between traveler choices and short-term factors. For example, consider a traveler's choice of highway route. One would expect that the traveler would be more likely to select a route between origin and destination that provides a shorter travel time. The travel time on various routes will be a function of route distance (highway infrastructure, long-term) and route traffic (higher traffic volumes reduce travel speed and increase travel time, as discussed in Chapter 5). But

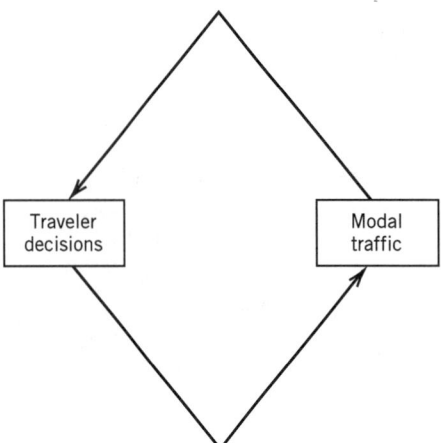

Figure 8.2 Interdependence of traveler decisions and traffic flow.

travelers' decisions to take specific routes ultimately determine the route traffic on which their route decisions are based. This interdependence between traveler decisions and modal traffic is schematically presented in Fig. 8.2. In addition to these short-term effects, persistently high traffic volumes may lead to a change in the highway infrastructure (construction of additional lanes and/or new highways to reduce congestion), again resulting in an interdependence. This interdependence creates the problem of equilibration, which is common to many modeling applications. Perhaps the most recognizable equilibration problem is determination of price in a classic model of economic supply and demand for a product. From a modeling perspective, as will be shown, equilibration adds yet another dimension of difficulty to an already complex travel demand and traffic forecasting problem. It is safe to say that no existing methodology has come close to accurately capturing the complexities involved in traveler decisions or fully addressing the issue of equilibration. However, within rather obvious limitations, the field of travel demand and traffic forecasting has, over the years, made progress toward more accurately modeling traveler decision complexities and equilibration concerns. This evolution of travel demand and traffic forecasting methodology has led to the popular approach of viewing traveler decisions as a sequence of three distinct decisions, as shown in Fig. 8.3, the result of which is forecasted traffic flow (a direct outgrowth of the highway route choice decision). Clearly, the sequential structure of traveler decisions is a considerable simplification of the actual decision-making process in which all trip-related decisions are considered simultaneously by the traveler. However, this sequential simplification permits the development of a sequence of mathematical models of traveler behavior that can be applied to forecast traffic flow. The following sections of this chapter present and discuss typical functional forms of the mathematical models used to forecast the three sequential traveler decisions shown in Fig. 8.3.

8.4 Trip Generation 275

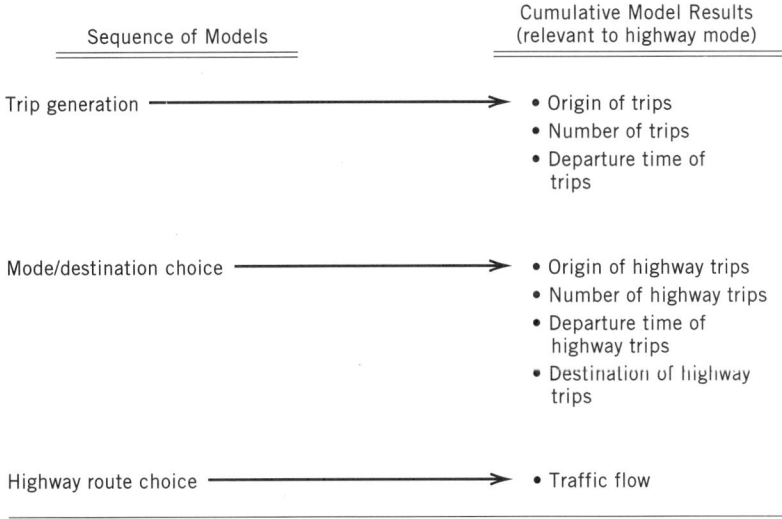

Figure 8.3 Overview of the sequential approach to traffic estimation.

8.4 TRIP GENERATION

The first traveler decision to be modeled in the sequential approach to travel demand and traffic forecasting is trip generation. The objective of trip generation modeling is to develop an expression that predicts exactly when a trip is to be made. This is an inherently difficult task due to the wide variety of trip types (working, social/recreational, shopping, etc.) and activities (eating lunch, exercising, visiting friends, etc.) undertaken by a traveler in a sample day, as is schematically shown in Fig. 8.4. To address the complexity of the trip generation decision, the following approach is typically taken:

1. *Aggregation of decision-making units.* Predicting trip generation behavior is simplified by considering the trip generation behavior of a household (a group of travelers sharing the same domicile) as opposed to the behavior of individual travelers. Such an aggregation of traveler decisions is justified on the basis of the comparatively homogeneous nature of household members (socially and economically) and household members' often intertwined trip-generating activities (joint shopping trips, etc.).

2. *Segmentation of trips by type.* Different types of trips have different characteristics that make them more or less likely to be taken at various times of the day. For example, work trips are more likely to be taken in the morning hours than are shopping trips, which are more likely to be taken during the evening hours. Also, it is more likely that a traveler will take multiple shopping trips during the course of a day as opposed to multiple work trips. To account for this, three distinct trip types are used: (1) work trips, including trips to and from work; (2) shopping trips; and (3) social/recreational trips, which include vacations, visiting friends, church meetings, sporting events, and so on.

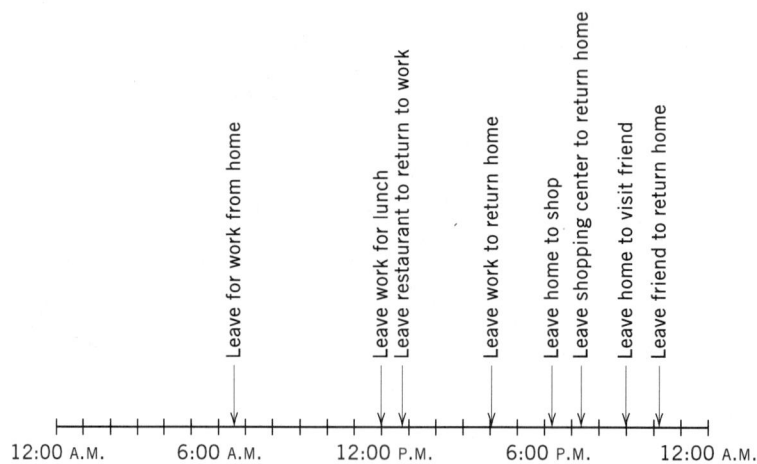

Figure 8.4 Weekday trip generation for a typical traveler.

3. *Temporal aggregation.* Although research has been undertaken to develop mathematical expressions that predict when a traveler is likely to make a trip (Hamed and Mannering 1993), trip generation more often focuses on the number of trips made over some period of time. Thus trips are aggregated temporally, and trip generation models seek to predict the number of trips per hour or per day.

8.4.1 Typical Trip Generation Models

Trip generation models generally assume a linear form, in which the number of vehicle-based (automobile, bus, or subway) trips is a function of various socioeconomic and/or distributional (residential and commercial) characteristics. An example of such a model, for a given trip type, is

$$T_i = b_0 + b_1 z_{1i} + b_2 z_{2i} + \cdots + b_k z_{ki} \qquad (8.1)$$

where

T_i = number of vehicle-based trips of a given type (shopping or social/recreational) in some specified time period made by household i,

b_k = coefficient estimated from traveler survey data and corresponding to characteristic k, and

z_{ki} = characteristic k (income, employment in neighborhood, number of household members) of household i.

The estimated coefficients (b's) are usually estimated by the method of least squares regression using data collected from traveler surveys. A brief description and example of this method are presented in Appendix 8A.

EXAMPLE 8.1

A simple linear regression model is estimated for shopping-trip generation during a shopping-trip peak hour. The model is

Number of peak-hour vehicle-based shopping trips per household
= 0.12 + 0.09(household size)
+ 0.011(annual household income in thousands of dollars)
− 0.15(employment in the household's neighborhood, in hundreds)

A particular household has six members and an annual income of $50,000. They currently live in a neighborhood with 450 retail employees, but are moving to a new home in a neighborhood with 150 retail employees. Calculate the predicted number of vehicle-based peak-hour shopping trips the household makes before and after the move.

SOLUTION

Note that the signs of the model coefficients (b's, +0.09, and +0.011) indicate that as household size and income increase, the number of shopping trips also increases. This is reasonable because wealthier, larger households can be expected to make more vehicle-based shopping trips. The negative sign of the employment coefficient (−0.15) indicates that as retail employment in a household's neighborhood increases, fewer vehicle-based shopping trips will be generated. This reflects the fact that larger retail employment in a neighborhood implies more shopping opportunities nearer to the household, thereby increasing the possibility that a shopping trip can be conducted without the use of a vehicle (a non–vehicle-based trip, such as walking).

Turning to the problem solution, before the household moves,

Number of vehicle trips = 0.12 + 0.09(6) + 0.011(50) − 0.15(4.5) = 0.535

After the household moves,

Number of vehicle trips = 0.12 + 0.09(6) + 0.011(50) − 0.15(1.5) = 0.985

Thus the model predicts that the move will result in 0.45 additional peak-hour vehicle-based shopping trips due to the decline in neighborhood shopping opportunities as reflected by the decline in neighborhood retail employment.

EXAMPLE 8.2

A model for social/recreational trip generation is estimated, with data collected during a major holiday, as

Number of peak-hour vehicle-based social/recreational trips per household
= 0.04 + 0.018(household size)
+ 0.009(annual household income in thousands of dollars)
+ 0.16(number of nonworking household members)

If the household described in Example 8.1 has one working member, how many peak-hour social/recreational trips are predicted?

SOLUTION

The positive signs of the model coefficients indicate that increasing household size, income, and number of nonworking household members result in more social/recreational trips. Again, wealthier and larger households can be expected to be involved in more vehicle-based trip-generating activities, and the larger the number of nonworking household members, the larger the number of people available at home to make peak-hour social/recreational trips.

The solution to this problem is

Number of vehicle trips = 0.04 + 0.018(6) + 0.009(50) + 0.16(5) = <u>1.398</u>

EXAMPLE 8.3

A neighborhood has 205 retail employees and 700 households that can be categorized into four types, with each type having characteristics as follows:

Type	Household size	Annual income	Number of nonworkers in the peak hour	Workers departing
1	2	$40,000	1	1
2	3	$50,000	2	1
3	3	$55,000	1	2
4	4	$40,000	3	1

There are 100 type 1, 200 type 2, 350 type 3, and 50 type 4 households. Assuming that shopping, social/recreational, and work vehicle-based trips all peak at the same time (for exposition purposes), determine the total number of peak-hour trips (work, shopping, social/recreational) using the generation models described in Examples 8.1 and 8.2.

SOLUTION

For vehicle-based shopping trips,

Type 1: 0.12 + 0.09(2) + 0.011(40) − 0.15(2.05) = 0.4325 trips/household
× 100 households
= 43.25 trips

Type 2: 0.12 + 0.09(3) + 0.011(50) − 0.15(2.05) = 0.6325 trips/household
× 200 households
= 126.5 trips

Type 3: 0.12 + 0.09(3) + 0.011(55) − 0.15(2.05) = 0.6875 trips/household
× 350 households
= 240.625 trips

Type 4: 0.12 + 0.09(4) + 0.011(40) − 0.15(2.05) = 0.6125 trips/household
× 50 households
= 30.625 trips

Therefore, there will be a total of 441 vehicle-based shopping trips.

For vehicle-based social/recreational trips,

Type 1: $0.04 + 0.018(2) + 0.009(40) + 0.16(1) = 0.596$ trips/household
$\times$ 100 households
= 59.6 trips

Type 2: $0.04 + 0.018(3) + 0.009(50) + 0.16(2) = 0.864$ trips/household
$\times$ 200 households
= 172.8 trips

Type 3: $0.04 + 0.018(3) + 0.009(55) + 0.16(1) = 0.749$ trips/household
$\times$ 350 households
= 262.15 trips

Type 4: $0.04 + 0.018(4) + 0.009(40) + 0.16(3) = 0.952$ trips/household
$\times$ 50 households
= 47.6 trips

Therefore, there will be a total of 542.15 vehicle-based social/recreational trips.

For vehicle-based work trips, there will be 100 generated from type 1 households (1×100), 200 from type 2 (1×200), 700 from type 3 (2×350), and 50 from type 4 (1×50), for a total of 1050 vehicle-based work trips. Summing the totals for the three trip types gives 2033 peak-hour vehicle-based trips.

It should be noted that the trip generation models used in Examples 8.1, 8.2, and 8.3 are a simplified representation of the actual trip generation decision-making process. First, there are many more traveler and household characteristics that affect trip-generating behavior (age, lifestyles, etc.), and second, the models have no variables to capture the equilibration concept discussed earlier. The equilibration concern is important, because if the highway system is heavily congested, travelers are likely to make fewer peak-hour trips as a result of either canceling trips or postponing them until a less congested time period. Unfortunately, such obvious model defects must often be accepted due to data and resource limitations.

8.4.2 Trip Generation with Count Data Models

Although linear regression has been a popular method for estimating trip generation models, there is a problem in that the estimated linear regression models can produce fractions of trips for a given time period. As an example, the model presented in Example 8.2 predicted that the household presented in Example 8.1 with one working member would produce 1.398 peak-hour social/recreational trips during the major holiday. Because fractions of trips are not realistic, a modeling approach that gives the probability of making a nonnegative-integer number of trips (0, 1, 2, 3, . . .) may be more appropriate [Washington et al. 2003]. One such model is the Poisson regression, which can be formulated for trip generation (for a given trip type) as

$$P(T_i) = \frac{e^{-\lambda_i} \lambda_i^{T_i}}{T_i!} \qquad (8.2)$$

where

T_i = number of vehicle-based trips of a given type (shopping or social/recreational) made in some specified time period by household i,

$P(T_i)$ = probability of household i making exactly T_i trips (where T_i is a nonnegative integer),

e = base of the natural logarithm (e = 2.718), and

λ_i = Poisson parameter for household i, which is equal to household i's expected number of vehicle-based trips in some specified time period, $E[T_i]$.

Poisson regressions are estimated by specifying the Poisson parameter λ_i (the expected number of trips of a specific type made by household i over some time period). The most common relationship between explanatory variables (variables that determine the Poisson parameter) and the Poisson parameter is the log-linear relationship

$$\lambda_i = e^{BZ_i} \qquad (8.3)$$

where

B = vector of estimable coefficients,

Z_i = vector of household characteristics determining trip generation, and

Other terms are as defined previously.

Note that the Poisson parameter λ_i (the expected number of trips of a specific type made by household i over some time period) is a real number (with fractions of trips) but when applied in Eq. 8.2 gives the probability of making a specified nonnegative-integer number of trips (T_i).

In Poisson regressions, the coefficient vector B is estimated by maximum-likelihood procedures. A brief description and example of this estimation procedure are presented in Appendix 8B.

EXAMPLE 8.4

Following Example 8.1, a Poisson regression is estimated for shopping-trip generation during a shopping-trip peak hour. The estimated coefficients are

BZ_i = −0.35 + 0.03(household size)
+ 0.004(annual household income in thousands of dollars)
− 0.10(employment in the household's neighborhood in hundreds)

Given that the household has six members, has an annual income of $50,000, and lives in their new neighborhood with a retail employment of 150, what is the expected number of peak-hour shopping trips and what is the probability that the household will not make a peak-hour shopping trip?

SOLUTION

For the expected number of peak-hour shopping trips (a real number),

$$E[T_i] = \lambda_i = e^{BZ_i} = e^{-0.35 + 0.03(6) + 0.004(50) - 0.1(1.5)} = \underline{\underline{0.887 \text{ vehicle trips}}}$$

For the probability of making zero peak-hour shopping trips (a nonnegative integer), Eq. 8.2 is used to give

$$P(0) = \frac{e^{-0.887} 0.887^0}{0!} = \underline{\underline{0.412}}$$

8.5 MODE AND DESTINATION CHOICE

Once the number of trips generated per unit time is known, the next step in the sequential approach to travel demand and traffic forecasting is to determine traveler mode and destination. As was the case with trip generation, trips are classified as work, shopping, and social/recreational. For both shopping and social/recreational trips, a traveler will have the option to choose a mode of travel (automobile, vanpool, or bus) as well as a destination (different shopping centers). In contrast, work trips offer only the mode option, because the choice of work location (destination) is usually a long-term decision that is beyond the time range of most traffic forecasts.

8.5.1 Methodological Approach

Following recent advances in the travel demand and traffic forecasting field, development of a model for mode/destination choice necessitates the use of some consistent theory of travelers' decision-making processes. Of the decision-making theories available, one that is based on the microeconomic concept of utility maximization has enjoyed widespread acceptance in mode/destination choice modeling. The basic assumption is that a traveler will select the combination of mode and destination that provides the most utility. The problem then becomes one of developing an expression for the utility provided by various mode and destination alternatives. Because it is unlikely that individual travelers' utility functions can ever be specified with certainty, the unspecifiable portion is assumed to be random. To illustrate this approach, consider a utility function of the following form:

$$V_{im} = \sum_i b_{mk} z_{imk} + \varepsilon_{im} \qquad (8.4)$$

where

V_{im} = total utility (specifiable and unspecifiable) provided by mode/destination alternative m to a traveler i,

b_{mk} = coefficient estimated from traveler survey data for mode/destination alternative m corresponding to mode/destination or traveler characteristic k,

z_{imk} = traveler or mode/destination characteristic k (income, travel time of mode, commercial floor space at destination, etc.) for mode/destination alternative m for traveler i, and

ε_{im} = unspecifiable portion of the utility of mode/destination alternative m for traveler i, which will be assumed to be random.

For notational convenience, define the specifiable nonrandom portion of utility V_{im} as

$$U_{im} = \sum_i b_{mk} z_{imk} \qquad (8.5)$$

With these definitions of utility, the probability that a traveler will choose some alternative, say m, is equal to the probability that the given alternative's utility is greater than the utility of all other possible alternatives. The probabilistic component arises from the fact that the unspecifiable portion of the utility expression is not known and is assumed to be a random variable. The basic probability statement is

$$P_{im} = \text{prob}\left[U_{im} + \varepsilon_{im} > U_{is} + \varepsilon_{is} \right] \quad \text{for all } s \neq m \qquad (8.6)$$

where

P_{im} = probability that traveler i will select alternative m,
prob[·] = notation for probability,
s = notation for available alternatives, and
Other terms are as defined previously.

With this basic probability and utility expression and an assumed random distribution of the unspecifiable components of utility (ε_{im}), a probabilistic choice model can be derived and the coefficients in the utility function (b_{mk}'s in Eqs. 8.4 and 8.5) can be estimated with data collected from traveler surveys, along the same lines as for the coefficients in the trip generation models. A popular approach to deriving such a probabilistic choice model is to assume that the random, unspecifiable component of utility (ε_{im} in Eq. 8.4) is generalized extreme value distributed. With this assumption, a rather lengthy and involved derivation gives rise to the logit model formulation [McFadden 1981],

8.5 Mode and Destination Choice

$$P_{im} = \frac{e^{U_{im}}}{\sum_{s} e^{U_{sm}}} \tag{8.7}$$

where e is the base of the natural logarithm ($e = 2.718$).

The coefficients that comprise the specifiable portion of utility (b_{mk}'s in Eq. 8.5) are estimated by the method of maximum likelihood (see Appendix 8B). For further information on logit model coefficient estimation and maximum-likelihood estimation techniques, refer to more specialized references [Washington et al. 2003; McFadden 1981].

8.5.2 Logit Model Applications

With the total number of vehicle-based trips made in specific time periods known (from trip generation models), the allocation of trips to vehicle-based modes and likely destinations can be undertaken by applying appropriate logit models. This process is best demonstrated by example.

EXAMPLE 8.5

A simple work-mode–choice model is estimated from data in a small urban area to determine the probabilities of individual travelers selecting various modes. The mode choices include automobile drive-alone (*DL*), automobile shared-ride (*SR*), and bus (*B*), and the utility functions are estimated as

$$U_{DL} = 2.2 - 0.2(\text{cost}_{DL}) - 0.03(\text{travel time}_{DL})$$
$$U_{SR} = 0.8 - 0.2(\text{cost}_{SR}) - 0.03(\text{travel time}_{SR})$$
$$U_{B} = -0.2(\text{cost}_{B}) - 0.01(\text{travel time}_{B})$$

where cost is in dollars and time is in minutes. Between a residential area and an industrial complex, 4000 workers (generating vehicle-based trips) depart for work during the peak hour. For all workers, the cost of driving an automobile is $4.00 with a travel time of 20 minutes, and the bus fare is 50 cents with a travel time of 25 minutes. If the shared-ride option always consists of two travelers sharing costs equally, how many workers will take each mode?

SOLUTION

Note that the utility function coefficients logically indicate that as modal costs and travel times increase, modal utilities decline and, consequently, so do modal selection probabilities (see Eq. 8.7). Substitution of cost and travel time values into the utility expressions gives

$$U_{DL} = 2.2 - 0.2(4) - 0.03(20) = 0.8$$
$$U_{SR} = 0.8 - 0.2(2) - 0.03(20) = -0.2$$
$$U_{B} = -0.2(0.5) - 0.01(25) = -0.35$$

Substituting these values into Eq. 8.7 yields

$$P_{DL} = \frac{e^{0.8}}{e^{0.8} + e^{-0.2} + e^{-0.35}} = \frac{2.226}{2.226 + 0.819 + 0.705} = \frac{2.226}{3.749} = 0.594$$

$$P_{SR} = \frac{0.819}{3.749} = 0.218$$

$$P_B = \frac{0.705}{3.749} = 0.188$$

Multiplying these probabilities by 4000 (the total number of workers departing in the peak hour) gives 2380 workers driving alone, 870 sharing a ride, and 750 using a bus.

EXAMPLE 8.6

A bus company is making costly efforts in an attempt to increase work-trip bus usage for the travel conditions described in Example 8.5. An exclusive bus lane is constructed that reduces bus travel time to 10 minutes.

   a. Determine the modal distribution of trips after the lane is constructed.
   b. If shared-ride vehicles are also permitted to use the facility, and travel time for bus and shared-ride modes is 10 min, determine the modal distribution.
   c. Given the conditions described in part (b), determine the modal distribution if the bus company offers free bus service.

SOLUTION

a. After the bus lane construction, the modal utilities of drive-alone and shared-ride are unchanged from those in Example 8.5. However, the bus modal utility becomes

$$U_B = -0.2(0.5) - 0.01(10) = -0.2$$

From Eq. 8.7 with 4000 work trips,

$$P_{DL} = \frac{e^{0.8}}{e^{0.8} + e^{-0.2} + e^{-0.2}} = \frac{2.226}{3.8635} = 0.576 \quad \text{and} \quad 0.576(4000) = 2304 \text{ trips}$$

$$P_{SR} = \frac{0.819}{3.8635} = 0.212 \quad \text{and} \quad 0.212(4000) = 848 \text{ trips}$$

$$P_B = \frac{0.819}{3.8635} = 0.212 \quad \text{and} \quad 0.212(4000) = 848 \text{ trips}$$

or an increase of 98 bus patrons from the prediction of Example 8.5.

b. With the bus lane opened to shared-ride vehicles, only the modal utility of shared rides will change from those in part (a):

$$U_{SR} = 0.8 - 0.2(2) - 0.03(10) = 0.1$$

From Eq. 8.7 with 4000 work trips,

$$P_{DL} = \frac{e^{0.8}}{e^{0.8} + e^{0.1} + e^{-0.2}} = \frac{2.226}{4.15} = 0.536 \text{ and } 0.536(4000) = \underline{2144 \text{ trips}}$$

$$P_{SR} = \frac{1.105}{4.15} = 0.267 \text{ and } 0.267(4000) = \underline{1068 \text{ trips}}$$

$$P_B = \frac{0.819}{4.15} = 0.197 \text{ and } 0.197(4000) = \underline{788 \text{ trips}}$$

or a loss of 60 bus patrons and a gain of 220 shared-ride users relative to part (a).

c. With free bus fare, the bus modal utility becomes [with other utilities unchanged from part (b)],

$$U_B = -0.2(0) - 0.01(10) = -0.1$$

From Eq. 8.7 with 4000 work trips,

$$P_{DL} = \frac{e^{0.8}}{e^{0.8} + e^{0.1} + e^{-0.1}} = \frac{2.226}{4.236} = 0.525 \text{ and } 0.525(4000) = \underline{2102 \text{ trips}}$$

$$P_{SR} = \frac{1.105}{4.236} = 0.261 \text{ and } 0.261(4000) = \underline{1043 \text{ trips}}$$

$$P_B = \frac{0.905}{4.236} = 0.214 \text{ and } 0.214(4000) = \underline{855 \text{ trips}}$$

or 67 more bus patrons compared with part (b).

EXAMPLE 8.7

Consider a residential area and two shopping centers that are possible destinations. From 7:00 to 8:00 P.M. on Friday night, 900 vehicle-based shopping trips leave the residential area for the two shopping centers. A joint shopping-trip mode-destination choice logit model (choice of either auto or bus) is estimated, giving the following coefficients:

Variable	Auto coefficient	Bus coefficient
Auto constant	0.6	0.0
Travel time in minutes	−0.3	−0.3
Commercial floor space (in thousands of ft^2)	0.012	0.012

Initial travel times to shopping centers 1 and 2 are as follows:

	By auto	By bus
Travel time to shopping center 1 (in minutes)	8	14
Travel time to shopping center 2 (in minutes)	15	22

If shopping center 2 has 400,000 ft² (37,160 m²) of commercial floor space and shopping center 1 has 250,000 ft² (23,225 m²), determine the distribution of Friday night shopping trips by destination and mode.

SOLUTION

The utility function coefficients indicate that as modal travel times increase, the likelihood of selecting the mode-destination combination declines. Also, as the destination's floor space increases, the probability of selecting that destination will increase, as suggested by the positive coefficient (+0.012). This reflects the fact that bigger shopping centers tend to have a greater variety of merchandise and hence are more attractive shopping destinations. Note that because this is a joint mode-destination choice model, there are four mode-destination combinations and four corresponding utility functions. Let U_{A1} be the utility of the auto mode to shopping center 1, U_{A2} the utility of the auto mode to shopping center 2, and U_{B1} and U_{B2} the utility of the bus mode to shopping centers 1 and 2, respectively. The utilities are

$$U_{A1} = 0.6 - 0.3(8) + 0.012(250) = 1.2$$
$$U_{B1} = -0.3(14) + 0.012(250) = -1.2$$
$$U_{A2} = 0.6 - 0.3(15) + 0.012(400) = 0.9$$
$$U_{B2} = -0.3(22) + 0.012(400) = -1.8$$

Substituting these values into Eq. 8.7 gives

$$P_{A1} = \frac{3.32}{6.246} = 0.532$$

$$P_{B1} = \frac{0.301}{6.246} = 0.048$$

$$P_{A2} = \frac{2.46}{6.246} = 0.394$$

$$P_{B2} = \frac{0.165}{6.246} = 0.026$$

Multiplying these probabilities by the 900 trips gives 479 trips by auto to shopping center 1, 43 trips by bus to shopping center 1, 355 trips by auto to shopping center 2, and 23 trips by bus to shopping center 2.

EXAMPLE 8.8

A joint mode-destination vehicle-based social/recreational trip logit model is estimated with the following coefficients:

Variable	Auto coefficient	Bus coefficient
Auto constant	0.9	0.0
Travel time in minutes	−0.22	−0.22
Population in thousands	0.16	0.16
Amusement floor space (in thousands of ft^2)	0.11	0.11

It is known that 500 social/recreational trips will depart from a residential area during the peak hour. There are three possible trip destinations with the following characteristics:

	Travel time (in minutes)		Population (in thousands)	Amusement floor space (in thousands of ft^2)
	Auto	Bus		
Destination 1	14	17	12.4	13.0
Destination 2	5	8	8.2	9.2
Destination 3	18	24	5.8	21.0

Determine the distribution of trips by mode and destination.

SOLUTION

As was the case for the shopping mode-destination model presented in Example 8.7, the signs of the coefficient estimates indicate that increasing travel time decreases an alternative's selection probability. Also, increasing population (reflecting an increase in social opportunities) and increasing amusement floor space (reflecting more recreational opportunities) both increase the probability of an alternative being selected. With two modes and three destinations, there are six alternatives, providing the following utilities (using the same subscripting notation as in Example 8.7):

$$U_{A1} = 0.9 - 0.22(14) + 0.16(12.4) + 0.11(13) = 1.234$$
$$U_{B1} = -0.22(17) + 0.16(12.4) + 0.11(13) = -0.326$$
$$U_{A2} = 0.9 - 0.22(5) + 0.16(8.2) + 0.11(9.2) = 2.124$$
$$U_{B2} = -0.22(8) + 0.16(8.2) + 0.11(9.2) = 0.564$$
$$U_{A3} = 0.9 - 0.22(18) + 0.16(5.8) + 0.11(21) = 0.178$$
$$U_{B3} = -0.22(24) + 0.16(5.8) + 0.11(21) = -2.042$$

Using Eq. 8.7 with 500 trips, the total number of trips to the six mode-destination alternatives are

$$P_{A1} = \frac{3.435}{15.607} = 0.220 \quad \text{and} \quad 0.220 \times 500 = \underline{110 \text{ trips}}$$

$$P_{B1} = \frac{0.722}{15.607} = 0.046 \quad \text{and} \quad 0.046 \times 500 = \underline{23 \text{ trips}}$$

$$P_{A2} = \frac{8.365}{15.607} = 0.536 \quad \text{and} \quad 0.536 \times 500 = \underline{268 \text{ trips}}$$

$$P_{B2} = \frac{1.76}{15.607} = 0.113 \quad \text{and} \quad 0.113 \times 500 = \underline{57 \text{ trips}}$$

$$P_{A3} = \frac{1.195}{15.607} = 0.077 \quad \text{and} \quad 0.077 \times 500 = \underline{38 \text{ trips}}$$

$$P_{B3} = \frac{0.13}{15.607} = 0.008 \quad \text{and} \quad 0.008 \times 500 = \underline{4 \text{ trips}}$$

EXAMPLE 8.9

Consider the situation described in Example 8.8. A labor dispute results in a bus union slowdown that increases travel times from the origin by 4, 2, and 8 minutes to destinations 1, 2, and 3, respectively. If the total number of trips remains constant, determine the resulting distribution of trips by mode and destination.

SOLUTION

The mode-destination utilities are computed as

$U_{A1} = 1.234$ (as in Example 8.8)

$U_{B1} = -0.22(21) + 0.16(12.4) + 0.11(13) = -1.206$

$U_{A2} = 2.124$ (as in Example 8.8)

$U_{B2} = -0.22(10) + 0.16(8.2) + 0.11(9.2) = 0.124$

$U_{A3} = 0.178$ (as in Example 8.8)

$U_{B3} = -0.22(32) + 0.16(5.8) + 0.11(21) = -3.802$

Applying Eq. 8.7 with 500 trips gives the following distribution of trips among mode-destination alternatives:

$$P_{A1} = \frac{3.435}{14.45} = 0.238 \quad \text{and} \quad 0.238 \times 500 = \underline{119 \text{ trips}}$$

$$P_{B1} = \frac{0.299}{14.45} = 0.021 \quad \text{and} \quad 0.021 \times 500 = \underline{10 \text{ trips}}$$

$$P_{A2} = \frac{8.365}{14.45} = 0.579 \quad \text{and} \quad 0.579 \times 500 = \underline{\underline{290 \text{ trips}}}$$

$$P_{B2} = \frac{1.132}{14.45} = 0.078 \quad \text{and} \quad 0.078 \times 500 = \underline{\underline{39 \text{ trips}}}$$

$$P_{A3} = \frac{1.195}{14.45} = 0.083 \quad \text{and} \quad 0.083 \times 500 = \underline{\underline{41 \text{ trips}}}$$

$$P_{B3} = \frac{0.022}{14.45} = 0.002 \quad \text{and} \quad 0.002 \times 500 = \underline{\underline{1 \text{ trip}}}$$

8.6 HIGHWAY ROUTE CHOICE

To summarize, the trip generation and mode-destination choice models give total highway traffic demand between a specified origin (the neighborhood from which trips originate) and a destination (the geographic area to which trips are destined), in terms of vehicles per some time period (usually vehicles per hour). With this information in hand, the final step in the sequential approach to travel demand and traffic forecasting—route choice—can be addressed. The result of the route choice decision will be traffic flow (generally in units of vehicles per hour) on specific highway routes, which is the desired output from the traffic forecasting process.

8.6.1 Highway Performance Functions

Route choice presents a classic equilibrium problem, because travelers' route choice decisions are primarily a function of route travel times, which are determined by traffic flow—itself a product of route choice decisions. This interrelationship between route choice decisions and traffic flow forms the basis of route choice theory and model development.

To begin modeling traveler route choice, a mathematical relationship between route travel time and route traffic flow is needed. Such a relationship is commonly referred to as a highway performance function. The most simplistic approach to formalizing this relationship is to assume a linear highway performance function in which travel time increases linearly with flow. An example of such a function is illustrated in Fig. 8.5. In this figure, the free-flow travel time refers to the travel time that a traveler would experience if no other vehicles were present to impede travel speed (as discussed in Chapter 5). This free-flow travel time is generally computed with the assumption that a vehicle travels at the posted speed limit of the route.

Although the linear highway performance function has the appeal of simplicity, it is not a particularly realistic representation of the travel time–traffic flow relationship. Recall that Chapter 5 presented a relationship between traffic speed and flow that is parabolic in nature, with significant reductions in travel speed occurring as the traffic flow approaches the roadway's capacity. This parabolic speed-flow relationship

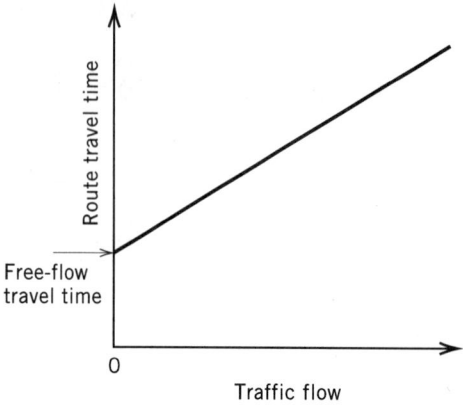

Figure 8.5 Linear travel time–traffic flow relationship.

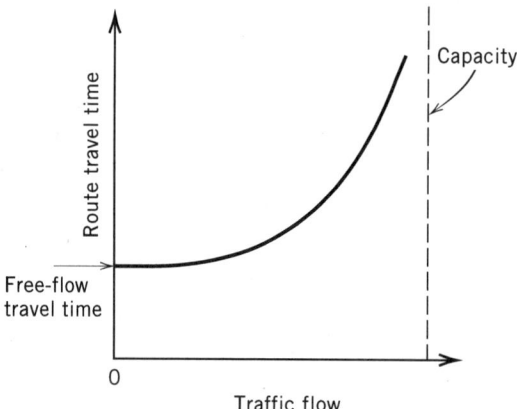

Figure 8.6 Nonlinear travel time–traffic flow relationship.

suggests a nonlinear highway performance function, such as that illustrated in Fig. 8.6. This figure shows route travel time increasing more quickly as traffic flow approaches capacity, which is consistent with the parabolic relationship presented in Chapter 5.

Both linear and nonlinear highway performance functions will be demonstrated, through example, using two theories of travel route choice: user equilibrium and system optimization. For other theories of route choice, refer to Sheffi [1985].

8.6.2 User Equilibrium

In developing theories of traveler route choice, two important assumptions are usually made. First, it is assumed that travelers will select routes between origins and destinations on the basis of route travel times only (they will tend to select the route with the shortest travel time). This assumption is not terribly restrictive, because travel time obviously plays the dominant role in route choice; however, other, more subtle factors that may influence route choice (scenery, pavement conditions, etc.) are not accounted for. The second assumption is that travelers know the travel times that would be encountered on all available routes between their origin and destination. This is potentially a strong assumption, because a traveler may not have actually traveled on all available routes between an origin and destination and may repeatedly (day after day) choose one route based only on the perception that travel times on alternative routes are higher. However, in support of this assumption, studies have shown that travelers' perceptions of alternative route travel times are reasonably close to actual observed travel times [Mannering 1989].

With these assumptions, the theory of user-equilibrium route choice can be made operational. The rule of choice underlying user equilibrium is that travelers will select a route so as to minimize their personal travel time between the origin and destination. User equilibrium is said to exist when individual travelers cannot improve their travel times by unilaterally changing routes. Stated differently [Wardrop 1952], user equilibrium can be defined as follows:

The travel time between a specified origin and destination on all used routes is the same and is less than or equal to the travel time that would be experienced by a traveler on any unused route.

EXAMPLE 8.10

Two routes connect a city and a suburb. During the peak-hour morning commute, a total of 4500 vehicles travel from the suburb to the city. Route 1 has a 60-mi/h (96.5-km/h) speed limit and is 6 mi (9.65 km) in length; route 2 is 3 mi (4.83 km) in length with a 45-mi/h (72.4-km/h) speed limit. Studies show that the total travel time on route 1 increases two minutes for every additional 500 vehicles added. Minutes of travel time on route 2 increase with the square of the number of vehicles, expressed in thousands of vehicles per hour. Determine user-equilibrium travel times.

SOLUTION

Determining free-flow travel times, in minutes, gives

$$\text{Route 1:} \quad 6 \text{ mi}/(60 \text{ mi/h}) \times 60 \text{ min/h} = 6 \text{ min}$$
$$\text{Route 2:} \quad 3 \text{ mi}/(45 \text{ mi/h}) \times 60 \text{ min/h} = 4 \text{ min}$$

With these data, the performance functions can be written as

$$t_1 = 6 + 4x_1$$
$$t_2 = 4 + 4x_2^2$$

where

t_1, t_2 = average travel times on routes 1 and 2 in minutes, and
x_1, x_2 = traffic flow on routes 1 and 2 in thousands of vehicles per hour.

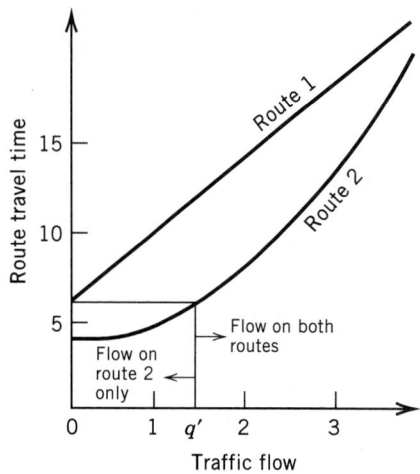

Figure 8.7 Illustration of performance curves for Example 8.10.

Also, the basic flow conservation identity is

$$q = x_1 + x_2 = 4.5$$

where q = total traffic flow between the origin and destination in thousands of vehicles per hour.

With Wardrop's definition of user equilibrium, it is known that the travel times on all used routes are equal. However, the first order of business is to determine whether or not both routes are used. Figure 8.7 gives a graphic representation of the two performance functions. Note that because route 2 has a lower free-flow travel time, any total origin-to-destination traffic flow less than q' (in Fig. 8.7) will result in only route 2 being used, because the travel time on route 1 would be greater even if only one vehicle used it. At flows of q' and above, route 2 is sufficiently congested, and its travel time sufficiently high, that route 1 becomes a viable alternative.

To check if the problem's flow of 4500 vehicles per hour exceeds q', the following test is conducted:

1. Assume that all traffic flow is on route 1. Substituting traffic flows of 4.5 and 0 into the performance functions gives $t_1(4.5) = 24$ min and $t_2(0) = 4$ min.
2. Assume that all traffic flow is on route 2, giving $t_1(0) = 6$ min and $t_2(4.5) = 24.25$ min.

Thus, because $t_1(4.5) > t_2(0)$ and $t_2(4.5) > t_1(0)$, both routes will be used. If $t_1(0)$ had been greater than $t_2(4.5)$, the 4500 vehicles would have been less than q' in Fig. 8.7, and only route 2 would have been used.

With both routes used, Wardrop's user-equilibrium definition gives

$$t_1 = t_2$$

or

$$6 + 4x_1 = 4 + x_2^2$$

From flow conservation, $x_1 + x_2 = 4.5$, so substituting,

$$6 + 4(4.5 - x_2) = 4 + x_2^2$$
$$x_2 = 2.899 \quad \text{or} \quad 2899 \text{ veh/h}$$
$$x_1 = 4.5 - x_2 = 4.5 - 2.899$$
$$= 1.601 \quad \text{or} \quad 1601 \text{ veh/h}$$

which gives average route travel times of

$$t_1 = 6 + 4(1.601) = \underline{\underline{12.4 \text{ min}}}$$
$$t_2 = 4 + (2.899)^2 = \underline{\underline{12.4 \text{ min}}}$$

EXAMPLE 8.11

Peak-hour traffic demand between an origin-destination pair is initially 3500 vehicles. The two routes connecting the pair have performance functions $t_1 = 2 + 3(x_1/c_1)$ and $t_2 = 4 + 2(x_2/c_2)$, where the t's are travel times in minutes, the x's are the peak-hour traffic volumes expressed in thousands, and the c's are the peak-hour route capacities expressed in thousands of vehicles per hour. Initially, the capacities of routes 1 and 2 are 2500 and 4000 veh/h, respectively. A reconstruction project reduces capacity on route 2 to 2000 veh/h. Assuming user equilibrium before and during reconstruction, what reduction in total peak-hour origin-destination traffic flow is needed to ensure that total travel times (summation of all $x_a t_a$'s, where a denotes route) during reconstruction are equal to those before reconstruction?

SOLUTION

First, focusing on the roads before reconstruction, a check to see if both routes are used gives (using performance functions)

$$t_1(3.5) = 6.2 \text{ min}, \quad t_2(0) = 4 \text{ min}$$
$$t_1(0) = 2 \text{ min}, \quad t_2(3.5) = 5.75 \text{ min}$$

which, because $t_1(3.5) > t_2(0)$ and $t_2(3.5) > t_1(0)$, indicates that both routes are used. Setting route travel times equal and substituting performance functions gives

$$2 + \frac{3}{2.5}(x_1) = 4 + \frac{2}{4}(x_2)$$

From conservation of flow, $x_2 = 3.5 - x_1$, so that

$$2 + 1.2x_1 = 4 + 0.5(3.5 - x_1)$$

Solving gives $x_1 = 2.206$ and $x_2 = 3.5 - 2.206 = 1.294$. For travel times,

$$t_1 = 2 + 1.2(2.206) = 4.647 \text{ min}$$
$$t_2 = 4 + 0.5(1.294) = 4.647 \text{ min}$$

The total peak-hour travel time before reconstruction will simply be the average route travel time multiplied by the number of vehicles:

$$\text{Total travel time} = 4.647(3500) = 16{,}264.5 \text{ veh-min}$$

During reconstruction, the performance function of route 1 is unchanged, but the performance function of route 2 is altered because of the reduction in capacity to

$$t_2 = 4 + \frac{2}{2}(x_2) = 4 + x_2$$

If it is assumed that both routes are used, $t_1 = t_2$. Also, it is known that the total travel time is

$$t_1(q) = t_2(q)$$
$$= 16{,}264.5 \text{ veh-min}$$

Using the performance function of route 2,

$$(4 + x_2)(q) = 16.2645 \text{ (thousands of vehicles)}$$
$$q = \frac{16.2645}{4 + x_2}$$

From $t_1 = t_2$, and $x_1 = q - x_2$ (flow conservation),

$$2 + 1.2x_1 = 4 + x_2$$
$$2 + 1.2(q - x_2) = 4 + x_2$$
$$q = 1.67 + 1.83x_2$$

Equating the two expressions for q gives

$$1.67 + 1.83x_2 = \frac{16.2645}{4 + x_2}$$
$$1.83x_2^2 + 8.99x_2 - 9.5845 = 0$$

which gives $x_2 = 0.901$, $q = 1.67 + 1.83(0.901) = 3.319$, and $x_1 = 3.319 - 0.901 = 2.418$. Because flow exists on both routes, the earlier assumption that both routes would be used is valid, and a reduction of <u>181 vehicles</u> (3500 − 3319) in peak-hour flow is needed to ensure equality of total travel times.

EXAMPLE 8.12

Two highways serve a busy corridor with a traffic demand that is fixed at 6000 vehicles during the peak hour. The performance functions for the two routes are $t_1 = 4 + 5(x_1/c_1)$ and $t_2 = 3 + 7(x_2/c_2)$, where t's are in minutes, and flows (x's) and capacities (c's) are in thousands of vehicles per hour. Initially, the capacities of routes 1 and 2 are 4400 veh/h and 5200 veh/h, respectively. If a highway reconstruction project cuts the capacity of

8.6 Highway Route Choice

route 2 to 2200 veh/h, how many additional vehicle hours of travel time will be added in the corridor assuming that user-equilibrium conditions hold?

SOLUTION

To determine the initial number of vehicle hours, first check to see if both routes are used:

$$t_1(6) = 10.82 \text{ min}, \quad t_2(0) = 3 \text{ min}$$
$$t_1(0) = 4 \text{ min}, \quad t_2(6) = 11.08 \text{ min}$$

Both routes are used, because $t_2(6) > t_1(0)$ and $t_1(6) > t_2(0)$. At user equilibrium, $t_1 = t_2$, so substituting performance functions gives

$$4 + \frac{5}{4.4}(x_1) = 3 + \frac{7}{5.2}(x_2)$$

With flow conservation, $x_2 = 6 - x_1$, so that

$$4 + 1.136(x_1) = 3 + 1.346(6 - x_1)$$
$$x_1 = 2.85$$

and

$$x_2 = 6 - 2.85$$
$$= 3.15$$

The total travel time in hours is $(t_1 x_1 + t_2 x_2)/60$ or, by substituting,

$$\frac{\{[4 + 1.136(2.85)]2850 + [3 + 1.346(3.15)]3150\}}{60} = 723.88 \text{ veh-h}$$

For the reduced-capacity case, the route usage check is

$$t_1(6) = 10.82 \text{ min}, \quad t_2(0) = 3 \text{ min}$$
$$t_1(0) = 4 \text{ min}, \quad t_2(6) = 22.09 \text{ min}$$

Again, both routes are used $[t_2(6) > t_1(0)$ and $t_1(6) > t_2(0)]$. Equating performance functions (because travel times are equal) and using flow conservation, $x_2 = 6 - x_1$,

$$4 + \frac{5}{4.4}(x_1) = 3 + \frac{7}{2.2}(x_2)$$
$$4 + 1.136 x_1 = 3 + 3.182(6 - x_1)$$
$$x_1 = 4.19$$

and

$$x_2 = 6 - 4.19 = 1.81$$

which gives a total travel time of $(t_1 x_1 + t_2 x_2)/60$ or, by substituting,

$$\frac{\{[4 + 1.136(4.19)]4190 + [3 + 3.182(1.81)]1810\}}{60} = 875.97 \text{ veh-h}$$

Thus the reduced capacity results in an additional 152.09 veh-h (875.97 − 723.88) of travel time.

8.6.3 Mathematical Programming Approach to User Equilibrium

Equating travel time on all used routes is a straightforward approach to user equilibrium, but can become cumbersome when many alternative routes are involved. The approach used to resolve this computational obstacle is to formulate the user equilibrium problem as a mathematical program. Specifically, user-equilibrium route flows can be obtained by minimizing the following function [Sheffi 1985]:

$$\min S(x) = \sum_n \int_0^{x_n} t_n(w)\,dw \tag{8.8}$$

where

n = a specific route, and
$t_n(w)$ = performance function corresponding to route n (w denotes flow, x_n's).

This function is subject to the constraints that the flow on all routes is greater than or equal to zero ($x_n \geq 0$) and that flow conservation holds (the flow on all routes between an origin and destination sums to the total number of vehicles, q, traveling between the origin and destination, $q = \sum_n x_n$).

Formulating the user equilibrium problem as a mathematical program allows an equilibrium solution to very complex highway networks (many origins and destinations) to be readily undertaken by computer. The reader is referred to Sheffi [1985] for an application of user-equilibrium principles to such a network.

EXAMPLE 8.13

Solve Example 8.10 by formulating user equilibrium as a mathematical program.

SOLUTION

From Example 8.10, the performance functions are

$$t_1 = 6 + 4x_1$$

$$t_2 = 4 + x_2^2$$

Substituting these into Eq. 8.8 gives

$$\min S(x) = \int_0^{x_1}(6 + 4w)\,dw + \int_0^{x_2}(4 + w^2)\,dw$$

The problem can be viewed in terms of x_2 only by noting that flow conservation implies $x_1 = 4.5 - x_2$. Substituting,

$$S(x) = \int_0^{4.5-x_2}(6+4w)dw + \int_0^{x_2}(4+w^2)dw$$

$$= 6w + 2w^2 \Big|_0^{4.5-x_2} + 4w + \frac{w^3}{3}\Big|_0^{x_2}$$

$$= 27 - 6x_2 + 40.5 - 18x_2 + 2x_2^2 + 4x_2 + \frac{x_2^3}{3}$$

To arrive at a minimum, the first derivative is set to zero, giving

$$\frac{dS(x)}{dx_2} = x_2^2 + 4x_2 - 20 = 0$$

which gives $x_2 = \underline{2899 \text{ veh/h}}$, the same value as found in Example 8.10. It can readily be shown that all other flows and travel times will also be the same as those computed in Example 8.10.

8.6.4 System Optimization

From an idealistic point of view, one can visualize a single route choice strategy that results in the lowest possible number of total vehicle hours of travel for some specified origin-destination traffic flow. Such strategy is known as a system-optimal route choice and is based on the choice rule that travelers will behave such that total system travel time will be minimized even though travelers may be able to decrease their own individual travel times by unilaterally changing routes. From this definition it is clear that system-optimal flows are not stable, because there will always be a temptation for travelers to switch to non–system-optimal routes in order to improve their travel times. Thus system-optimal flows are generally not a realistic representation of actual traffic. Nevertheless, system-optimal flows often provide useful comparisons with the more realistic user-equilibrium traffic forecasts.

The system-optimal route choice rule is made operational by the following mathematical program:

$$\min S(x) = \sum_n x_n t_n(x_n) \tag{8.9}$$

This program is subject to the constraints of flow conservation ($q = \sum_n x_n$) and nonnegativity ($x_n \geq 0$).

EXAMPLE 8.14

Determine the system-optimal travel time for the situation described in Example 8.10.

SOLUTION

Using Eq. 8.9 and substituting the performance functions for routes 1 and 2,

$$S(x) = x_1(6 + 4x_1) + x_2\left(4 + x_2^2\right)$$

$$= 6x_1 + 4x_1^2 + 4x_2 + x_2^3$$

From flow conservation, $x_1 = 4.5 - x_2$; therefore,

$$S(x) = 6(4.5 - x_2) + 4(4.5 - x_2)^2 + 4x_2 + x_2^3$$

$$= x_2^3 + 4x_2^2 - 38x_2 + 108$$

To find the minimum, the first derivative is set to zero, giving

$$\frac{dS(x)}{dx_2} = 3x_2^2 + 8x_2 - 38 = 0$$

which gives $x_2 = 2.467$ and $x_1 = 4.5 - 2.467 = 2.033$. For system-optimal travel times,

$$t_1 = 6 + 4(2.033) = 14.13 \text{ min}$$

$$t_2 = 4 + (2.467)^2 = 10.08 \text{ min}$$

which are not user-equilibrium travel times, because t_1 is not equal to t_2. In Example 8.10, the total user-equilibrium travel time is computed as 930 veh-h [4500(12.4)/60]. For the system-optimal total travel time [$(t_1 x_1 + t_2 x_2)/60$],

$$\frac{[2033(14.13) + 2467(10.08)]}{60} = \underline{\underline{893.2 \text{ veh-h}}}$$

Therefore, the system-optimal solution results in a systemwide travel time savings of 36.8 veh-h.

EXAMPLE 8.15

Two roads begin at a gate entrance to a park and take different scenic routes to a single main attraction in the park. The park manager knows that 4000 vehicles arrive during the peak hour, and he distributes these vehicles among the two routes so that an equal number of vehicles take each route. The performance functions for the routes are $t_1 = 10 + x_1$ and $t_2 = 5 + 3x_2$, with the x's expressed in thousands of vehicles per hour and the t's in minutes. How many vehicle-hours would have been saved had the park manager distributed the vehicular traffic so as to achieve a system-optimal solution?

SOLUTION

For the number of vehicle hours, assuming an equal distribution of traffic among the two routes,

$$\text{Route 1:} \quad \frac{x_1 t_1}{60} = \frac{2000[10 + (2)]}{60} = 400 \text{ veh-h}$$

$$\text{Route 2:} \quad \frac{x_2 t_2}{60} = \frac{2000[5 + 3(2)]}{60} = 366.67 \text{ veh-h}$$

for a total of 766.67 veh-h. With the system-optimal traffic distribution, the performance functions are substituted into Eq. 8.9, giving

$$S(x) = (10 + x_1)x_1 + (5 + 3x_2)x_2$$

With flow conservation, $x_1 = 4.0 - x_2$, so that

$$S(x) = 4x_2^2 - 13x_2 + 56$$

Setting the first derivative equal to zero,

$$\frac{dS(x)}{dx_2} = 8x_2 - 13 = 0$$

gives $x_2 = 1.625$ and $x_1 = 4 - 1.625 = 2.375$. The total travel times are

$$\text{Route 1:} \quad \frac{x_1 t_1}{60} = \frac{2375[10 + 2.375]}{60} = 489.84 \text{ veh-h}$$

$$\text{Route 2:} \quad \frac{x_2 t_2}{60} = \frac{1625[5 + 3(1.625)]}{60} = 267.45 \text{ veh-h}$$

which gives a total system travel time of 757.27 veh-h or a savings of 9.38 veh-h (766.67 − 757.29) over the equal distribution of traffic to the two routes.

EXAMPLE 8.16

During the peak hour, an urban freeway segment has a traffic flow of 4000 veh/h (2000 vehicles with one occupant and 2000 vehicles with two occupants). The freeway has five lanes, four of which are unrestricted (open to all vehicles regardless of vehicle occupancy) and one that is restricted for use by vehicles with two occupants. The performance functions for the length of this freeway segment are $t_u = 4 + 0.5x_u$ for the unrestricted lanes (all four combined) and $t_r = 4 + 2x_r$ for the restricted lane (t's are in minutes and x's in thousands of vehicles per hour). Determine the distribution of traffic among the lanes such that the total number of person hours is minimized, and compare the savings in person hours relative to a user equilibrium solution (assume that compliance is perfect and that no single-occupant vehicles use the restricted lane).

SOLUTION

As stated in the problem, the 2000 single-occupant vehicles must use the unrestricted lanes. Begin by determining the distribution of traffic that will minimize total person hours. Using the subscripts r for restricted lane, $u1$ for single-occupant vehicles using the unrestricted lanes, and $u2$ for two-occupant vehicles using the unrestricted lanes, total person hours can be written as

$$S(x) = x_r t_r \times 2 + x_{u2} t_u \times 2 + x_{u1} t_u \times 1$$

where

x_r = flow on the restricted lane (two-occupant vehicles only),
t_r = travel time on the restricted lane,
x_{u2} = flow of two-occupant vehicles on the unrestricted lanes,
t_u = travel time on the unrestricted lanes, and
x_{u1} = flow of single-occupant vehicles on the unrestricted lanes.

It is given that $t_u = 4 + 0.5x_u$, where $x_u = x_{u1} + x_{u2}$. And because $x_{u1} = 2.0$, $t_u = 4 + 0.5(2.0 + x_{u2})$. Substituting gives

$$S(x) = 2x_r(4 + 2x_r) + 2x_{u2}[4 + 0.5(2.0 + x_{u2})] + 2[4 + 0.5(2.0 + x_{u2})]$$
$$= 8x_r + 4x_r^2 + 10x_{u2} + x_{u_2}^2 + 10 + x_{u2}$$

The total number of two-occupant vehicles is 2000, so $x_r + x_{u2} = 2.0$. Substituting,

$$S(x) = 8(2 - x_{u2}) + 4(2 - x_{u2})^2 + 10x_{u2} + x_{u_2}^2 + 10 + x_{u2} = 5x_{u_2}^2 - 13x_{u2} + 42$$

Taking the first derivative,

$$\frac{dS(x)}{dx_{u2}} = 10x_{u2} - 13 = 0$$

which gives $x_{u2} = 1.3$, and so $x_r = 2.0 - 1.3 = 0.7$. With this, total person hours is [with $t_r = 4 + 2(0.7) = 5.4$ and $t_u = 4 + 0.5(3.3) = 5.65$]

$$2[5.4(700)] + 2[5.65(1300)] + 2000(5.65) = 33{,}550 \text{ person-min or } \underline{559.167 \text{ person-h}}$$

For the user-equilibrium solution, with 2000 vehicles on the unrestricted lanes, t_u can be written as

$$t_u = 4 + 0.5(2.0 + x_{u2}) = 5 + 0.5x_{u2}$$

To check if both two-occupant lane choices are used by two-occupant vehicles, note that when $x_{u2} = 2$ and $x_r = 0$, $t_u = 6$ and $t_r = 4$. And when $x_{u2} = 0$ and $x_r = 2$, $t_u = 5$ and $t_r = 8$, so both lane choices might be used by two-occupant vehicles. Equating travel times ($t_u = t_r$) gives

$$5 + 0.5x_{u2} = 4 - 2x_r$$

and with $x_r = 2 - x_{u2}$,

$$5 + 0.5x_{u2} = 4 - 2(2 - x_{u2})$$

Solving gives $x_{u2} = 1.2$ and $x_r = 2 - 1.2 = 0.8$. This produces user-equilibrium travel times $t_u = t_r = 5.6$. Total person hours for the user-equilibrium solution is then

$$2[5.6(2000)] + 5.6(2000) = 33{,}600 \text{ person-min or } \underline{560 \text{ person-h}}$$

So the savings is $\underline{0.833 \text{ person-h}}$ $(560 - 559.167)$ when person-hours are minimized relative to the user-equilibrium solution.

8.7 THE STATE OF TRAVEL DEMAND AND TRAFFIC FORECASTING IN PRACTICE

As mentioned earlier, the approach to travel demand and traffic forecasting presented in this chapter provides an excellent exposition of the principles underlying the problem. In practice, the most widely used approach to travel demand and traffic forecasting is a four-step procedure: trip generation, mode choice, destination choice, and traffic assignment. This differs from the three-step procedure presented in this chapter in that mode and destination choices are modeled separately. Separating the mode and destination choice is theoretically incorrect, but it tends to simplify the modeling process considerably and makes model implementation easier. Although a logit formulation is often used for the mode choice decision in practice, the destination choice is often determined by using the gravity model [Meyer and Miller 2001]. The gravity model is based on the gravitational modeling principles covered in physics (the gravitational forces of planets), where the likelihood of a trip going to a destination is a function of the distance from the trip origin and some measure of attractiveness (the equivalent of mass in gravitational theory) of the destination. The use of the gravity model is a relic that remains from the earliest travel demand and traffic forecasting efforts, before notions of utility maximization were fully implementable. Its continued use is not necessarily an indication of resistance in the field to more theoretically consistent modeling approaches (although some resistance to more advanced approaches certainly exists), but more a reflection of the complexity of the underlying travel decision problem and the methodological and computational barriers that have made it difficult for the profession to evolve beyond the trip generation, mode choice, destination choice, and route choice stepped process.

Currently, the travel demand and traffic forecasting profession is attempting to deal with some of the obvious limitations of the stepped process. These limitations include the difficulty of modeling trip chaining (the combining of trips, such as shopping on the way to or from work), the temporal problem (the ability of trip makers to delay their trips in response to congestion), equilibrium concerns (interaction among the models of the stepped modeling process), the use of the household as the decision-making unit instead of individual household members, the long-term effects of traveler decisions on residential and commercial locations (see Fig. 8.1), and the stability of model coefficients over time (as one forecasts into the future, it is a matter of some debate as to whether or not the estimated logit and regression model coefficients can be assumed to be the same as they were when the model was estimated). All of these limitations suggest that the stepped process is in need of considerable revision. However, the recent theoretical advances made in the field have proven difficult to implement, and, consequently, the stepped process remains the most practical approach to the travel demand and traffic forecasting problem.

APPENDIX 8A LEAST SQUARES ESTIMATION

Least squares regression is a popular method for developing mathematical relationships from empirical data. As mentioned earlier, it is a method that is well suited to the estimation of trip generation models. To illustrate the least squares approach, consider the hypothetical trip generation data presented in Table 8A.1, which could have been gathered from a typical survey of travelers.

To begin formalizing a mathematical expression, note that the objective is to predict the number of shopping trips made on a Saturday for each household, i; this number is referred to as the dependent variable (Y_i). This prediction is to be a function of the number of people in household i (z_i), which is referred to as the independent variable. A simple linear relationship between Y_i and z_i is

$$Y_i = b_0 + b_1 z_i \tag{8A.1}$$

where

Y_i = number of shopping trips made by household i,
b_0, b_1 = coefficients to be determined by estimation, and
z_i = number of people in household i.

Ideally, one wants to determine the b's in Eq. 8A.1 that will give predictions of the number of shopping trips (Y_i's) that are as close as possible to the actual observed number of shopping trips (Y_i's, as shown in Table 8A.1). The difference or deviation between the observed and predicted number of shopping trips can be expressed mathematically as

$$\text{Deviation} = Y_i - (b_0 + b_1 z_i) \tag{8A.2}$$

Table 8A.1 Example of Shopping Trip Generation Data

Household number, i	Number of shopping trips made all day Saturday, Y_i	People in household i, z_i
1	3	4
2	1	2
3	1	3
4	5	4
5	3	2
6	2	4
7	6	8
8	4	6
9	5	6
10	2	2

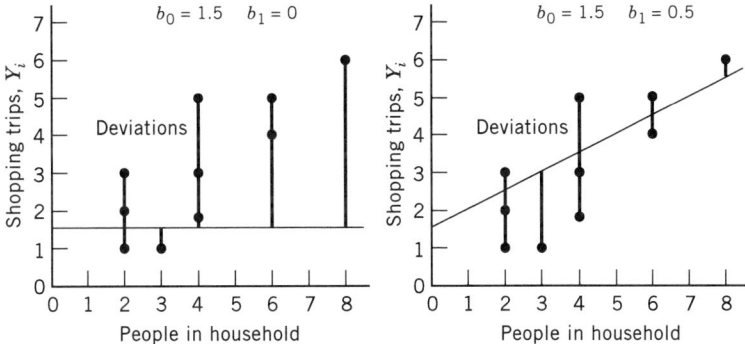

Figure 8A.1 Illustration of deviations.

Graphically, such deviations are illustrated in Fig. 8A.1 for two groups of b_0 and b_1 values. In the first illustration in this figure, $b_0 = 1.5$ and $b_1 = 0$, which implies that the number of household members does not affect the number of shopping trips made. The second illustration has $b_0 = 1.5$ and $b_1 = 0.5$ and, as can readily be seen, the deviations (differences between the points representing the observed number of shopping trips and the line representing the equation $b_0 + b_1 z_i$) are reduced relative to the first illustration. These two illustrations suggest the need for some method of determining the values of b_0 and b_1 that produce the smallest possible deviations relative to observed data. Such a method can be solved by a mathematical program whose objective is to minimize the sum of the square of deviations, or

$$\min S(b_0 + b_1) = \sum_i (Y_i - b_0 - b_1 z_i)^2 \tag{8A.3}$$

The minimization is accomplished by setting partial derivatives equal to zero:

$$\frac{\partial S}{\partial b_0} = -2 \sum_i (Y_i - b_0 - b_1 z_i) = 0 \tag{8A.4}$$

$$\frac{\partial S}{\partial b_1} = -2 \sum_i z_i (Y_i - b_0 - b_1 z_i) = 0 \tag{8A.5}$$

Solving these equations using gives

$$\sum_i Y_i - n b_0 - b_1 \sum_i z_i = 0 \tag{8A.6}$$

$$\sum_i z_i Y_i - b_0 \sum_i X_i - b_1 \sum_i z_i^2 = 0 \tag{8A.7}$$

where

n = number of households used to estimate the coefficients, and
Other terms are as defined previously.

Solving these equations simultaneously for b_0 and b_1 gives

$$b_1 = \frac{\sum_i (z_i - \bar{z})(Y_i - \bar{Y})}{\sum_i (z_i - \bar{z})^2} \quad (8A.8)$$

$$b_0 = \bar{Y} - b_1 \bar{z} \quad (8A.9)$$

where

$\bar{Y}$ = average number of shopping trips (averaged over all households, n),
$\bar{z}$ = average household income (averaged over all households, n), and
Other terms are as defined previously.

This approach to determining the values of estimable coefficients (b's) is referred to as least squares regression, and it can be shown that for the data values given in Table 8A.1, the smallest deviations between the number of predicted and actual shopping trips is given by the equation

$$Y_i = 0.33 + 0.7 z_i \quad (8A.10)$$

When many coefficient values (b's) must be determined, a matrix representation of the least squares solution is appropriate:

$$\mathbf{B} = (\mathbf{Z'Z})^{-1} \mathbf{Z'Y} \quad (8A.11)$$

where

$\mathbf{B}$ is an $n \times 1$ vector of coefficients (with n being the number of households),
$\mathbf{Z}$ is an $n \times k$ matrix of variables determining Y (where k is the number of variables, such as household income, number of people in the household, etc.),
$\mathbf{Z'}$ is the $k \times n$ transpose matrix of $\mathbf{Z}$, and
$\mathbf{Y}$ is an $n \times 1$ vector of the dependent variable (number of shopping trips in this example).

For additional information on least squares regression, refer to Washington, Karlaftis, and Mannering [2003].

APPENDIX 8B MAXIMUM-LIKELIHOOD ESTIMATION

Maximum-likelihood estimation is used extensively in the statistical analysis of traffic data [Washington et al. 2003]. The idea underlying maximum-likelihood estimation is that different statistical distributions generate different samples, and any one sample is more likely to come from some distributions than from others. To illustrate this, suppose there is a sample of six randomly drawn numbers, $Y_1, Y_2, \ldots, Y_6$, and there are two possible distributions that could generate these numbers as shown in Fig. 8B.1.

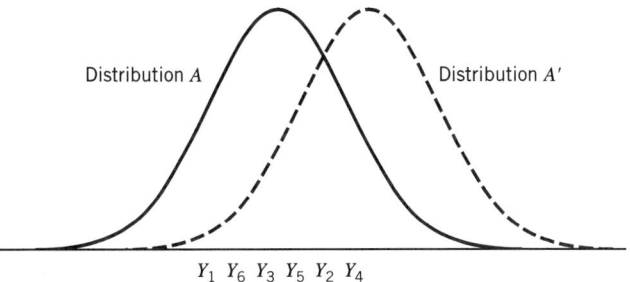

Figure 8B.1 Illustration of randomly drawn numbers and possible source distributions.

It is clear from Fig. 8B.1 that distribution A is much more likely to generate these six numbers than distribution A'. The objective of maximum-likelihood estimation is to estimate a coefficient vector, say B, that defines a distribution that is most likely to generate some observed data. To show how this is done, consider the Poisson regression of trip generation discussed in Section 8.4.2. The maximum-likelihood function can be written as a simple product of the probabilities of a Poisson distribution with coefficients B generating observed household trip generation. This is for a given trip type,

$$L(B) = \prod_i P(T_i) \qquad (8B.1)$$

where

B = vector of estimable coefficients,
$L(B)$ = likelihood function,
T_i = number of vehicle-based trips of a specific type (shopping, social/recreational, etc.) in some specified time period by household i, and
$P(T_i)$ = probability of household i making T trips.

Using the Poisson equation (Eq. 8.2), Eq. 8B.1 becomes

$$L(B) = \prod_i \frac{e^{-\lambda_i} \lambda_i^{T_i}}{T_i!} \qquad (8B.2)$$

where

λ_i = Poisson parameter for household i, which is equal to household i's expected number of vehicle-based trips in some specified time period, $E[T_i]$, and
Other terms are as defined previously.

With $\lambda_i = e^{BZ_i}$ as in Eq. 8.3,

$$L(B) = \prod_i \frac{e^{-e^{BZ_i}}\left(e^{BZ_i}\right)^{T_i}}{T_i!} \tag{8B.3}$$

where

Z_i = vector of household i characteristics determining trip generation for a given trip type, and

Other terms are as defined previously.

The problem then becomes one of finding the vector B that maximizes this function (maximizes the product of probabilities as shown in Eq. 8B.1). To do this, the natural logarithm is used to transform the likelihood function into a log-likelihood function (this does not affect the maximization process). In the Poisson regression case, this log transformation of Eq. 8B.3 gives

$$LL(B) = \sum_i \left[-e^{BZ_i} + T_i BZ_i - \ln(T_i!)\right] \tag{8B.4}$$

where

$LL(B)$ = log-likelihood function, and

Other terms are as defined previously.

Maximization of this expression with respect to B is undertaken by setting the first derivative to zero such that

$$\frac{\partial LL(B)}{\partial B} = \sum_i \left[-e^{BZ_i} + T_i\right] Z_i = 0 \tag{8B.5}$$

Equation 8B.5 can be solved numerically using standard software packages [Washington et al. 2003]. Using such a software package with the data in Table 8A.1, the estimated maximum-likelihood values of B give

$$\lambda_i = e^{BZ_i} = e^{-0.206 + 0.2z_i} \tag{8B.6}$$

where

z_i = people in household i (see Table 8A.1), and

Other terms are as defined previously.

Refer to Washington, Karlaftis, and Mannering [2003] for an extensive discussion of maximum-likelihood estimation.

NOMENCLATURE FOR CHAPTER 8

b_k estimated coefficients
B   vector of estimable coefficients
$L(\cdot)$ likelihood function
$LL(\cdot)$ log-likelihood function
P probability of an alternative being selected
$P(T_i)$ probability of T_i (a nonnegative integer) trips being generated by household i
q total origin-to-destination traffic flow
$S(\cdot)$ mathematical objective function
s notation for the set of available alternatives
T_i number of household trips generated per unit time for household i
t_n travel time on route n
U specifiable portion of an alternative's utility
V total utility for an alternative
w route flow operative for x_n
x_n traffic flow on route n
Y_i dependent variable for household i
z traveler, household, or alternative characteristic
Z_i vector of household i's characteristics determining trip generation
λ_i Poisson parameter for household i
ε unspecifiable portion of an alternative's utility (assumed to be a random variable)

REFERENCES

Hamed, M., and F. Mannering. "Modeling Travelers' Post-Work Activity Involvement: Toward a New Methodology." *Transportation Science*, vol. 17, no. 4, 1993.

Mannering, F. "Poisson Analysis of Commuter Flexibility in Changing Route and Departure Times." *Transportation Research*, vol. 23B, no. 1, 1989.

McFadden, D. "Econometric Models of Probabilistic Choice." In *Structural Analysis of Discrete Data with Econometric Applications*, edited by C. Manski and D. McFadden. Cambridge, MA: MIT Press, 1981.

Meyer, M., and E. Miller. *Urban Transportation Planning: A Decision-Oriented Approach*. New York: McGraw-Hill, 2001.

Sheffi, Y. *Urban Transportation Networks: Equilibrium Analysis with Mathematical Programming Models*. Englewood Cliffs, NJ: Prentice-Hall, 1985.

Wardrop, F. "Some Theoretical Aspects of Road Traffic Research." *Proceedings, Institution of Civil Engineers II*, vol. 1, 1952.

Washington, S., M. Karlaftis, and F. Mannering. *Statistical and Econometric Methods for Transportation Data Analysis*. Boca Raton, FL: Chapman & Hall/CRC, 2003.

PROBLEMS

8.1 A large retirement village has a total retail employment of 100. All 1700 of the households in this village consist of two nonworking family members with household income of $20,000. Assuming that shopping and social/recreational trip rates both peak during the same hour (for exposition purposes), predict the total number of peak-hour trips generated by this village using the trip generation models of Examples 8.1 and 8.2.

8.2 Consider the retirement village described in Problem 8.1. Determine the amount of additional retail employment (in the village) necessary to reduce the total predicted number of peak-hour shopping trips to 100.

8.3 A large residential area has 1500 households with an average household income of $15,000, an average household size of 5.2, and, on average, 1.2 working members. Using the model described in Example 8.2 (assuming it was estimated using zonal averages instead of individual households), predict the change in the number of peak-hour social/recreational trips if employment in the area increases by 20% and household income by 10%.

8.4 Consider the Poisson trip generation model in Example 8.4. Suppose that a household has four members with an annual income of $150,000 and lies in a neighborhood with a retail employment of 300. What is the expected number of peak-hour shopping trips, and what is the probability that the household will make more than one peak-hour shopping trip?

8.5 Consider a Poisson regression model for the number of social/recreational trips generated during a peak-hour period that is estimated by (see Eq. 8.3) $BZ_i = -0.75 + 0.025$(household size) $+ 0.008$(annual household income, in thousands of dollars) $+ 0.10$(number of nonworking household members). Suppose a household has five members (three of whom work) and an annual income of \$100,000. What is the expected number of peak-hour social/recreational trips, and what is the probability that the household will not make a peak-hour social/recreational trip?

8.6 If small express buses leave the origin described in Example 8.5 and all are filled to their capacity of 15 travelers, how many work trip vehicles leave from origin to destination in Example 8.5 during the peak hour?

8.7 Consider the conditions described in Example 8.5. If an energy crisis doubles the cost of the auto modes (drive-alone and shared-ride) and bus costs are not affected, how many workers will use each mode?

8.8 It is known that 4000 automobile trips are generated in a large residential area from noon to 1:00 P.M. on Saturdays for shopping purposes. Four major shopping centers have the following characteristics:

Shopping center	Distance from residential area (mi)	Commercial floor space (thousands of ft^2)
1	2.5	200
2	5.5	150
3	5.0	300
4	8.7	600

If a logit model is estimated with coefficients of -0.455 for distance and 0.0172 for commercial space (in thousands of ft^2), how many shopping trips will be made to each of the four shopping centers?

8.9 Consider the shopping trip situation described in Problem 8.8. Suppose that shopping center 3 goes out of business and shopping center 2 is expanded to 500,000 ft^2 of commercial space. What would be the new distribution of the 4000 Saturday afternoon shopping trips?

8.10 If shopping center 3 is closed (see Problem 8.9), how much commercial floor space is needed in shopping centers 1 and 2 to ensure that each of them has the same probability of being selected as shopping center 4?

8.11 Consider the situation described in Example 8.7. If the construction of a new freeway lowers auto and transit travel times to shopping center 2 by 20%, determine the new distribution of shopping trips by destination and mode.

8.12 Consider the conditions described in Example 8.7. Heavily congested highways have caused travel times to shopping center 2 to increase by 4 min for both auto and transit modes (travel times to shopping center 1 are not affected). In order to attract as many total trips (auto and transit) as it did before the congestion, how much commercial floor space must be added to shopping center 2 (given that the total number of departing shopping trips remains at 900)?

8.13 A total of 700 auto-mode social/recreational trips are made from an origin (residential area) during the peak hour. A logit model estimation is made, and three factors were found to influence the destination choice: (1) population at the destination, in thousands (coefficient $= 0.2$); (2) distance from origin to destination, in miles (coefficient $= -0.24$); and (3) square feet of amusement floor space (movie theaters, video game centers, etc.), in thousands (coefficient $= 0.09$). Four possible destinations have the following characteristics:

	Population (thousands)	Distance from origin (mi)	Amusement space (thousands of ft^2)
Destination 1	15.5	7.5	5
Destination 2	6.0	5	10
Destination 3	0.8	2	8
Destination 4	5.0	7	15

Determine the distribution of trips among possible destinations.

8.14 Consider the situation described in Problem 8.13. If a new 15,000-ft^2 arcade center is built at destination 3, determine the distribution of the 700 peak-hour social/recreational trips.

8.15 Note that with the situation described in Example 8.8, 26.6% [$(110 + 23)/500$] of all social/recreational trips are to destination 1. If the total number of trips remains constant, how much additional amusement floor space would have to be added to destination 1 to have it capture 40.0% of the total social/recreational trips?

8.16 Consider the situation described in Problem 8.13. Destination 2 currently attracts 148 of the 700 social/recreational trips. If the total number of trips remains constant, determine the amount of amusement floor space that must be added to destination 2 to attract a total of 250 social/recreational trips.

8.17 Two routes connect an origin and a destination, and the flow is 15,000 veh/h. Route 1 has a performance function $t_1 = 4 + 3x_1$, and route 2 has a function of $t_2 = b + 6x_2$, with the x's expressed in thousands of vehicles per hour and the t's in minutes.

(a) If the user equilibrium flow on route 1 is 9780 veh/h, determine the free-flow travel time on route 2 (b) and equilibrium travel times.

(b) If population declines reduce the number of travelers at the origin, and the total origin-destination flow is reduced to 7000 veh/h, determine user equilibrium travel times and flows.

8.18 An origin-destination pair is connected by a route with a performance function $t_1 = 8 + x_1$, and another with a function $t_2 = 1 + 2x_2$ (with x's in thousands of vehicles per hour and t's in minutes). If the total origin-destination flow is 4000 veh/h, determine user equilibrium and system-optimal route travel times, total travel time (in vehicle-minutes), and route flows.

8.19 Because of the great increase in vehicle-hours caused by the reconstruction project in Example 8.12, the state transportation department decides to regulate the flow of traffic on the two routes (until reconstruction is complete) to achieve a system-optimal solution. How many vehicle-hours will be saved during each peak-hour period if this strategy is implemented and travelers are not permitted to achieve a user equilibrium solution?

8.20 For Example 8.11, what reduction in peak-hour traffic demand is needed to ensure an equality of total vehicle travel time (in vehicle-minutes) assuming a system-optimal solution before and during reconstruction?

8.21 Two routes connect an origin and a destination. Routes 1 and 2 have performance functions $t_1 = 2 + x_1$ and $t_2 = 1 + x_2$, where the t's are in minutes and the x's are in thousands of vehicles per hour. The travel times on the routes are known to be in user equilibrium. If an observation for route 1 finds that the gaps between 40% of the vehicles are less than 5 seconds, estimate the volume and average travel times for the two routes. (*Hint:* Assume a Poisson distribution of vehicle arrivals as discussed in Chapter 5.)

8.22 Three routes connect an origin and a destination with performance functions $t_1 = 8 + 0.5x_1$, $t_2 = 1 + 2x_2$, and $t_3 = 3 + 0.75x_3$, with the x's expressed in thousands of vehicles per hour and the t's expressed in minutes. If the peak-hour traffic demand is 3000 vehicles, determine user equilibrium traffic flows.

8.23 Two routes connect a suburban area and a city, with route travel times (in minutes) given by the expressions $t_1 = 6 + 8(x_1/c_1)$ and $t_2 = 10 + 3(x_2/c_2)$, where the x's are expressed in thousands of vehicles per hour and the c's are the route capacities in thousands of vehicles per hour. Initially, the capacities of routes 1 and 2 are 4000 and 2000 veh/h, respectively. A reconstruction project on route 1 reduces the capacity to 3000 veh/h, but total traffic demand is unaffected. Observational studies note a 35.28-second increase in average travel time on route 1 and a 68.5% increase in flow on route 2 after reconstruction begins. User equilibrium conditions exist before and during reconstruction. If both routes are always used, determine equilibrium flows and travel times before and after reconstruction begins.

8.24 Three routes connect an origin and destination with performance functions $t_1 = 2 + 0.5x_1$, $t_2 = 1 + x_2$, and $t_3 = 4 + 0.2x_3$ (with t's in minutes and x's in thousands of vehicles per hour). Determine user equilibrium flows if the total origin-to-destination demand is (a) 10,000 veh/h and (b) 5000 veh/h.

8.25 For the routes described in Problem 8.24, what is the minimum origin-to-destination traffic demand (in vehicles per hour) that will ensure that all routes are used (assuming user equilibrium conditions)?

8.26 A multilane highway has two northbound lanes. Each lane has a capacity of 1200 vehicles per hour. Currently, northbound traffic consists of 2500 vehicles with 1 occupant, 500 vehicles with 2 occupants, 300 vehicles with 3 occupants, and 20 buses with 50 occupants each. The highway's performance function is $t = t_0[1 + 1.15(x/c)^{6.87}]$, where t is in minutes, t_0 is equal to 15 minutes, and x and c are volumes and capacities in vehicles per hour. An additional lane is being added (with 1200 veh/h capacity). What will the total person hours of travel be if the lane is (a) open to all traffic, (b) open to vehicles with 2 or more occupants only, and (c) open to vehicles with 3 or more occupants only? (Assume that all qualified higher-occupancy vehicles use only the new lane, no unqualified vehicles use the new lane, and there is no mode shift.)

8.27 Consider the new lane addition in Problem 8.26. First, suppose 500 one-occupant vehicle travelers take 10 buses (50 on each bus), and the new lane is open to vehicles with two or more occupants. What would the total person hours be? Second, referring back to part (a) of Problem 8.26, what is the minimum mode shift from one-occupant vehicles to buses (with 50 persons each) needed to ensure that the person hours of travel time on the highway with the new lane (which is restricted to vehicles with two or more occupants) is as low if all three lanes (the two existing lanes and the new lane) were open to all traffic? (Set up the equation, and solve to the nearest 100 one-occupant vehicles.)

8.28 Two routes connect an origin and destination with performance functions $t_1 = 5 + 3x_1$ and $t_2 = 7 + x_2$, with t's in minutes and x's in thousands of vehicles per hour. Total origin-destination demand is 7000 vehicles in the peak hour. What are user equilibrium and system-optimal route flows and total travel times?

8.29 Consider the conditions in Problem 8.28. What is the value of the derivative of the user equilibrium math program evaluated at the system-optimal solution with respect to x_1 (with x_1 equal to the system-optimal solution)?

8.30 Two routes connect an origin and a destination. Their performance functions are $t_1 = 3 + 1.5(x_1/c_1)^2$ and $t_2 = 5 + 4(x_2/c_2)$, with t's in minutes and x's and c's being route flows and capacities, respectively. The origin-destination demand is 6000 vehicles per hour, and c_1 and c_2 are equal to 2000 and 1500 vehicles per hour, respectively. Proposed capacity improvements will increase c_2 by 1000 vehicles per hour. It is known that the routes are currently in user equilibrium, and it is estimated that each 1-minute reduction in route travel time will attract an additional 500 vehicles per hour (from latent travel demand and mode shifts). What will the user equilibrium flows and total hourly origin-destination demand be after the capacity improvement?

8.31 Two routes connect an origin-destination pair with performance functions $t_1 = 5 + 4x_1$ and $t_2 = 7 + 2x_2$, with t's in minutes and x's in thousands of vehicles per hour. Assuming both routes are used, can user equilibrium and system-optimal solutions be equal at some feasible value of total origin-destination demand (q)? (Prove your answer.)

8.32 Three routes connect an origin-destination pair with performance functions $t_1 = 5 + 1.5x_1$, $t_2 = 12 + 3x_2$, and $t_3 = 2 + 0.2 x_3^2$ (with t's in minutes and x's in thousands of vehicles per hour). Determine user equilibrium flows if $q = 4000$ veh/h.

8.33 Two routes connect an origin-destination pair with performance functions $t_1 = 6 + 4x_1$ and $t_2 = 2 + 0.5x_2^2$ (with t's in minutes and x's in thousands of vehicles per hour). The origin-destination demand is 4000 veh/h at a travel time of 2 minutes, but for each additional minute beyond these 2 minutes, 100 fewer vehicles depart. Determine user equilibrium route flows and total vehicle travel time.

8.34 A freeway has six lanes, four of which are unrestricted (open to all vehicle traffic), and two of which are restricted lanes that can be used only by vehicles with two or more occupants. The performance function for the highway is $t = 12 + (2/NL)x$ (with t in minutes, NL being the number of lanes, and x in thousands of vehicles). During the peak hour, 3000 vehicles with one occupant and 4000 vehicles with two occupants depart for the destination. Determine the distribution of traffic between restricted and unrestricted lanes such that total person hours are minimized.

8.35 Two routes connect an origin-destination pair with performance functions $t_1 = 5 + (x_1/2)^2$ and $t_2 = 7 + (x_2/4)^2$ (with t's in minutes and x's in thousands of vehicles per hour). It is known that at user equilibrium, 75% of the origin-destination demand takes route 1. What percentage would take route 1 if a system-optimal solution were achieved, and how much travel time would be saved?

8.36 Two routes connect an origin-destination pair with performance functions $t_1 = 5 + 3.5x_1$ and $t_2 = 1 + 0.5x_2^2$ (with t's in minutes and x's in thousands of vehicles per hour). It is known that at $x_2 = 3$, the difference between the first derivatives of the system-optimal and user equilibrium math programs, evaluated with respect to x_2, is 7 $[dS(x)_{SO}/dx_2 - dS(x)_{UE}/dx_2 = 7]$. Determine the difference in total vehicle travel times (in vehicle minutes) between user equilibrium and system-optimal solutions.

Appendix A

Metric Example Problems

EXAMPLE 2.1

An 11.0-kN car is driven at sea level ($\rho = 1.2256$ kg/m^3) on a level paved surface. The car has $C_D = 0.38$ and 2.0 m^2 of frontal area. It is known that at maximum speed, 38 kW is being expended to overcome rolling and aerodynamic resistance. Determine the car's maximum speed.

SOLUTION

It is known that at maximum speed (V_m),

$$\text{available power} = R_a V_m + R_{rl} V_m$$

or

$$\text{available power} = \frac{\rho}{2} C_D A_f V_m^3 + f_{rl} W V_m$$

Substituting,

$$38{,}000 = \frac{1.2256}{2}0.38 \cdot 2.0 V_m^3 + 0.01\left(1 + \frac{V_m}{44.73}\right)11{,}000 V_m$$

or

$$38{,}000 = 0.4657 V_m^3 + 2.459 V_m^2 + 110 V_m$$

Solving for V_m gives

$$V_m = \underline{39.95 \text{ m/s}} \quad \text{or} \quad \underline{143.82 \text{ km/h}}$$

EXAMPLE 2.2

An 8.9-kN car has $C_D = 0.40$, $A_f = 2.0$ m^2, and an available tractive effort of 1135 N. If the car is traveling at an elevation of 1500 m ($\rho = 1.0567$ kg/m^3) on a paved surface at a speed of 110 km/h, what is the maximum grade that this car could ascend and maintain the 110-km/h speed?

SOLUTION

To maintain the speed, the available tractive effort will be exactly equal to the summation of resistances. Thus no tractive effort will remain for vehicle acceleration ($ma = 0$). Therefore, Eq. 2.2 can be written as

$$F = R_a + R_{rl} + R_g$$

For grade resistance (using Eq. 2.9),
$$R_g = WG = 8900G$$
For aerodynamic resistance (using Eq. 2.3),
$$R_a = \frac{\rho}{2}C_D A_f V^2$$
$$= \frac{1.0567}{2}(0.4)(2)(110 \times 1000/3600)^2$$
$$= 394.63 \text{ N}$$

and for rolling resistance (using Eq. 2.6),
$$R_{rl} = f_{rl}W$$
$$= 0.01\left(1 + \frac{110 \times 1000/3600}{44.73}\right) \times 8900$$
$$= 149.8$$

Therefore,
$$F = 1135 = 394.63 + 149.8 + 8900G$$
$$G = \underline{0.0664} \quad \text{or a} \quad \underline{6.64\% \text{ grade}}$$

EXAMPLE 2.3

An 11.0-kN car is designed with a 3.05-m wheelbase. The center of gravity is located 550 mm above the pavement and 1.00 m behind the front axle. If the coefficient of road adhesion is 0.6, what is the maximum tractive effort that can be developed if the car is (a) front-wheel drive and (b) rear-wheel drive?

SOLUTION

For the front-wheel–drive case Eq. 2.15 is used:
$$F_{max} = \frac{\mu W(l_r + f_{rl}h)/L}{1 + \mu h/L}$$

and, from Eq. 2.5, $f_{rl} = 0.01$ because $V = 0$ m/s, so
$$F_{max} = \frac{[0.6 \times 11{,}000 \times (2.050 + 0.01(0.550))]/3.050}{1 + (0.6 \times 0.550)/3.050}$$
$$= \underline{4013.70 \text{ N}}$$

For the rear-wheel–drive case, Eq. 2.14 can be used:
$$F_{max} = \frac{[0.6 \times 11{,}000 \times (1.000 - 0.01(0.550))]/3.050}{1 - (0.6 \times (0.550))/3.050}$$
$$= \underline{2413.13 \text{ N}}$$

EXAMPLE 2.4

It is known that an experimental engine has a torque curve of the form $M_e = an_e - bn_e^2$, where M_e is engine torque in N-m, n_e is engine speed in revolutions per second, and a and b are unknown parameters. If the engine develops a maximum torque of 125 N-m at 3200 rev/min (revolutions per minute), what is the engine's maximum power?

SOLUTION

At maximum torque, $n_e = 53.33$ rev/s (3200/60) and

$$\frac{dM_e}{dn_e} = 0 = a - 2bn_e$$

$$a = 2(53.33)b = 106.67b$$

Also, at maximum torque,

$$M_e = an_e - bn_e^2$$

$$125 = a(53.33) - b(53.33)^2$$

Using these two equations to solve for the two unknowns (a and b), we find that $b = 0.044$ and $a = 4.692$. Using Eq. 2.16 and $M_e = an_e - bn_e^2$,

$$P_e = \frac{2\pi(an_e - bn_e^2)n_e}{1000}$$

$$= \frac{2\pi(4.692n_e - 0.044n_e^2)n_e}{1000}$$

The first derivative of the power equation is used to solve for the engine speed at maximum power:

$$\frac{dP_e}{dn_e} = 0 = (0.00628)(9.384n_e - 0.132n_e^2)$$

$$n_e = 71.09 \text{ rev/s}$$

so the engine's maximum power is

$$P_e = \frac{2\pi(4.692n_e - 0.044n_e^2)n_e}{1000}$$

$$= \frac{2\pi[4.692(71.09) - 0.044(71.09)^2]71.09}{1000}$$

$$= \underline{49.64 \text{ kW}}$$

EXAMPLE 2.5

A car is traveling at 16 km/h on a roadway covered with hard-packed snow (coefficient of road adhesion of 0.20). The car has $C_D = 0.30$, $A_f = 2.0$ m², and $W = 13.3$ kN. The wheelbase is 3.05 m, and the center of gravity is 500 mm above the roadway surface and 1.27 m behind the front axle. The air density is 1.0567 kg/m³. The car's engine is producing 130 N-m of torque and is in a gear that gives an overall gear reduction ratio of 4.5 to 1, the wheel radius is 360 mm, and the mechanical efficiency of the driveline is 80%. If the driver needs to accelerate quickly to avoid an accident, what would the acceleration be if the car is (a) front-wheel drive and (b) rear-wheel drive?

SOLUTION

We begin by computing the resistances, tractive effort generated by the engine, and mass factor because all of these factors will be the same for both front- and rear-wheel drive.

The air resistance is (from Eq. 2.3)

$$R_a = \frac{\rho}{2} C_D A_f V^2$$

$$= \frac{1.0567}{2}(0.3)(2)(16 \times 1000/3600)^2$$

$$= 6.26 \text{ N}$$

The rolling resistance is (from Eq. 2.6)

$$R_{rl} = f_{rl} W$$

$$= 0.01 \left(1 + \frac{16 \times 1000/3600}{44.73}\right) \times 13{,}300$$

$$= 146.22 \text{ N}$$

The engine-generated tractive effort is (from Eq. 2.17)

$$F_e = \frac{M_e \varepsilon_0 \eta_d}{r}$$

$$= \frac{130(4.5)(0.8)}{0.36}$$

$$= 1300 \text{ N}$$

The mass factor is (from Eq. 2.20)

$$\gamma_m = 1.04 + (0.0025)\varepsilon_0^2$$

$$= 1.04 + (0.0025)(4.5)^2$$

$$= 1.091$$

Recall that to determine acceleration, we need the resistances (already computed) and the available tractive effort, F, which is the lesser of F_e and F_{max}. For the case of the front-wheel–drive car, Eq. 2.15 can be applied to determine F_{max}:

$$F_{max} = \frac{\mu W (l_r + f_{rl} h)/L}{1 + \mu h/L}$$

$$= \frac{[0.2 \times 13{,}300 \times (1.780 + 0.011(0.500))]/3.050}{1 + (0.2 \times 0.500)/3.050}$$

$$= 1507.76 \text{ N}$$

Thus for a front-wheel–drive car $F = 1300$ N (the lesser of 1300 and 1507.76), and the acceleration is (from Eq. 2.19)

$$F - \sum R = \gamma_m ma$$

$$a = \frac{F - \sum R}{\gamma_m m} = \frac{1300 - 152.48}{1.091(13,300/9.807)} = \underline{0.776 \text{ m/s}^2}$$

For the case of the rear-wheel–drive car, Eq. 2.14 can be applied to determine F_{max}:

$$F_{max} = \frac{[0.2 \times 13,300 \times (1.270 - 0.011(0.500))]/3.050}{1 - (0.2 \times 0.500)/3.050} = 1140.19 \text{ N}$$

Thus for a rear-wheel–drive car $F = 1140.19$ N (the lesser of 1300 and 1140.19), and the acceleration is (from Eq. 2.19)

$$a = \frac{F - \sum R}{\gamma_m m} = \frac{1140.19 - 152.48}{1.091(13,300/9.807)} = \underline{0.668 \text{ m/s}^2}$$

EXAMPLE 2.6

A car has a wheelbase of 2.5 m and a center of gravity that is 1.0 m behind the front axle at a height of 600 mm. If the car is traveling at 130 km/h on a road with poor pavement that is wet, determine the percentage of braking forces that should be allocated to the front and rear brakes (by the vehicle's braking system) to ensure that maximum braking forces are developed.

SOLUTION

The coefficient of rolling resistance is

$$f_{rl} = 0.01\left(1 + \frac{130 \times 0.2778}{44.73}\right) = 0.0181$$

and $\mu = 0.6$ from Table 2.4 (maximum because we want the tires to be at the point of impending slide). Applying Eq. 2.30 gives

$$BFR_{f/r\ max} = \frac{l_r + h(\mu + f_{rl})}{l_f - h(\mu + f_{rl})}$$

$$= \frac{1.500 + 0.600(0.6 + 0.0181)}{1.000 - 0.600(0.6 + 0.0181)}$$

$$= 2.97$$

Using Eq. 2.31, the percentage of the force allocated to the front brakes should be

$$PBF_f = 100 - \frac{100}{1 + BFR_{f/r\ max}}$$

$$= 100 - \frac{100}{1 + 2.97}$$

$$= \underline{74.81\%}$$

and using Eq. 2.32 (or simply $100 - PBF_f$), the percentage of the force allocated to the rear brakes should be

$$PBF_r = \frac{100}{1 + BFR_{f/r\ max}}$$

$$= \frac{100}{1 + 2.97}$$

$$= \underline{25.19\%}$$

EXAMPLE 2.7

A new experimental 11-kN car, with $C_D = 0.25$ and $A_f = 2.0\ m^2$, is traveling at 145 km/h down a 10% grade. The coefficient of road adhesion is 0.7 and the air density is 1.2256 kg/m³. The car has an advanced antilock braking system that gives it a braking efficiency of 100%. Determine the theoretical minimum stopping distance for the case where aerodynamic resistance is considered and the case where aerodynamic resistance is ignored.

SOLUTION

With aerodynamic resistance considered, Eq. 2.42 can be applied with $\gamma_b = 1.04$, $\theta_g = 5.71°$, and

$$f_{rl} = 0.01\left[1 + \frac{\frac{145 \times 1000/3600 + 0}{2}}{44.73}\right] = 0.0145$$

$$K_a = \frac{1.2256}{2}(0.25)(2) = 0.3064$$

Then

$$S = \frac{1.04(11,000)}{2(9.807)(0.3064)}$$

$$\times \ln\left[1 + \frac{0.3064(145 \times 1000/3600)^2}{(1.0)(0.7)(11,000) + (0.0145)(11,000) - 11,000\sin(5.71°)}\right]$$

$$= \underline{134.97\ m}$$

With aerodynamic resistance excluded, Eq. 2.43 is used:

$$S = \frac{1.04(145 \times 1000/3600)^2}{2(9.807)(0.7 + 0.0145 - \sin(5.71°))}$$

$$= \underline{139.87\ m}$$

EXAMPLE 2.8

A car is traveling at 130 km/h and has a braking efficiency of 80%. The brakes are applied to miss an object that is 45 m from the point of brake application, and the coefficient of road adhesion is 0.85. Ignoring aerodynamic resistance and assuming theoretical minimum stopping distance, estimate how fast the car will be going when it strikes the object if (a) the surface is level and (b) the surface is on a 5% upgrade.

SOLUTION

In both cases rolling resistance will be approximated as

$$f_{rl} = 0.01\left(1 + \frac{\frac{130 \times 1000/3600 + V_2}{2}}{44.73}\right) = 0.014 + 0.0001118V_2$$

Applying Eq. 2.43 for the level grade with $\gamma_b = 1.04$, $\theta_g = 0°$,

$$S = \frac{\gamma_b\left(V_1^2 - V_2^2\right)}{2g(\eta_b\mu + f_{rl} \pm \sin\theta_g)}$$

$$45 = \frac{1.04\left[(130 \times 1000/3600)^2 - V_2^2\right]}{2(9.807)\left[0.8(0.85) + (0.014 + 0.0001118V_2) \pm 0\right]}$$

$V_2 = \underline{26.70 \text{ m/s}}$ or $\underline{96.11 \text{ km/h}}$

On a 5% grade with $\theta_g = 2.86°$,

$$45 = \frac{1.04\left[(130 \times 1000/3600)^2 - V_2^2\right]}{2(9.807)\left[0.8(0.85) + (0.014 + 0.0001118V_2) + 0.05\right]}$$

$V_2 = \underline{25.89 \text{ m/s}}$ or $\underline{93.20 \text{ km/h}}$

EXAMPLE 2.9

A car ($W = 9.8$ kN, $C_D = 0.25$, $A_f = 2.0$ m^2) has an antilock braking system that gives it a braking efficiency of 100%. The car's stopping distance is tested on a level roadway with poor, wet pavement (with tires at the point of impending skid), and $\rho = 1.227$ kg/m^3. How inaccurate will the stopping distance predicted by the practical-stopping-distance equation be compared with the theoretical stopping distance, assuming the car is initially traveling 100 km/h? How inaccurate will the practical-stopping-distance equation be if the same car has a braking efficiency of 85%?

SOLUTION

First, to calculate the theoretical minimum stopping distance, Eq. 2.42 is applied with $\gamma_b = 1.04$, $\theta_g = 0°$, $\mu = 0.60$ (maximum for poor, wet pavement, from Table 2.4), and

$$f_{rl} = 0.01\left(1 + \frac{\frac{(100 \times 1000/3600 + 0)}{2}}{44.73}\right) = 0.0131$$

$$K_a = \frac{1.227}{2}(0.25)(2.0) = 0.3068$$

so from Eq. 2.42,

$$S = \frac{1.04(9800)}{2(9.807)(0.3068)} \ln\left[1 + \frac{0.3068(100 \times 1000/3600)^2}{(1.0)(0.60)(9800) + (0.0131)(9800) \pm 0}\right]$$

$$= 65.45 \text{ m}$$

For the same conditions but with a vehicle braking efficiency of 85%, Eq. 2.42 gives

$$S = \frac{1.04(9800)}{2(9.807)(0.3064)} \ln\left[1 + \frac{0.3064(100 \times 1000/3600)^2}{(0.85)(0.60)(9800) + (0.0131)(9800) \pm 0}\right]$$

$$= 76.46 \text{ m}$$

Now applying Eq. 2.46 (since $G = 0$) for the practical stopping distance, we find

$$d = \frac{V_1^2}{2a} = \frac{(100 \times 1000/3600)^2}{2(3.4)} = 113.47 \text{ m}$$

In the first case, the error is 48.02 m. In the case of 85% braking efficiency, the error is 37.01 m. Rearranging Eq. 2.47 to solve for a, we find that the stopping distances of 65.45 m and 76.46 m correspond to deceleration rates of 5.90 m/s^2 and 5.046 m/s^2, respectively. Studies [Fambro et al. 1997] have shown that most drivers decelerate at rates of 5.6 m/s^2 or greater for emergency stopping situations. Thus, this range of theoretical values is consistent with observed distances for situations in which minimum stopping distances are trying to be achieved. Comparing these theoretical values with the AASHTO recommended deceleration rate of 3.4 m/s^2, it is readily apparent that a considerable level of conservatism is built into the deceleration rate for practical stopping sight distance.

EXAMPLE 2.10

Two drivers each have a reaction time of 2.5 seconds. One is obeying a 90-km/h speed limit, and the other is traveling illegally at 120 km/h. How much distance will each of the drivers cover while perceiving/reacting to the need to stop?

SOLUTION

For the driver traveling at 90 km/h,

$$d_r = V_1 \times t_r = (90 \times 1000/3600)(2.5) = \underline{62.5 \text{ m}}$$

For the driver traveling at 120 km/h,

$$d_r = V_1 \times t_r = (120 \times 1000/3600)(2.5) = \underline{83.33 \text{ m}}$$

Therefore, driving at 120 km/h increases the distance covered during perception/reaction by a substantial 20.83 m.

EXAMPLE 3.1

A 200-m equal-tangent sag vertical curve has the PVC at station 3 + 700.000 and elevation 321 m. The initial grade is -3.5% and the final grade is $+0.5\%$. Determine the stationing and elevation of the PVI, the PVT, and the lowest point on the curve.

SOLUTION

Since the curve is equal tangent, the *PVI* will be 100 m (measured in a horizontal plane) from the *PVC*, and the *PVT* will be 200 m from the *PVC*. Therefore, the stationing of the *PVI* and *PVT* are 3 + 800.000 and 3 + 900.000, respectively. For the elevations of the *PVI* and *PVT*, it is known that a −3.5% grade can be equivalently written as −0.035 m/m. Since the *PVI* is 100 m from the *PVC*, which is known to be at elevation 321 m, the elevation of the *PVI* is

$$321 - 0.035 \text{ m/m} \times (100 \text{ m}) = \underline{317.5 \text{ m}}$$

Similarly, with the *PVI* at elevation 317.5 m, the elevation of the *PVT* is

$$317.5 + 0.005 \text{ m/m} \times (100 \text{ m}) = \underline{318.0 \text{ m}}$$

It is clear from the values of the initial and final grades that the lowest point on the vertical curve will occur when the first derivative of the parabolic function (Eq. 3.1) is zero because the initial and final grades are opposite in sign. When initial and final grades are not opposite in sign, the low (or high) point on the curve will not be where the first derivative is zero because the slope along the curve will never be zero. For example, a sag curve with an initial grade of −2.0% and a final grade of −1.0% will have its lowest elevation at the *PVT*, and the first derivative of Eq. 3.1 will not be zero at any point along the curve. However, in our example problem the derivative will be equal to zero at some point, so the low point will occur when

$$\frac{dy}{dx} = 2ax + b = 0$$

From Eq. 3.3 we have

$$b = G_1 = -0.035$$

and from Eq. 3.6,

$$a = \frac{0.005 - (-0.035)}{2(200)} = 0.0001$$

Substituting for *a* and *b* gives

$$\frac{dy}{dx} = 2(0.0001)x + (-0.035) = 0$$

$$x = 175 \text{ m}$$

This gives the stationing of the low point at $\underline{3 + 875.000}$ (175 m from the *PVC*). For the elevation of the lowest point on the vertical curve, the values of *a*, *b*, *c* (elevation of the *PVC*), and *x* are substituted into Eq. 3.1, giving

$$y = 0.0001(175)^2 + (-0.035)(175) + 321$$
$$= \underline{317.94 \text{ m}}$$

EXAMPLE 3.2

An equal-tangent vertical curve is to be constructed between grades of −2.0% (initial) and +1.0% (final). The *PVI* is at station 3 + 350.000 and at elevation 130 m. Due to a street crossing the roadway, the elevation of the roadway at station 3 + 415.000 must be at 131 m. Design the curve.

SOLUTION

The design problem is one of determining the length of the curve required to ensure that station 3 + 415.000 is at elevation 131 m. To begin, we use Eq. 3.1:

$$y = ax^2 + bx + c$$

From Eq. 3.3,

$$b = G_1 = -0.02$$

and from Eq. 3.6,

$$a = \frac{G_2 - G_1}{2L}$$

Substituting $G_1 = -0.02$ and $G_2 = 0.01$, we have

$$a = \frac{G_2 - G_1}{2L} = \frac{0.01 - (-0.02)}{2L} = \frac{0.015}{L}$$

Now note that c (the elevation of the PVC) in Eq. 3.1 will be equal to the elevation of the PVI plus $G_1 \times 0.5L$ (this is simply using the slope of the initial grade to determine the elevation difference between the PVI and PVC). With G_1 in m/m and the curve length L in meters, we have

$$c = 130 + 0.02(0.5L) = 130 + 0.01L$$

Finally, the value of x to be used in Eq. 3.1 will be $0.5L + 65$ because the point of interest (station 3 + 415.000) is 65 meters from the PVI (which is at station 3 + 350.000). Substituting $b = -0.02$, the expressions for a, c, and x, and $y = 131$ m (the given elevation) into Eq. 3.1 gives

$$131 = (0.015/L)(0.5L + 65)^2 + (-0.02)(0.5L + 65) + (130 + 0.01L)$$

$$0 = 0.00375L^2 - 1.325L + 63.375$$

Solving this quadratic equation gives $L = 57.04$ m (which is not feasible because we know that the point of interest is 65 m beyond the PVI, so the curve length must be longer than 57.04 m) or $L = 296.30$ m (which is the only feasible solution). This means that the curve must be 296.30 m long. Using this value of L,

elevation of PVC = $c = 130 + 0.01L = 130 + 2.963 = 132.96$ m

stationing of PVC = 3 + 350.000 − (296.30 / 2 = 3 + 210.850

elevation of PVT = elevation of PVI + (0.5L)G_2 = 130 + [0.5(296.30)](0.01)

= 131.48 m

stationing of PVT = 3 + 350.000 + (296.30) / 2 = 3 + 498.150

and

$$x = 0.5L + 65 = 148.15 + 65 = 213.15 \text{ m from the PVC}$$

To check the elevation of the curve at station 3 + 415.000, we apply Eq. 3.1 with $x = 213.15$:

$$y = ax^2 + bx + c$$
$$= \frac{0.03}{2(296.30)}(215.15)^2 + (-0.02)(213.15) + 132.96$$
$$= 131.00 \text{ m}$$

Therefore, all calculations are correct.

EXAMPLE 3.3

A curve has initial and final grades of +3% and −4%, respectively, and is 210 m long. The *PVC* is at elevation 100 m. Graph the vertical curve elevations and the slope of the curve against the length of the curve. Compute the *K*-value and use it to locate the high point of the curve (distance from the *PVC*).

SOLUTION

Recall that to find the slope at any point on the curve, we take the derivative of Eq. 3.1, which gives Eq. 3.2. To apply this equation, a and b need to be determined. From Eq. 3.6,

$$a = \frac{-0.04 - (0.03)}{2(210)} = -0.000166\overline{6}$$

and from Eq. 3.3,

$$b = G_1 = 0.03$$

The results of applying Eq. 3.2 and solving for the slope at all points along the curve, as well as a profile view of the curve itself (by application of Eq. 3.1), are shown graphically in Fig. A3.5 (exaggerating the vertical scale). Figure A3.5 shows the constant rate of change of the slope along the length of the curve. The circular points on the slope-of-curve line correspond to changes in grade of 1%, and these points occur at equal intervals of 30 m.

To show that this is consistent with the *K*-value, Eq. 3.10 gives

$$K = \frac{L}{A} = \frac{210}{|3 - (-4)|} = \underline{\underline{30 \text{ m}}}$$

This indicates that there should be a change in grade of 1% for every 30 m of curve length (measured in the horizontal plane), and this is consistent with Fig. A3.5. Applying Eq. 3.11 with the *K*-value of 30 m gives the high point at $\underline{90 \text{ m}}$ from the beginning of the curve ($x_{hl} = 30 \times 3 = 90$ m). This is shown in Fig. A3.5, where the slope of the curve at 90 m is zero (the same result obtained by setting the derivative of Eq. 3.2 equal to zero and solving for x). This result can also be explained conceptually based on the definition of the *K*-value. The *K*-value gives the horizontal distance required to effect a 1% change in the slope of the curve, and for this curve, that value is 30 m. Thus, to go from an initial grade (G_1) of 3% to a grade of 0% (the high point), it takes a horizontal distance equal to $K \times 3$, or 90 m.

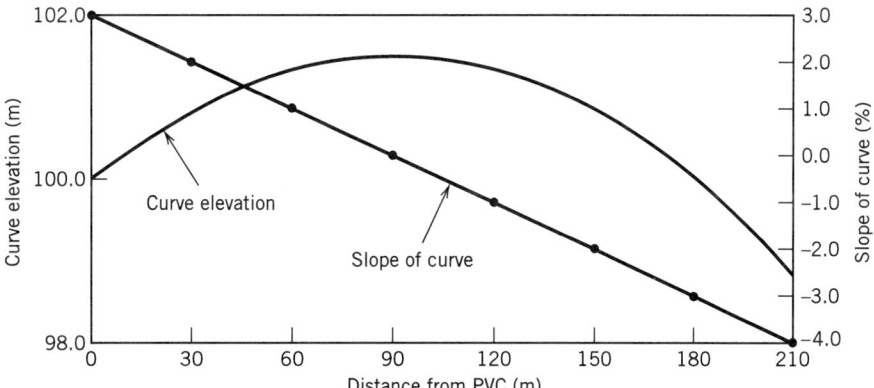

Figure A3.5 Profile view of vertical curve for Example 3.3 with graph of slope at all points along curve overlayed.

EXAMPLE 3.4

A vertical curve crosses a 1-m–diameter pipe at right angles. The pipe is located at station 3 + 420 and its centerline is at elevation 333 m. The *PVI* of the vertical curve is at station 3 + 400 and elevation 335 m. The vertical curve is equal tangent, 180 m long, and connects an initial grade of +1.20% and a final grade of −1.08%. Using offsets, determine the depth, below the surface of the curve, of the top of the pipe and determine the station of the highest point on the curve.

SOLUTION

The *PVC* is at station 3 + 310 (3 + 400 minus 0 + 090, which is half of the curve length), so the pipe is 110 m (3 + 420 minus 3 + 310) from the beginning of the curve (*PVC*). The elevation of the *PVC* will be the elevation of the *PVI* minus the drop in grade over one-half the curve length,

$$335 - (90 \text{ m} \times 0.012 \text{ m/m}) = 333.92 \text{ m}$$

Using this, the elevation of the initial tangent above the pipe is

$$333.92 + (110 \text{ m} \times 0.012 \text{ m/m}) = 335.24 \text{ m}$$

Using Eq. 3.7 to determine the offset above the pipe at $x = 110$ m (the distance of the pipe from the *PVC*), we have

$$Y = \frac{A}{200L}x^2$$

$$Y = \frac{|1.2 - (-1.08)|}{200(180)}(110)^2 = 0.77 \text{ m}$$

Thus the elevation of the curve above the pipe is 334.47 m (335.24 − 0.77). The elevation of the top of the pipe is 333.5 m (elevation of the centerline plus one-half of the pipe's diameter), so the pipe is 0.97 m below the surface of the curve (334.47 − 333.5).

To determine the location of the highest point on the curve, we find *K* from Eq. 3.10 as

$$K = \frac{180}{|1.2 - (-1.08)|} = 78.95$$

and the distance from the *PVC* to the highest point is (from Eq. 3.11)

$$x = K \times |G_1| = 78.95 \times 1.2 = 94.73 \text{ m}$$

This gives the station of the highest point at 3 + 404.73 (3 + 310 plus 0 + 094.73). Note that this example could also be solved by applying Eq. 3.1, setting Eq. 3.2 equal to zero (for determining the location of the highest point on the curve), and following the procedure used in Example 3.1.

EXAMPLE 3.5

A highway is being designed to AASHTO guidelines with a 120-km/h design speed, and at one section, an equal-tangent vertical curve must be designed to connect grades of +1.0% and −2.0%. Determine the minimum length of curve necessary to meet SSD requirements.

SOLUTION

If we ignore the effect of grades ($Gs = 0$), the SSD can be read directly from Table 3.1. In this case, the SSD corresponding to a speed of 120 km/h is 250 m. If we assume that $L >$ SSD (an assumption that is typically made), Eq. 3.15 gives

$$L_m = \frac{A \times SSD^2}{658} = \frac{3 \times 250^2}{658} = \underline{\underline{284.95 \text{ m}}}$$

Since $284.95 > 250$, the assumption that $L >$ SSD was correct.

EXAMPLE 3.6

Solve Example 3.5 using the K-values in Table 3.2.

SOLUTION

From Example 3.5, $A = 3$. For a 120-km/h design speed, $K = 95.0$ (from Table 3.2). Therefore, application of Eq. 3.17 gives

$$L_m = KA = 95.0(3) = \underline{\underline{285 \text{ m}}}$$

which is almost identical to the 284.95 m obtained in Example 3.5. This difference is due to rounding. In this example the rounded K of 95 was used as opposed to the calculated K of 94.98. The rounded values are typically used in design for computational convenience. Note, however, that fractional calculated values are always rounded up to the nearest integer value, to be conservative.

EXAMPLE 3.7

If the grades in Example 3.5 intersect at station $3 + 000$, determine the stationing of the PVC, PVT, and curve high point for the minimum curve length based on SSD requirements.

SOLUTION

Using the curve length from Example 3.6, $L = 285$ m. Since the curve is equal tangent (as are virtually all curves used in practice), one-half of the curve will occur before the PVI and one-half after, so that

$$PVC \text{ is at } 3 + 000 - L/2 = 3 + 000 \text{ minus } 0 + 142.5 = \underline{\underline{2 + 857.5}}$$

$$PVT \text{ is at } 3 + 000 + L/2 = 3 + 000 \text{ plus } 0 + 142.5 = \underline{\underline{3 + 142.5}}$$

For the stationing of the high point, Eq. 3.11 is used:

$$x_{hl} = K \times |G_1| = 95(1) = 95 \text{ m}$$

or

$$\text{station } 2 + 857.5 \text{ plus } 0 + 095 = \underline{\underline{2 + 952.5}}$$

EXAMPLE 3.8

An existing tunnel needs to be connected to a newly constructed bridge with sag and crest vertical curves. The profile view of the tunnel and bridge is shown in Fig. A3.8. Develop a vertical alignment to connect the tunnel and bridge by determining the highest possible common design speed for the sag and crest (equal-tangent) vertical curves needed. Compute the stationing and elevations of *PVC*, *PVI*, and *PVT* curve points.

SOLUTION

From left to right (see Fig. A3.8), a sag vertical curve (with subscript s) and a crest vertical curve (with subscript c) are needed to connect the tunnel and bridge. From given information, it is known that $G_{1s} = 0\%$ (the initial slope of the sag vertical curve) and $G_{2c} = 0\%$ (the final slope of the crest vertical curve). To obtain the highest possible design speed, we want to use all of the horizontal distance available. This means we want to connect the curve such that the *PVT* of the sag curve (PVT_s) will be the *PVC* of the crest curve (PVC_c). If this is the case, $G_{2s} = G_{1c}$, and since $G_{1s} = G_{2c} = 0$, $A_s = A_c = A$, the common algebraic difference in the grades.

Since 310 m separates the tunnel and bridge,

$$L_s + L_c = 310$$

Also, the summation of the end-of-curve offset for the sag curve and the beginning-of-curve offset (relative to the final grade) for the crest curve must equal 12 m. Using the equation for the final offset, Eq. 3.9, we have

$$\frac{AL_s}{200} + \frac{AL_c}{200} = 12$$

Rearranging,

$$\frac{A}{200}(L_s + L_c) = 12$$

and since $L_s + L_c = 310$,

$$\frac{A}{200} 310 = 12$$

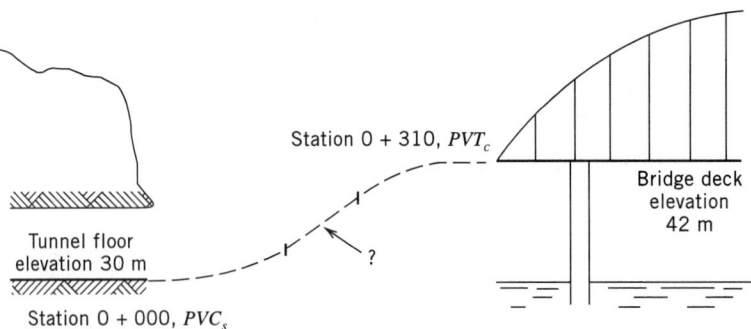

Figure A3.8 Profile view (vertical alignment diagram) for Example 3.8.

Solving for A gives 7.742%. The problem now becomes one of finding K-values that allow $L_s + L_c = 310$. Since $L = KA$ (Eq. 3.17), we can write

$$K_s A + K_c A = 310$$

Substituting $A = 7.742$,

$$K_s + K_c = 40$$

To find the highest possible design speed, Tables 3.2 and 3.3 are used to arrive at K-values to solve $K_s + K_c = 40$. From Tables 3.2 and 3.3 it is apparent that the highest possible design speed is 70 km/h, at which speed $K_c = 17$ and $K_s = 23$ (the summation of K's is 40).

To arrive at the stationing of curve points, we first determine curve lengths as

$$L_s = K_s A = 23(7.742) = 178 \text{ m}$$

$$L_c = K_c A = 17(7.742) = 132 \text{ m}$$

Since the station of PVC_s is $\underline{0 + 000}$ (given), it is clear that the $PVI_s = \underline{0 + 089}$, $PVT_s = PVC_c = \underline{0 + 178}$, $PVI_c = \underline{0 + 244}$, and $\overline{PVT_c = 0 + 310}$. For elevations, $PVC_s = PVI_s = \underline{30 \text{ m}}$ and $PVI_c = PVT_c = \underline{42 \text{ m}}$. Finally, the elevation of PVT_s and PVC_c can be computed as

$$30 + \frac{AL_s}{200} = 30 + \frac{7.742(178)}{200} = \underline{\underline{36.9 \text{ m}}}$$

EXAMPLE 3.9

Consider the conditions described in Example 3.8. Suppose a design speed of only 60 km/h is needed. Determine the lengths of curves required to connect the bridge and tunnel while keeping the connecting grade as small as possible.

SOLUTION

It is known that the 310 m separating the tunnel and bridge are more than enough to connect a 60-km/h alignment because Example 3.8 showed that 70 km/h is possible. Therefore, to connect the tunnel and bridge and keep the connecting grade as small as possible, we will place a constant-grade section between the sag and crest curves (as shown in Fig. A3.9).

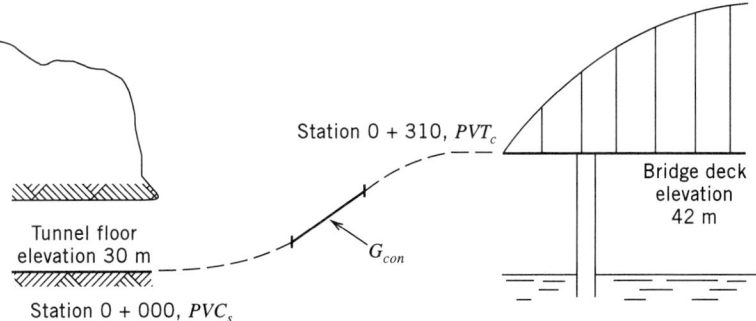

Figure A3.9 Profile view (vertical alignment diagram) for Example 3.9.

The elevation change will be the final offsets of the sag and crest curves plus the change in elevation resulting from the constant-grade section connecting the two curves. Let G_{con} be the grade of the constant-grade section. This means that $G_{2s} = G_{1c} = G_{con}$ and, since $G_{1s} = G_{2c} = 0$ (as in Example 3.8), $G_{con} = A_s = A_c = A$. The equation that will solve the vertical alignment for this problem is

$$\frac{AL_s}{200} + \frac{AL_c}{200} + \frac{A(310 - L_s - L_c)}{100} = 12$$

where the third term accounts for the elevation difference attributable to the constant-grade section connecting the sag and crest curves (the 100 in the denominator of this term converts A from percent to m/m). Using $L = KA$, we have

$$\frac{A^2 K_s}{200} + \frac{A^2 K_c}{200} + \frac{A(310 - K_s A - K_c A)}{100} = 12$$

From Table 3.2, $K_c = 11$, and from Table 3.3, $K_s = 18$. Putting these values in the above equation gives

$$0.145A^2 + 3.10A - 0.29A^2 = 12$$

$$-0.145A^2 + 3.10A - 12 = 0$$

Solving this gives $A = 5.076$ and $A = 16.303$; $A = 5.076$ is chosen because we want to minimize the grade. For this value of A, the curve lengths are

$$L_s = K_s A = 18(5.076) = \underline{91.368 \text{ m}}$$

$$L_c = K_c A = 11(5.076) = \underline{55.836 \text{ m}}$$

and the length of the constant-grade section will be 162.796 m. This means that about 8.264 m of the elevation difference will occur in the constant-grade section, with the remainder of the elevation difference attributable to the final curve offsets.

EXAMPLE 3.10

Two sections of highway are separated by 700 m, as shown in Fig. A3.10. Determine the curve lengths required for a 100-km/h vertical alignment to connect these two highway segments while keeping the connecting grade as small as possible.

SOLUTION

Let Y_{fc} and Y_{fs} be the final offsets of the crest and sag curves, respectively. Let G_{con} be the slope of a constant-grade section connecting the crest and sag curves (we will assume that the horizontal distance is sufficient to connect the highway with a 100-km/h alignment; if this assumption is incorrect, the following equations will produce an obviously erroneous answer and a lower design speed will have to be chosen). Finally, let Δy_{con} be the change in elevation over the constant-grade section, and let Δy_c and Δy_s be the changes in elevation due to the extended curve tangents. The elevation equation is then (see Fig. A3.10)

$$Y_{fc} + Y_{fs} + \Delta y_{con} + \Delta y_s = 10 + \Delta y_c$$

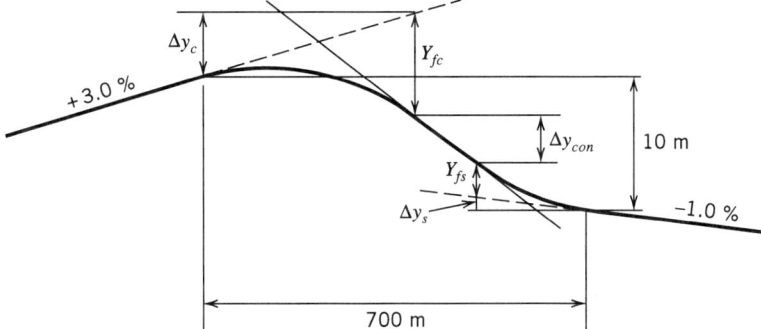

Figure A3.10 Profile view (vertical alignment diagram) for Example 3.10.

Substituting offset equations and equations for elevation changes (with subscripts c for crest and s for sag),

$$\frac{A_c L_c}{200} + \frac{A_s L_s}{200} + \frac{G_{con}(700 - L_c - L_s)}{100} + \frac{1.0 L_s}{100} = 10 + \frac{3.0 L_c}{100}$$

Using $L = KA$, this equation becomes

$$\frac{A_c^2 K_c}{200} + \frac{A_s^2 K_s}{200} + \frac{G_{con}(700 - K_c A_c - K_s A_s)}{100} + \frac{1.0 K_s A_s}{100} = 10 + \frac{3.0 K_c A_c}{100}$$

From Tables 3.2 and 3.3, $K_c = 52$ and $K_s = 45$. Substituting and defining A's (and arranging the equation such that G_{con} will be positive, and assuming G_{con} will be greater than 1%) gives

$$\frac{(3 + G_{con})^2 52}{200} + \frac{(G_{con} - 1)^2 45}{200} + \frac{G_{con}[700 - 52(3 + G_{con}) - 45(G_{con} - 1)]}{100}$$
$$+ \frac{1.0[45(G_{con} - 1)]}{100} = 10 + \frac{3.0[52(3 + G_{con})]}{100}$$

or

$$-0.485 G_{con}^2 + 5.890 G_{con} - 12.565 = 0$$

which gives $G_{con} = 2.761$ (the other possible solution is 9.383, which is rejected because we want to minimize the grade). Using $L = KA$ gives $L_c = \underline{299.572 \text{ m}}$ (52 × 5.761) and $L_s = \underline{79.245 \text{ m}}$ (45 × 1.761). Accordingly, the length of the constant-grade section is 321.183 m $(700 - L_c - L_s)$. Elevations and the locations of curve points can be readily computed with this information.

EXAMPLE 3.11

An equal-tangent crest vertical curve is 1200 m long and connects a +2.5% and a −1.5% grade. If the design speed of the roadway is 90 km/h, does this curve have adequate passing sight distance?

SOLUTION

To determine the length of curve required to provide adequate passing sight distance at a design speed of 90 km/h, we use $L = KA$ with $K = 438$ (as read from Table 3.4). This gives

$$L = 438(4.0) = 1752 \text{ m}$$

Since the curve is only 1200 m long, it is not long enough to provide adequate passing sight distance. Alternatively, the K-value for the existing design can be compared with that required for a PSD-based design. The K-value for the existing design is

$$K = \frac{1200}{4} = 300$$

Since the K-value of 300 for the existing curve design is less than 438, this curve does not provide adequate PSD for a 90-km/h design speed.

EXAMPLE 3.12

An equal-tangent sag curve has an initial grade of -4.0%, a final grade of $+3.0\%$, and a length of 385 m. An overhead guide sign is being placed directly over the PVI of this curve. At what height above the roadway should the bottom of this sign be placed?

SOLUTION

For this situation, Eq. 3.29 or 3.30 must be used to solve for the necessary clearance height based on stopping sight distance. Thus, the required SSD must be determined for the given sag curve specifications, based on the design speed. The design speed for the curve can be determined from the K-value by applying Eq. 3.10, as follows:

$$K = \frac{L}{A} = \frac{385}{|-4 - 3|} = 55$$

From Table 3.3, this K-value corresponds approximately to a design speed of 110 km/h ($K = 55$). For a 110-km/h design speed, the required stopping sight distance is 220 m. Since the curve length is greater than the required SSD (385 > 220), Eq. 3.29 applies:

$$L = \frac{A \times \text{SSD}^2}{800(H_c - 1.5)}$$

Rearranging this equation to solve for the clearance height, H_c, and substituting $A = 7\%$, SSD = 220 m, and $L = 385$ m gives

$$H_c = \frac{A \times \text{SSD}^2}{800L} + 1.5$$

$$= \frac{7 \times 220^2}{800(385)} + 1.5$$

$$= 2.6 \text{ m}$$

Although only 2.6 m is needed for SSD requirements, AASHTO 2001 recommends a minimum clearance height of 4.4 m to account for maximum vehicle heights. Thus, the bottom of the sign should be placed at least 4.4 m above the roadway surface (at the *PVI*), but desirably at a height of 5 m according to AASHTO 2001.

EXAMPLE 3.13

A roadway is being designed for a speed of 110 km/h. At one horizontal curve, it is known that the superelevation is 8.0% and the coefficient of side friction is 0.10. Determine the minimum radius of curve (measured to the traveled path) that will provide for safe vehicle operation.

SOLUTION

The application of Eq. 3.34 gives [with 0.2778 (1000/3600) converting from km/h to m/s]

$$R_v = \frac{V^2}{g\left(f_s + \dfrac{e}{100}\right)} = \frac{(110 \times 0.2778)^2}{9.807(0.10 + 0.08)} = \underline{\underline{528.982 \text{ m}}}$$

This value is the minimum radius, because radii smaller than 528.982 m will generate centripetal forces higher than those capable of being safely supported by the superelevation and the side frictional force.

EXAMPLE 3.14

A horizontal curve is designed with a 610-m radius. The curve has a tangent length of 120 m and the *PI* is at station 3 + 140. Determine the stationing of the *PT*.

SOLUTION

Equation 3.36 is applied to determine the central angle, Δ.

$$T = R \tan \frac{\Delta}{2}$$

$$120 = 610 \tan \frac{\Delta}{2}$$

$$\Delta = 22.26°$$

So, from Eq. 3.39, the length of the curve is

$$L = \frac{\pi}{180} R \Delta$$

$$L = \frac{3.1416}{180} 610(22.26) = 236.99 \text{ m}$$

Given that the tangent length is 120 m,

stationing $PC = 3 + 140$ minus $0 + 120 = 3 + 020$

Since horizontal curve stationing is measured along the alignment of the road,

$$\text{stationing } PT = \text{stationing } PC + L$$
$$= 3 + 020 \text{ plus } 0 + 236.99 = \underline{3 + 256.99}$$

EXAMPLE 3.15

A horizontal curve on a two-lane highway is designed with a 610-m radius, 3.6-m lanes, and a 100-km/h design speed. Determine the distance that must be cleared from the inside edge of the inside lane to provide a sufficient stopping sight distance.

SOLUTION

Because the curve radius is usually taken to the centerline of the roadway, $R_v = R - 3.6/2 = 610 - 1.8 = 608.2$ m, which gives the radius to the middle of the inside lane (the critical driver location). From Table 3.1, the SSD for a 100-km/h design speed is 185 m, so applying Eq. 3.42 gives

$$M_s = R_v \left(1 - \cos\frac{90(\text{SSD})}{\pi R_v}\right)$$

$$= 608.2 \left(1 - \cos\frac{90(185)}{\pi(608.2)}\right) = \underline{7.021 \text{ m}}$$

Therefore, 7.021 m must be cleared, as measured from the center of the inside lane, or 5.221 m as measured from the inside edge of the inside lane.

EXAMPLE 3.16

A two-lane highway (two 3.6-m lanes) has a posted speed limit of 80 km/h and, on one section, has both horizontal and vertical curves, as shown in Fig. A3.15. A recent daytime crash (driver traveling eastbound and striking a stationary roadway object) resulted in a fatality and a lawsuit alleging that the 80-km/h posted speed limit is an unsafe speed for the curves in question and was a major cause of the crash. Evaluate and comment on the roadway design.

SOLUTION

Begin with an assessment of the horizontal alignment. Two concerns must be considered: the adequacy of the curve radius and the superelevation, and the adequacy of the sight distance on the eastbound (inside) lane. For the curve radius, note from Fig. A3.15 that

$$L = \text{station of } PT - \text{station of } PC$$
$$L = 4 + 600 \text{ minus } 4 + 160 = 440 \text{ m}$$

Rearranging Eq. 3.39, we get

$$R = \frac{180}{\pi \Delta} L = \frac{180}{\pi(80)} (440) = 315.126 \text{ m}$$

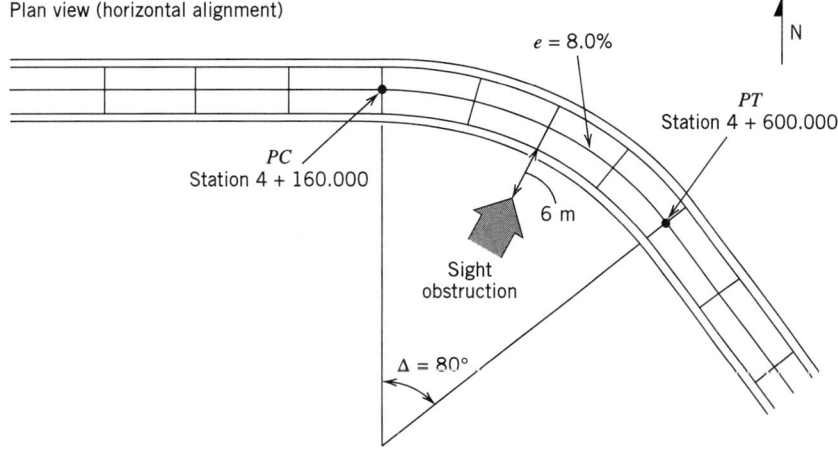

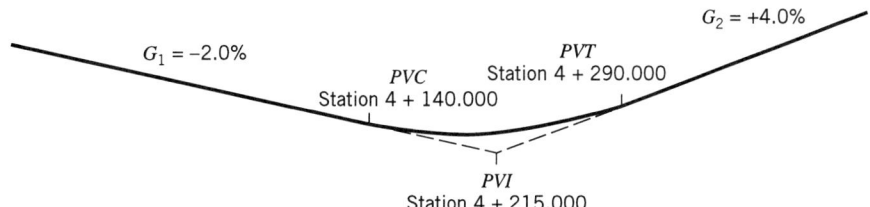

Figure A3.15 Horizontal and vertical alignment for Example 3.16.

Using the posted speed limit of 80 km/h with $e = 8.0\%$, we find that Eq. 3.34 can be rearranged to give (with the vehicle traveling in the middle of the inside lane, $R_v = R - $ half the lane width, or $R_v = 315.126 - 1.8 = 313.326$ m)

$$f_s = \frac{V^2}{gR_v} - e = \frac{(80 \times 0.2778)^2}{9.807(313.326)} - 0.08 = 0.0807$$

From Table 3.5, the maximum f_s for 80 km/h is 0.14. Since 0.0807 does not exceed 0.14, the radius and superelevation are sufficient for the 80-km/h design speed.

For sight distance, the available M_s is 6 m plus the 1.8-m distance to the center of the eastbound (inside) lane, or 7.8 m. Application of Eq. 3.43 gives

$$\text{SSD} = \frac{\pi R_v}{90}\left[\cos^{-1}\left(\frac{R_v - M_s}{R_v}\right)\right]$$

$$= \frac{\pi(313.326)}{90}\left[\cos^{-1}\left(\frac{313.326 - 7.8}{313.326}\right)\right]$$

$$= 140.119 \text{ m}$$

From Table 3.1, the required SSD at 80 km/h is 130 m, so the 140.119 m of SSD provided is sufficient. Turning to the sag vertical curve, the length of curve is

$$L = \text{station of } PVT - \text{station of } PVC$$

$$L = 4 + 290 \text{ minus } 4 + 140 = 150 \text{ m}$$

Using $A = 6\%$ (From Fig. A3.15) and applying Eq. 3.10, we obtain

$$K = \frac{L}{A} = \frac{150}{6} = 25$$

For the 80-km/h design speed, Table 3.3 indicates a necessary K-value of 30. Thus the K-value of 25 reveals that the curve is inadequate for the 80-km/h speed. However, because the crash occurred in daylight and sight distances on sag vertical curves are governed by nighttime conditions, this design did not contribute to the crash.

EXAMPLE 4.1

A tire with 690-kPa air pressure distributes a load over an area with a circular contact radius, a, of 125 mm. The pavement was constructed with a material that has a modulus of elasticity of 345,000 kPa and a Poisson ratio of 0.45. Calculate the radial-horizontal stress and deflection at a point on the pavement surface under the center of the tire load. Also, calculate the radial-horizontal stress and deflection at a point at a depth of 500 mm and a radial distance of 250 mm from the center of the tire load. (Use Ahlvin and Ulery equations.)

SOLUTION

With $z = 0$ mm and $r = 0$ mm,

$$\frac{z}{a} = \frac{0}{125} = 0 \quad \text{and} \quad \frac{r}{a} = \frac{0}{125} = 0$$

From Table 4.1, for the above values,

$$A = 1.0, \quad B = 0, \quad C = 0, \quad F = 0.5, \quad \text{and} \quad H = 2.0$$

The radial-horizontal stress as calculated from Eq. 4.5 is

$$\sigma_r = p[2\mu A + C + (1 - 2\mu)F]$$
$$= 690\{2(0.45)(1.0) + [1 - 2(0.45)](0.5)\}$$
$$= \underline{655 \text{ kPa}}$$

The deflection as calculated from Eq. 4.6 is

$$\Delta_z = \frac{p(1 + \mu)a}{E}\left[\frac{z}{a}A + (1 - \mu)H\right]$$

$$= \frac{690(1 + 0.45)125}{345,000}[0(1) + (1 - 0.45)2.0]$$

$$= \underline{0.399 \text{ mm}}$$

With $z = 500$ mm and $r = 250$ mm,

$$\frac{z}{a} = \frac{500}{125} = 4 \quad \text{and} \quad \frac{r}{a} = \frac{250}{125} = 2$$

From Table 4.1, for the above values,

$$A = 0.02193, \quad B = 0.03066, \quad C = -0.00956, \quad F = 0.00912, \quad \text{and} \quad H = 0.22188$$

The radial-horizontal stress is

$$\sigma_r = p[2\mu A + C + (1 - 2\mu)F]$$
$$= 690\{2(0.45)(0.02193) + (-0.00956) + [1 - 2(0.45)](0.00912)\}$$
$$= \underline{\underline{7.65 \text{ kPa}}}$$

The deflection is

$$\Delta_z = \frac{p(1 + \mu)a}{E}\left[\frac{z}{a}A + (1 - \mu)H\right]$$

$$= \frac{690(1 + 0.45)125}{345{,}000}[4(0.02193) + (1 - 0.45)(0.22188)]$$

$$= \underline{\underline{0.076 \text{ mm}}}$$

EXAMPLE 4.2

A pavement is to be designed to last 10 years. The initial PSI is 4.2 and the TSI (the final PSI) is determined to be 2.5. The subgrade has a soil resilient modulus of 105,000 kPa. Reliability is 95% with an overall standard deviation of 0.4. For design, the daily car, pickup truck, and light van traffic is 30,000, and the daily truck traffic consists of 1000 passes of single-unit trucks with two single axles and 350 passes of tractor semi-trailer trucks with single, tandem, and triple axles. The axle weights are

$$\text{cars, pickups, light vans} = \text{two 9-kN single axles}$$
$$\text{single-unit truck} = \text{36-kN steering, single axle}$$
$$= \text{100-kN drive, single axle}$$
$$\text{tractor semi-trailer truck} = \text{45-kN steering, single axle}$$
$$= \text{70-kN drive, tandem axle}$$
$$= \text{200-kN trailer, triple axle}$$

M_2 and M_3 are equal to 1.0 for the materials in the pavement structure. One hundred millimeters of hot-mix asphalt is to be used as the wearing surface and 250 mm of crushed stone as the subbase. Determine the thickness required for the base if soil cement is the material to be used.

SOLUTION

Because the axle-load equivalency factors presented in Tables 4.2, 4.3, and 4.4 are a function of the structural number (SN), we have to assume an SN to start the problem (later we will arrive at a structural number and check to make sure that it is consistent with our assumed value). A typical assumption is to let SN = 4. Given this, the 80-kN–equivalent single-axle load (80-kN ESAL) for cars, pickups, and light vans is

$$\text{9-kN single-axle equivalent} = 0.0002 \text{ (Table 4.2)}$$

This gives an 80-kN ESAL total of 0.0004 for each vehicle. For single-unit trucks,

$$36\text{-kN single-axle equivalent} = 0.041 \text{ (Table 4.2)}$$
$$100\text{-kN single-axle equivalent} = 2.090 \text{ (Table 4.2)}$$

This gives an 80-kN ESAL total of 2.131 for single-unit trucks. For tractor semi-trailer trucks,

$$45\text{-kN single-axle equivalent} = 0.102 \text{ (Table 4.2)}$$
$$70\text{-kN tandem-axle equivalent} = 0.057 \text{ (Table 4.3)}$$
$$200\text{-kN triple-axle equivalent} = 0.769 \text{ (Table 4.4)}$$

This gives an 80-kN ESAL total of 0.928 for tractor semi-trailer trucks. Note the comparatively small effect of cars and other light vehicles in terms of the 80-kN ESAL. This small effect underscores the nonlinear relationship between axle loads and pavement damage. For example, looking at Table 4.2 with SN = 4, a 160-kN single-axle load has 14.4 times the impact on pavement as an 80-kN single-axle load (twice the weight has 14.4 times the impact).

Given the computed 80-kN ESAL, the daily traffic on this highway produces an 80-kN ESAL total of 2467.8 (0.0004 × 30,000 + 2.131 × 1000 + 0.928 × 350). Traffic (total axle accumulations) over the 10-year design period will be

$$2467.8 \times 365 \times 10 = 9{,}007{,}470 \text{ 80-kN ESAL}$$

With an initial PSI of 4.2 and a TSI of 2.5, $\Delta PSI = 1.7$. Solving Eq. 4.7 for SN (using an equation solver on a calculator or computer) with $Z_R = -1.645$ (which corresponds to $R = 95\%$, as shown in Table 4.5) gives SN = 3.94 (Fig. 4.7 can also be used to arrive at an approximate solution for SN). Note that this is very close to the value that was assumed (SN = 4.0) to get the load equivalency factors from Tables 4.2, 4.3, and 4.4. If Eq. 4.7 gave SN = 5, we would go back and recompute total axle accumulations using the SN of 5 to read the axle-load equivalency factors in Tables 4.2, 4.3, and 4.4. Usually one iteration of this type is all that is needed. Later, Example 4.5 will provide a demonstration of this type of iteration.

Given that SN = 3.94, Eq. 4.9 can be applied with $a_1 = 0.44$ (surface course, hot-mix asphalt, Table 4.6), $a_2 = 0.20$ (base course, soil cement, Table 4.6), and $a_3 = 0.11$ (subbase, crushed stone, Table 4.6), $M_2 = 1.0$ (given), $M_3 = 1.0$ (given), $D_1 = 100$ mm (given), and $D_3 = 250$ mm (given).

$$SN = a_1 D_1 + a_2 D_2 M_2 + a_3 D_3 M_3$$
$$3.94 = 0.01732(100) + 0.00787 D_2(1.0) + 0.00433(250)(1.0)$$

Solving for D_2 gives $D_2 = 143$ mm. Using $D_2 = \underline{150 \text{ mm}}$ would be a conservative estimate and allow for variations in construction. Rounding up to the nearest 10 mm is a safe practice.

EXAMPLE 4.3–4.6

The current AASHTO design equations are based only on U.S. Customary units, so a metric solution is not presented here.

EXAMPLE 5.1

The speeds of five vehicles were measured (with radar) at the midpoint of a 0.8-kilometer section of roadway. The speeds for vehicles 1, 2, 3, 4, and 5 were 71, 68, 82, 79, and 74 km/h, respectively. Assuming all vehicles were traveling at constant speed over this roadway section, calculate the time-mean and space-mean speeds.

SOLUTION

For the time-mean speed, Eq. 5.5 is applied, giving

$$\bar{u}_t = \frac{\sum_{i=1}^{n} u_i}{n}$$

$$= \frac{71 + 68 + 82 + 79 + 74}{5}$$

$$= \underline{74.8 \text{ km/h}}$$

For the space-mean speed, Eq. 5.9 will be applied. This equation is based on travel time; however, because it is known that the vehicles were traveling at constant speed, we can rearrange this equation to utilize the measured speed, knowing that distance, l, divided by travel time, t, is equal to speed ($l/t_i = u$).

$$\bar{u}_s = \frac{1}{\frac{1}{n}\sum_{i=1}^{n}\left[\frac{1}{(l/t_i)}\right]} = \frac{1}{\frac{1}{n}\sum_{i=1}^{n}\left[\frac{1}{u_i}\right]}$$

$$= \frac{1}{\frac{1}{5}\left(\frac{1}{71} + \frac{1}{68} + \frac{1}{82} + \frac{1}{79} + \frac{1}{74}\right)}$$

$$= \frac{1}{0.013431}$$

$$= \underline{74.45 \text{ km/h}}$$

Note that the space-mean speed will always be lower than the time-mean speed, unless all vehicles are traveling at the exact same speed, in which case the two measures will be equal.

EXAMPLE 5.2

Vehicle time headways and spacings were measured at a point along a highway, from a single lane, over the course of an hour. The average values were calculated as 2.5 s/veh for headway and 60 m/veh for spacing. Calculate the average speed of the traffic.

SOLUTION

To calculate the average speed of the traffic, the fundamental relationship in Eq. 5.14 is used. To begin, the flow and density need to be calculated from the headway and spacing data. Flow is determined from Eq. 5.4 as

$$q = \frac{1}{2.5 \text{ s/veh}}$$

$$= 0.40 \text{ veh/s}$$

or, because the data were collected for an hour,

$$q = 0.40 \text{ veh/s} \times 3600 \text{ s/h}$$
$$= 1440 \text{ veh/h}$$

Density is determined from Eq. 5.13 as

$$k = \frac{1}{60 \text{ m/veh}}$$
$$= 0.0167 \text{ veh/m}$$

or, applying this spacing over the course of one kilometer,

$$k = 0.0167 \text{ veh/m} \times 1000 \text{ m/km}$$
$$= 16.7 \text{ veh/km}$$

Now applying Eq. 5.14, after rearranging to solve for speed, gives

$$u = \frac{q}{k}$$
$$= \frac{1440 \text{ veh/h}}{16.7 \text{ veh/km}}$$
$$= \underline{86.4 \text{ km/h}}$$

Note that the average speed of traffic can be determined directly from the average headway and spacing values, as follows:

$$u = \frac{\bar{s}}{\bar{h}}$$
$$= \frac{60 \text{ m/veh}}{2.5 \text{ s/veh}}$$
$$= \underline{24 \text{ m/s } (86.4 \text{ km/h})}$$

EXAMPLE 5.3

A section of highway is known to have a free-flow speed of 90 km/h and a capacity of 3300 veh/h. In a given hour, 2100 vehicles were counted at a specified point along this highway section. If the linear speed-density relationship shown in Eq. 5.15 applies, what would you estimate the space-mean speed of these 2100 vehicles to be?

SOLUTION

The jam density is first determined from Eq. 5.20 as

$$k_j = \frac{4 q_{cap}}{u_f}$$
$$= \frac{4 \times 3300}{90}$$
$$= 146.7 \text{ veh/km}$$

Rearranging Eq. 5.22 to solve for u,

$$\frac{k_j}{u_f}u^2 - k_j u + q = 0$$

Substituting,

$$\frac{146.7}{90}u^2 - 146.7u + 2100 = 0$$

which gives $u = \underline{72.14 \text{ km/h}}$ or $\underline{17.86 \text{ km/h}}$. Both of these speeds are feasible, as shown in Fig. 5.3.

EXAMPLE 5.4–5.13

Same as in Chapter 5.

EXAMPLE 6.1

A six-lane urban freeway (three lanes in each direction) is on rolling terrain with 3.4-m lanes, obstructions 0.6 m from the right edge of the traveled pavement, and 0.9 interchanges per kilometer. The traffic stream consists of primarily commuters. A directional weekday peak-hour volume of 2200 vehicles is observed, with 700 vehicles arriving in the most congested 15-min period. If the traffic stream has 15% large trucks and buses and no recreational vehicles, determine the level of service.

SOLUTION

Determine the free-flow speed according to Eq. 6.2.

$$FFS = BFFS - f_{LW} - f_{LC} - f_N - f_{ID}$$

with

$BFFS = 110$ km/h (urban freeway)
$f_{LW} = 2.1$ km/h (Table 6.3)
$f_{LC} = 2.6$ km/h (Table 6.4)
$f_N = 4.8$ km/h (Table 6.5)
$f_{ID} = 8.1$ km/h (Table 6.6)

$$FFS = 110 - 2.1 - 2.6 - 4.8 - 8.1 = 92.4 \text{ km/h}$$

Determine the flow rate according to Eq. 6.3:

$$v_p = \frac{V}{PHF \times N \times f_{HV} \times f_p}$$

with

$PHF = \dfrac{2200}{700 \times 4} = 0.786$
$N = 3$ (given)
$f_p = 1.0$ (commuters)
$E_T = 2.5$ (rolling terrain, Table 6.7)

From Eq. 6.5 we obtain

$$f_{HV} = \frac{1}{1 + 0.15(2.5 - 1)} = 0.816$$

So,

$$v_p = \frac{2200}{0.786 \times 3 \times 0.816 \times 1.0} = 1143.4 \rightarrow 1144 \text{ pc/h/ln}$$

Obtaining average passenger car speed from Fig. 6.2 (metric) for a flow rate of 1144 and FFS of 92.4 km/h yields 92.4 km/h. In this case the average speed is the same as FFS because the flow rate is low enough that it is still on the linear/flat part of the speed-flow curve.

Now density can be calculated with Eq. 6.6:

$$D = \frac{1144}{92.4} = 12.4 \text{ pc/km/ln}$$

From Table 6.1, it can be seen that this corresponds to LOS C [11.0 (maximum density for LOS B) < 12.4 < 16.0 (maximum density for LOS C)]. Thus, the freeway segment operates at level of service C.

This problem can also be solved graphically by applying Fig. 6.2 (metric). Using this figure, draw a vertical line up from 1144 pc/h/ln (on the figure's x-axis) and find that this line intersects the 92.4-km/h free-flow speed curve (between the 90 km/h and the 100 km/h curves) in the LOS C density region (the dashed diagonal lines).

EXAMPLE 6.2

Consider the freeway and traffic conditions in Example 6.1. At some point further along the roadway there is a 6% upgrade that is 2.4 km long. All other characteristics are the same as in Example 6.1. What is the level of service along this portion of the roadway, and how many vehicles can be added before the roadway reaches capacity (assuming that the proportion of vehicle types and the peak-hour factor remain constant)?

SOLUTION

To determine the LOS of this section of the freeway, we note that all adjustment factors are the same as those in Example 6.1 except f_{HV}, which must now be determined using an equivalency factor, E_T, drawn from specific-upgrade tables (in this case Table 6.8). From Table 6.8, $E_T = 3.5$, which gives

$$f_{HV} = \frac{1}{1 + 0.15(3.5 - 1)} = 0.727$$

So,

$$v_p = \frac{2200}{0.786 \times 3 \times 0.727 \times 1.0} = 1283.3 \rightarrow 1284 \text{ pc/h/ln}$$

The average passenger car speed remains 92.4 km/h; thus

$$D = \frac{1284}{92.4} = 13.9 \text{ pc/km/ln}$$

which still gives LOS C from Table 6.1.

To determine how many vehicles can be added before capacity is reached, the hourly volume at capacity must be computed. Recall that capacity corresponds to a volume-to-capacity ratio of 1.0 (the threshold between LOS E and LOS F). The maximum service flow rate that can be accommodated for a free-flow speed of 92.4 km/h is 2262 pc/h/ln (by linear interpolation). Equation 6.3 is rearranged and used to solve for the hourly volume based upon the maximum service flow rate:

$$V = v_p \times PHF \times N \times f_{HV} \times f_p \Rightarrow V = 2262 \times 0.786 \times 3 \times 0.727 \times 1.0$$

which gives $V = 3878$ veh/h. This means that about $\underline{1678 \text{ vehicles}}$ (3878 − 2200) can be added during the peak hour before capacity is reached. It should be noted that the assumption that the peak-hour factor will remain constant as the roadway approaches capacity is not very realistic. In practice it is observed that as a roadway approaches capacity, PHF gets closer to 1. This implies that the flow rate over the peak hour becomes more uniform. This uniformity is the result of, among other factors, motorists adjusting their departure and arrival times to avoid congested periods within the peak hour.

EXAMPLE 6.3

A four-lane undivided highway has 3.4-m lanes, with 1.2-m shoulders on the right side. There are 4.5 access points per kilometer, and the posted speed limit is 80 km/h. What is the estimated free-flow speed?

SOLUTION

This problem can be solved by direct application of Eq. 6.7 to arrive at an estimated free-flow speed.

$$FFS = BFFS - f_{LW} - f_{LC} - f_M - f_A$$

with

$BFFS = 88$ km/h (assume FFS = posted speed + 8 km/h)
$f_{LW} = 2.1$ km/h (from Table 6.3)
$f_{LC} = 0.6$ km/h (from Table 6.13 with $TLC = 1.2 + 1.8 = 3.0$, from Eq. 6.8, with $LC_L = 1.8$ m because the highway is undivided)
$f_M = 2.6$ km/h (from Table 6.14)
$f_A = 3.0$ km/h (from Table 6.15, by interpolation)

Substitution gives

$$FFS = 88 - 2.1 - 0.6 - 2.6 - 3.0 = \underline{79.7 \text{ km/h}}$$

which means that the more restrictive roadway characteristics relative to the base conditions result in a reduction in free-flow speed of 8.3 km/h.

EXAMPLE 6.4

A six-lane divided highway is on rolling terrain with 1.2 access points per kilometer and has 3.0-m lanes, with a 1.5-m shoulder on the right side and a 0.9-m shoulder on the left side. The peak-hour factor is 0.80, and the directional peak-hour volume is 3000 vehicles per hour. There are 6% large trucks, 2% buses, and 2% recreational vehicles. A significant percentage of nonfamiliar roadway users are in the traffic stream (the driver population adjustment factor is estimated as 0.95). No speed studies are available, but the posted speed limit is 90 km/h. Determine the level of service.

SOLUTION

We begin by determining FFS by applying Eq. 6.7:

$$FFS = BFFS - f_{LW} - f_{LC} - f_M - f_A$$

with

$BFFS = 98$ km/h (assume FFS = posted speed + 8 km/h)
$f_{LW} = 10.6$ km/h (from Table 6.3)
$f_{LC} = 1.5$ km/h (from Table 6.13, with $TLC = 1.5 + 0.9 = 2.4$, from Eq 6.8)
$f_M = 0.0$ km/h (from Table 6.14)
$f_A = 0.8$ km/h (from Table 6.15, by interpolation)

Substitution gives

$$FFS = 98 - 10.6 - 1.5 - 0.0 - 0.8 = 85.1 \text{ km/h}$$

Next we determine the analysis flow rate using Eq. 6.3:

$$v_p = \frac{V}{PHF \times N \times f_{HV} \times f_p}$$

with

$V = 3000$ veh/h (given) $f_P = 0.95$ (given)
$PHF = 0.80$ (given) $E_T = 2.5$ (Table 6.7)
$N = 3$ (given) $E_R = 2.0$ (Table 6.7)

From Eq. 6.5,

$$f_{HV} = \frac{1}{1 + 0.08(2.5 - 1) + 0.02(2 - 1)} = 0.877$$

Substitution gives

$$v_p = \frac{3000}{0.80 \times 3 \times 0.877 \times 0.95} = 1500.3 \text{ pc/h/ln}$$

Using Fig. 6.4 (metric), construct a speed-flow curve for $FFS = 85.1$ km/h (drawing it parallel to the curves for 80 km/h and 90 km/h) and note that the 1500.3-pc/h/ln flow rate intersects this curve in the LOS D density region. Therefore, this highway is operating at <u>LOS D</u>.

EXAMPLE 6.5

A local manufacturer wishes to open a factory near the section of highway described in Example 6.4. How many large trucks can be added to the peak-hour directional volume before capacity is reached? (Add only trucks and assume that the *PHF* remains constant.)

SOLUTION

Note that *FFS* will remain unchanged at 85.1 km/h. Table 6.11 shows that capacity for *FFS* = 90 km/h is 2100 pc/h/ln and for *FFS* = 80 km/h is 2000 pc/h/ln, so a linear interpolation gives us a capacity of 2051 pc/h/ln for *FFS* = 85.1 km/h. The current number of large trucks and buses in the peak-hour traffic stream is 240 (0.08 × 3000) and the current number of recreational vehicles is 60 (0.02 × 3000). Let us denote the number of new trucks as V_{nt}, so the combination of Eqs. 6.3 and 6.5 gives

$$v_p = \frac{V + V_{nt}}{(PHF)(N)\left[\dfrac{1}{1 + \left(\dfrac{240 + V_{nt}}{V + V_{nt}}\right)(E_T - 1) + \left(\dfrac{60}{V + V_{nt}}\right)(E_R - 1)}\right](f_p)}$$

with

v_p = 2051 pc/h/ln $\qquad f_p$ = 0.95 (Example 6.4)
V = 3000 veh/h (Example 6.4) $\qquad E_T$ = 2.5 (Example 6.4)
PHF = 0.80 (Example 6.4) $\qquad E_R$ = 2.0 (Example 6.4)
N = 3 (Example 6.4)

$$2051 = \frac{3000 + V_{nt}}{(0.80)(3)\left[\dfrac{1}{1 + \left(\dfrac{240 + V_{nt}}{3000 + V_{nt}}\right)(2.5 - 1) + \left(\dfrac{60}{3000 + V_{nt}}\right)(2 - 1)}\right](0.95)}$$

which gives (V_{nt} = 502.5) (≈502), which is the number of trucks that can be added to the peak-hour volume before capacity is reached.

EXAMPLE 6.6

Same as in Chapter 6.

EXAMPLE 6.7

The two-lane highway segment in Example 6.6 has the following additional characteristics: 3.4-m lanes, 0.6-m shoulders, access frequency of 6 per km, 50% no-passing zones, base *FFS* of 90 km/h, and a directional traffic split of 60/40. Using the analysis flow rates for *ATS* and *PTSF* from Example 6.6, determine the level of service for this two-lane highway segment.

SOLUTION

We begin by checking whether the highway segment is over capacity.

The analysis flow rates of 612 and 586 for *ATS* and *PTSF*, respectively, are both well below the two-way capacity of 3200 pc/h. Because both two-way flow rates are also below the directional capacity of 1700 pc/h, it is clear just by inspection that the directional capacity is also not exceeded.

Since the facility is not over capacity, we can proceed with the LOS determination. We first estimate the free-flow speed using Eq. 6.10.

$$FFS = BFFS - f_{LS} - f_A$$

with

$BFFS = 90$ km/h (given)
$f_{LS} = 4.9$ km/h (Table 6.16)
$f_A = 4.0$ km/h (from Table 6.15)

Substituting these values into Eq. 6.10 gives

$$FFS = 90 - 4.9 - 4.0 = 81.1 \text{ km/h}$$

The average travel speed will be calculated first, using Eq. 6.12.

$$ATS = FFS - 0.0125 v_p - f_{np}$$

with

$FFS = 81.1$ km/h (from previous calculation)
$v_p = 612$ pc/h (from Example 6.6)
$f_{np} = 4.3$ km/h (from Table 6.19, by linear interpolation for v_p and percent no-passing zones)

Substituting these values into Eq. 6.12 gives

$$ATS = 81.1 - 0.0125(612) - 4.3 = 69.15 \text{ km/h}$$

The percent time spent following is calculated next, using Eq. 6.13.

$$PTSF = BPTSF + f_{d/np}$$

BPTSF is calculated from Eq. 6.14.

$$BPTSF = 100(1 - e^{-0.000879 v_p})$$

Substituting the v_p value of 586 pc/h determined from Example 6.6 gives

$$BPTSF = 100(1 - e^{-0.000879(586)}) = 40.3\%$$

$f_{d/np}$ is found to be 17.0% from Table 6.20, again by linear interpolation of flow rate and percent no-passing zones. Note that a three-way linear interpolation is possible with this table if the directional split does not equal one of the five predefined categories. Substituting these values into Eq. 6.13 gives

$$PTSF = 40.3 + 17.0 = 57.3\%$$

We now determine the LOS for the calculated ATS and PTSF values from Table 6.21, because it is a Class I highway. From this table, the LOS is D. Although PTSF falls within the LOS C category, ATS falls within the LOS D category; thus, ATS governs the level of service for this two-lane highway under these roadway and traffic conditions.

EXAMPLE 6.8

A freeway is to be designed as a passenger-car–only facility for an AADT of 35,000 vehicles per day. It is estimated that the freeway will have a free-flow speed of 110 km/h. The design will be for commuters, and the peak-hour factor is estimated to be 0.85 with 65% of the peak-hour traffic traveling in the peak direction. Assuming Fig. 6.7 applies, determine the number of lanes required to provide at least LOS C using the highest annual hourly volume and the 30th highest annual hourly volume.

SOLUTION

By inspection of Fig. 6.7, the highest annual hourly volume has $K_1 = 0.148$. Application of Eq. 6.16 gives

$$\text{DDHV} = K_1 \times D \times \text{AADT}$$
$$= 0.148 \times 0.65 \times 35{,}000 = 3367 \text{ veh/h}$$

The next step is to determine the maximum service flow rate that can be accommodated at LOS C for $FFS = 110$ km/h. From Table 6.1, we see that this value is 1740 pc/h/ln. Thus, we must provide enough lanes such that the per-lane traffic flow is less than or equal to this value.

We can use Eq. 6.3 to find v_p, based on an assumed number of lanes. Comparing the calculated value of v_p to the maximum service flow rate of 1740 will determine whether we have an adequate number of lanes.

Assuming a four-lane freeway, Eq. 6.3 gives

$$v_p = \frac{3367}{0.85 \times 2 \times 1.0 \times 1.0} = 1980.6 \text{ pc/h/ln}$$

with

$V = 3367$ (DDHV from above)

$f_{HV} = 1.0$ (no heavy vehicles)

$f_p = 1.0$ (commuters)

This value is higher than 1740, so we need to provide more lanes. The calculation is repeated, this time with an assumed six-lane freeway:

$$v_p = \frac{3367}{0.85 \times 3 \times 1.0 \times 1.0} = 1320.4 \text{ pc/h/ln}$$

Since this value is less than 1740, a six-lane freeway (three lanes in each direction) is necessary to provide LOS C operation for the design traffic flow rate.

For the 30th highest hourly annual volume, Fig. 6.7 gives $K_{30} = K = 0.12$, which when used in Eq. 6.16 gives

$$\text{DDHV} = K \times D \times \text{AADT}$$
$$= 0.12 \times 0.65 \times 35{,}000 = 2730 \text{ veh/h}$$

Again applying Eq. 6.3, with an assumed four-lane freeway, yields

$$v_p = \frac{2730}{0.85 \times 2 \times 1.0 \times 1.0} = 1605.9 \text{ pc/h/ln}$$

This value is less than 1740, so a <u>four-lane freeway</u> (two lanes in each direction) is adequate for this design traffic flow rate.

This example demonstrates the impact that the chosen design traffic flow rate has on roadway design. Only a four-lane freeway is necessary to provide LOS C for the 30th highest annual hourly volume, as opposed to a six-lane freeway needed to satisfy the level-of-service conditions for the highest annual hourly volume.

EXAMPLE 7.1–7.11

Same as in Chapter 7.

EXAMPLE 7.12

Determine the yellow and all-red times for vehicles traveling on Vine and Maple Streets as shown in Fig A7.9.

SOLUTION

For the Vine Street phasing (applying Eqs. 7.23 and 7.24),

$$Y = 1.0 + \frac{(55 \times 1000/3600)}{2(3.05)}$$

$$= 3.51 \quad \rightarrow \quad \underline{4.0 \text{ s}} \text{ (rounding up to the nearest 0.5 s)}$$

$$AR = \frac{18.3 + 6}{55 \times 1000/3600}$$

$$= 1.59 \quad \rightarrow \quad \underline{2.0 \text{ s}} \text{ (rounding up to the nearest 0.5 s)}$$

For the Maple Street phasing (applying Eqs. 7.23 and 7.24),

$$Y = 1.0 + \frac{(65 \times 1000/3600)}{2(3.05)}$$

$$= 3.96 \quad \rightarrow \quad \underline{4.0 \text{ s}} \text{ (rounding up to the nearest 0.5 s)}$$

$$AR = \frac{11 + 6}{65 \times 1000/3600}$$

$$= 0.94 \quad \rightarrow \quad \underline{1.0 \text{ s}} \text{ (rounding up to the nearest 0.5 s)}$$

Note that separate calculations are usually required for exclusive left-turn phases, as vehicle approach speeds are often lower than for through vehicles and intersection crossing distances may be longer (due to the width of the opposing direction and the circular travel path).

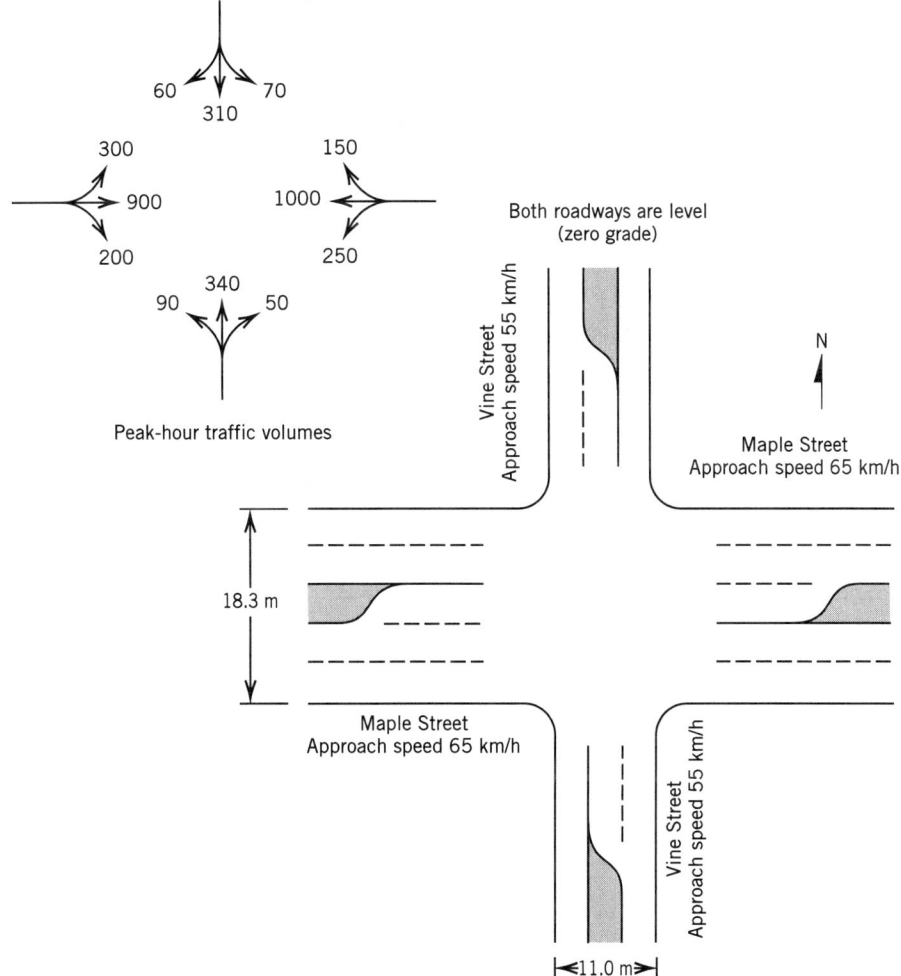

Figure A7.9 Intersection geometry and peak traffic volumes for example problems.

EXAMPLE 7.13

Determine the minimum amount of pedestrian green time required for the intersection of Vine and Maple streets. Assume a maximum of 15 pedestrians crossing either street during any one phase and a crosswalk width of 2.5 m.

SOLUTION

A pedestrian who crosses Maple Street will cross while Vine Street has a green interval. The minimum pedestrian green time needed on Vine Street is (using Eq. 7.25, as the effective crosswalk width is less than or equal to 3.05 m)

$$G_p = 3.2 + \frac{11}{1.2} + (0.27 \times 15) = \underline{\underline{16.42 \text{ s}}}$$

In Example 7.11, Vine Street was assigned 15.8 seconds of effective green time [13.8 seconds of displayed green time (from Eq. 7.3)]. This amount of time is insufficient for pedestrians crossing Maple Street. Therefore, the green time for this phase will have to be increased to accommodate crossing pedestrians, and the overall signal timing plan adjusted accordingly (although we will continue to use the previously computed green time in subsequent examples). The minimum pedestrian green time needed on Maple Street (for the through/right-turn phase, when pedestrian movement would be permitted) is

$$G_p = 3.2 + \frac{18.3}{1.2} + (0.27 \times 15) = \underline{\underline{22.50 \text{ s}}}$$

In Example 7.11, Maple Street was assigned 24.7 seconds of effective green time (23.7 seconds of displayed green time) for this phase, so this green time is adequate for pedestrians crossing Vine Street.

EXAMPLE 7.14–7.15

Same as in Chapter 7.

EXAMPLE 8.1–8.6

Same as in Chapter 8.

EXAMPLE 8.7

Consider a residential area and two shopping centers that are possible destinations. From 7:00 to 8:00 P.M. on Friday night, 900 vehicle-based shopping trips leave the residential area for the two shopping centers. A joint shopping-trip mode-destination choice logit model (choice of either auto or bus) is estimated, giving the following coefficients:

Variable	Auto coefficient	Bus coefficient
Auto constant	0.6	0.0
Travel time in minutes	−0.3	−0.3
Commercial floor space (in thousands of m^2)	0.12	0.12

Initial travel times to shopping centers 1 and 2 are as follows:

	By auto	By bus
Travel time to shopping center 1 (in minutes)	8	14
Travel time to shopping center 2 (in minutes)	15	22

If shopping center 2 has 40,000 m^2 of commercial floor space and shopping center 1 has 25,000 m^2, determine the distribution of Friday night shopping trips by destination and mode.

SOLUTION

The utility function coefficients indicate that as modal travel times increase, the likelihood of selecting the mode-destination combination declines. Also, as the destination's floor space increases, the probability of selecting that destination will increase, as suggested by the positive coefficient ($+0.12$). This reflects the fact that bigger shopping centers tend to have a greater variety of merchandise and hence are more attractive shopping destinations. Note that because this is a joint mode-destination choice model, there are four mode-destination combinations and four corresponding utility functions. Let U_{A1} be the utility of the auto mode to shopping center 1, U_{A2} the utility of the auto mode to shopping center 2, and U_{B1} and U_{B2} the utility of the bus mode to shopping centers 1 and 2, respectively. The utilities are

$$U_{A1} = 0.6 - 0.3(8) + 0.12(25) = 1.2$$
$$U_{B1} = -0.3(14) + 0.12(25) = -1.2$$
$$U_{A2} = 0.6 - 0.3(15) + 0.12(40) = 0.9$$
$$U_{B2} = -0.3(22) + 0.12(40) = -1.8$$

Substituting these values into Eq. 8.7 gives

$$P_{A1} = \frac{3.32}{6.246} = 0.532$$

$$P_{B1} = \frac{0.301}{6.246} = 0.048$$

$$P_{A2} = \frac{2.46}{6.246} = 0.394$$

$$P_{B2} = \frac{0.165}{6.246} = 0.026$$

Multiplying these probabilities by the 900 trips gives 479 trips by auto to shopping center 1, 43 trips by bus to shopping center 1, 355 trips by auto to shopping center 2, and 23 trips by bus to shopping center 2.

EXAMPLE 8.8

A joint mode-destination vehicle-based social/recreational trip logit model is estimated with the following coefficients:

Variable	Auto coefficient	Bus coefficient
Auto constant	0.9	0.0
Travel time in minutes	−0.22	−0.22
Population in thousands	0.16	0.16
Amusement floor space (in thousands of m^2)	1.1	1.1

It is known that 500 social/recreational trips will depart from a residential area during the peak hour. There are three possible trip destinations with the following characteristics:

	Travel time (in minutes)		Population (in thousands)	Amusement floor space (in thousands of m^2)
	Auto	Bus		
Destination 1	14	17	12.4	1.3
Destination 2	5	8	8.2	0.92
Destination 3	18	24	5.8	2.1

Determine the distribution of trips by mode and destination.

SOLUTION

As was the case for the shopping mode-destination model presented in Example 8.7, the signs of the coefficient estimates indicate that increasing travel time decreases an alternative's selection probability. Also, increasing population (reflecting an increase in social opportunities) and increasing amusement floor space (reflecting more recreational opportunities) both increase the probability of an alternative being selected. With two modes and three destinations, there are six alternatives, providing the following utilities (using the same subscripting notation as in Example 8.7):

$$U_{A1} = 0.9 - 0.22(14) + 0.16(12.4) + 1.1(1.3) = 1.234$$

$$U_{B1} = -0.22(17) + 0.16(12.4) + 1.1(1.3) = -0.326$$

$$U_{A2} = 0.9 - 0.22(5) + 0.16(8.2) + 1.1(0.92) = 2.124$$

$$U_{B2} = -0.22(8) + 0.16(8.2) + 1.1(0.92) = 0.564$$

$$U_{A3} = 0.9 - 0.22(18) + 0.16(5.8) + 1.1(2.1) = 0.178$$

$$U_{B3} = -0.22(24) + 0.16(5.8) + 1.1(2.1) = -2.042$$

Using Eq. 8.7 with 500 trips, the total number of trips to the six mode-destination alternatives are

$$P_{A1} = \frac{3.435}{15.607} = 0.220 \quad \text{and} \quad 0.220 \times 500 = \underline{110 \text{ trips}}$$

$$P_{B1} = \frac{0.722}{15.607} = 0.046 \quad \text{and} \quad 0.046 \times 500 = \underline{23 \text{ trips}}$$

$$P_{A2} = \frac{8.365}{15.607} = 0.536 \quad \text{and} \quad 0.536 \times 500 = \underline{268 \text{ trips}}$$

$$P_{B2} = \frac{1.76}{15.607} = 0.113 \quad \text{and} \quad 0.113 \times 500 = \underline{57 \text{ trips}}$$

$$P_{A3} = \frac{1.195}{15.607} = 0.077 \quad \text{and} \quad 0.077 \times 500 = \underline{38 \text{ trips}}$$

$$P_{B3} = \frac{0.13}{15.607} = 0.008 \quad \text{and} \quad 0.008 \times 500 = \underline{4 \text{ trips}}$$

EXAMPLE 8.9

Consider the situation described in Example 8.8. A labor dispute results in a bus union slowdown that increases travel times from the origin by 4, 2, and 8 minutes to destinations 1, 2, and 3, respectively. If the total number of trips remains constant, determine the resulting distribution of trips by mode and destination.

SOLUTION

The mode-destination utilities are computed as

$$U_{A1} = 1.234 \quad \text{(as in Example 8.8)}$$
$$U_{B1} = -0.22(21) + 0.16(12.4) + 1.1(1.3) = -1.206$$
$$U_{A2} = 2.124 \quad \text{(as in Example 8.8)}$$
$$U_{B2} = -0.22(10) + 0.16(8.2) + 1.1(0.92) = 0.124$$
$$U_{A3} = 0.178 \quad \text{(as in Example 8.8)}$$
$$U_{B3} = -0.22(32) + 0.16(5.8) + 1.1(2.1) = -3.802$$

Applying Eq. 8.7 with 500 trips gives the following distribution of trips among mode-destination alternatives:

$$P_{A1} = \frac{3.435}{14.45} = 0.238 \quad \text{and} \quad 0.238 \times 500 = \underline{\underline{119 \text{ trips}}}$$

$$P_{B1} = \frac{0.299}{14.45} = 0.021 \quad \text{and} \quad 0.021 \times 500 = \underline{\underline{10 \text{ trips}}}$$

$$P_{A2} = \frac{8.365}{14.45} = 0.579 \quad \text{and} \quad 0.579 \times 500 = \underline{\underline{290 \text{ trips}}}$$

$$P_{B2} = \frac{1.132}{14.45} = 0.078 \quad \text{and} \quad 0.078 \times 500 = \underline{\underline{39 \text{ trips}}}$$

$$P_{A3} = \frac{1.195}{14.45} = 0.083 \quad \text{and} \quad 0.083 \times 500 = \underline{\underline{41 \text{ trips}}}$$

$$P_{B3} = \frac{0.022}{14.45} = 0.002 \quad \text{and} \quad 0.002 \times 500 = \underline{\underline{1 \text{ trip}}}$$

EXAMPLE 8.10

Two routes connect a city and a suburb. During the peak-hour morning commute, a total of 4500 vehicles travel from the suburb to the city. Route 1 has a 100-km/h speed limit and is 10 km in length; route 2 is 5 km in length with a 75-km/h speed limit. Studies show that the total travel time on route 1 increases two minutes for every additional 500 vehicles added. Minutes of travel time on route 2 increase with the square of the number of vehicles, expressed in thousands of vehicles per hour. Determine user-equilibrium travel times.

SOLUTION

Determining free-flow travel times, in minutes, gives

Route 1: 10 km/(100 km/h) × 60 min/h = 6 min

Route 2: 5 km/(75 km/h) × 60 min/h = 4 min

With these data, the performance functions can be written as

$$t_1 = 6 + 4x_1$$

$$t_2 = 4 + x_2^2$$

where

t_1, t_2 = average travel times on routes 1 and 2 in minutes, and

x_1, x_2 = traffic flow on routes 1 and 2 in thousands of vehicles per hour.

Also, the basic flow conservation identity is

$$q = x_1 + x_2 = 4.5$$

where q = total traffic flow between the origin and destination in thousands of vehicles per hour.

With Wardrop's definition of user equilibrium, it is known that the travel times on all used routes are equal. However, the first order of business is to determine whether or not both routes are used. Figure A8.7 gives a graphic representation of the two performance functions. Note that because route 2 has a lower free-flow travel time, any total origin-to-destination traffic flow less than q' (in Fig. A8.7) will result in only route 2 being used, because the travel time on route 1 would be greater even if only one vehicle used it. At flows of q' and above, route 2 is sufficiently congested, and its travel time sufficiently high, that route 1 becomes a viable alternative.

To check if the problem's flow of 4500 vehicles per hour exceeds q', the following test is conducted:

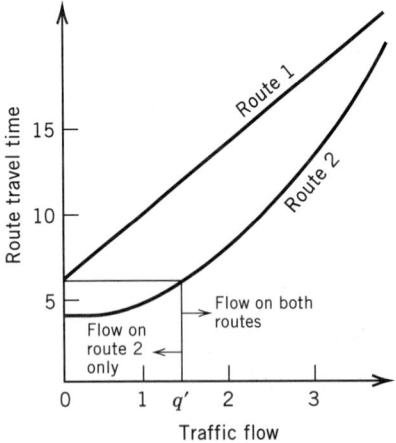

Figure A8.7 Illustration of performance curves for Example 8.10.

1. Assume that all traffic flow is on route 1. Substituting traffic flows of 4.5 and 0 into the performance functions gives $t_1(4.5) = 24$ min and $t_2(0) = 4$ min.
2. Assume that all traffic flow is on route 2, giving $t_1(0) = 6$ min and $t_2(4.5) = 24.25$ min.

Thus, because $t_1(4.5) > t_2(0)$ and $t_2(4.5) > t_1(0)$, both routes will be used. If $t_1(0)$ had been greater than $t_2(4.5)$, the 4500 vehicles would have been less than q' in Fig. A8.7, and only route 2 would have been used.

With both routes used, Wardrop's user-equilibrium definition gives

$$t_1 = t_2$$

or

$$6 + 4x_1 = 4 + x_2^2$$

From flow conservation, $x_1 + x_2 = 4.5$, so substituting,

$$6 + 4(4.5 - x_2) = 4 + x_2^2$$

$$x_2 = 2.899 \quad \text{or} \quad 2899 \text{ veh/h}$$

$$x_1 = 4.5 - x_2 = 4.5 - 2.899$$

$$= 1.601 \quad \text{or} \quad 1601 \text{ veh/h}$$

which gives average route travel times of

$$t_1 = 6 + 4(1.601) = \underline{\underline{12.4 \text{ min}}}$$

$$t_2 = 4 + (2.899)^2 = \underline{\underline{12.4 \text{ min}}}$$

EXAMPLE 8.11–8.16

Same as in Chapter 8.

Appendix B

Metric End-of-Chapter Problems

CHAPTER 2

2.1 A new sports car has a drag coefficient of 0.29 and a frontal area of 1.9 m², and is traveling at 160 km/h. How much power is required to overcome aerodynamic drag if $\rho = 1.2256$ kg/m³?

2.2 A vehicle manufacturer is considering an engine for a new sedan ($C_D = 0.30$, $A_f = 2$ m²). The car is being designed to achieve a top speed of 160 km/h on a paved surface at sea level ($\rho = 1.2256$ kg/m³). The car currently weighs 9.3 kN, but the designer initially selected an underpowered engine because he did not account for aerodynamic and rolling resistances. If 9 N of additional vehicle weight is added for each kilowatt needed to overcome the neglected resistance, what will be the final weight of the car if it is to achieve the 160-km/h top speed?

2.3 For Example 2.3, how far back from the front axle would the center of gravity have to be to ensure that the maximum tractive effort developed for front- and rear-wheel–drive options is equal (assume that all other variables are unchanged)?

2.4 A rear-wheel–drive 13.3-kN drag race car has a 5.10-m wheelbase and a center of gravity 500 mm above the pavement and 3.55 m behind the front axle. The owners wish to achieve an initial acceleration from rest of 4.6 m/s² on a level paved surface. What is the minimum coefficient of road adhesion needed to achieve this acceleration? (Assume $\gamma_m = 1.00$.)

2.5 If the race car in Problem 2.4 has a center of gravity 0.9 m above the roadway and is run on a pavement with a coefficient of road adhesion of 1.0, how far back from the front axle would the center of gravity have to be to develop a maximum acceleration from rest of 1.0 g (9.807 m/s²)? (Assume $\gamma_m = 1.00$.)

2.6 A rear-wheel–drive car weighs 12 kN, has 355-mm–radius wheels, a driveline efficiency of 95%, and an engine that develops 400 N-m of torque. Its wheelbase is 2.5 m, and the center of gravity is 450 mm above the road surface and 1 m behind the front axle. What is the lowest gear reduction ratio that would allow this car to achieve the highest possible acceleration from rest on good, dry pavement?

2.7 A newly designed car has a 2.8-m wheelbase, is rear-wheel drive, and has a center of gravity 450 mm above the road and 1.3 m behind the front axle. The car weighs 10.9 kN, the mechanical efficiency of the driveline is 90%, and the wheel radius is 355 mm. The base engine develops 250 N-m of torque, and a modified version of the engine develops 290 N-m of torque. If the overall gear reduction ratio is 9 to 1, what is the maximum acceleration from rest for the car with the 250-N-m engine and for the car with the 290–N-m engine? (It is on good, dry, and level pavement.)

2.8 A 13.3-kN car with $C_D = 0.35$, $A_f = 2$ m², and $\rho = 1.2256$ kg/m³ is traveling on a paved road. Its engine is running at 3000 rev/min and is producing 340 N-m of torque. The car's gear reduction ratio is 3.5 to 1, driveline efficiency is 90%, driveline slippage is 3.5%, and the road-wheel radius is 380 mm. What will the car's maximum acceleration rate be under these conditions on a level road? (Assume that the available tractive effort is the engine-generated tractive effort.)

2.9 A 9.5-kN car is traveling at sea level at a constant speed. Its engine is running at 4500 rev/min and is producing 200 N-m of torque. It has a driveline efficiency of 90%, a driveline slippage of 2%, 380-mm–radius wheels, and an overall gear reduction ratio of 3 to 1. If the car's frontal area is 1.8 m², what is the drag coefficient?

2.10 An 11-kN car has a maximum speed (at sea level, and on a level, paved surface) of 240 km/h with 355-mm–radius wheels, a gear reduction of 3 to 1, and a driveline efficiency of 90%. It is known that at the car's top speed the engine is producing 270 N-m of torque. If the car's frontal area is 2.3 m², what is its drag coefficient?

352

2.11 An 11-kN car ($C_D = 0.35$, $A_f = 2.3$ m², $\rho = 1.2256$ kg/m³) has 355-mm–radius wheels, a driveline efficiency of 90%, an overall gear reduction ratio of 3.2 to 1, and driveline slippage of 3.5%. The engine develops a maximum torque of 270 N-m at 3500 rev/min. What is the maximum grade this vehicle could ascend, on a paved surface while the engine is developing maximum torque? (Assume that the available tractive effort is the engine-generated tractive effort.)

2.12 A 12-kN car is traveling in third gear (overall gear reduction ratio of 2.5 to 1) on a level road at its top speed of 200 km/h. The air density is 1.06 kg/m³. The car has a frontal area of 1.8 m², a drag coefficient of 0.28, a wheel radius of 320 mm, a driveline slippage of 0.03, and a driveline efficiency of 90%. At this vehicle speed, what torque is the engine producing and what is the engine speed (in revolutions per minute)?

2.13 A rear-wheel–drive car weighs 11 kN, has a 2.03-m wheelbase, a center of gravity 500 mm above the roadway surface and 760 mm behind the front axle, a driveline efficiency of 75%, 355-mm–radius wheels, and an overall gear reduction of 11 to 1. The car's torque/engine speed curve is given by

$$M_e = 6n_e - 0.045n_e^2$$

If the car is on a paved, level roadway surface with a coefficient of adhesion of 0.75, determine its maximum acceleration from rest.

2.14 Consider the car in Problem 2.13. If it is known that the car achieves maximum speed at an overall gear reduction ratio of 2 to 1 with a driveline slippage of 3.5%, how fast would the car be going if it could achieve its maximum speed when its engine is producing maximum power?

2.15 Consider the situation described in Example 2.5. If the vehicle is redesigned with 330-mm–radius wheels (assume that the mass factor is unchanged) and a center of gravity located at the same height but at the midpoint of the wheelbase, determine the acceleration for front- and rear-wheel–drive options.

2.16 An engineer designs a rear-wheel–drive car (without an engine) that weighs 8.9 kN and has a 2.55-m wheelbase, a driveline efficiency of 80%, 355-mm–radius wheels, an overall gear reduction ratio of 10 to 1, and a center of gravity (without engine) that is 550 mm above the roadway surface and 1.40 m behind the front axle. An engine that weighs 13 N for each N-m of developed torque is to be placed in the front portion of the car. Calculations show that for every 89 N of engine weight added, the car's center of gravity moves 25 mm closer to the front axle (but stays at the same height above the roadway surface). If the car is starting from rest on a level paved roadway with a coefficient of adhesion of 0.8, select an engine size (weight and associated torque) that will result in the highest possible available tractive effort.

2.17 If the car in Example 2.7 had $C_D = 0.45$ and $A_f = 2.3$ m², what would have been the difference in minimum theoretical stopping distances with and without aerodynamic resistance considered (all other factors the same as in Example 2.7)?

2.18 A 15.6-kN vehicle ($C_D = 0.40$, $A_f = 2.4$ m², $\rho = 1.2256$ kg/m³) is driven on a surface with a coefficient of adhesion of 0.5, and the coefficient of rolling friction is approximated as 0.015 for all speeds. Assuming minimum theoretical stopping distances, if the vehicle comes to a stop 76 m after brake application on a level surface and has a braking efficiency of 0.78, what was its initial speed (a) if aerodynamic resistance is considered and (b) if aerodynamic resistance is ignored?

2.19 A level test track has a coefficient of road adhesion of 0.75, and a car being tested has a coefficient of rolling friction that is approximated as 0.018 for all speeds. The vehicle is tested unloaded and achieves the theoretical minimum stop in 61 m (from brake application). The initial speed was 100 km/h. Ignoring aerodynamic resistance, what is the unloaded braking efficiency?

2.20 A driver is traveling at 180 km/h down a 3% grade on good, wet pavement. An accident investigation team noted that braking skid marks started 180 m before a parked car was hit at an estimated 90 km/h. Ignoring air resistance, and using theoretical stopping distance, what was the braking efficiency of the car?

2.21 A small truck is to be driven down a 4% grade at 120 km/h. The coefficient of road adhesion is 0.95, and it is known that the braking efficiency is 80% when the truck is empty and it decreases by one percentage point for every 445 N of cargo added. Ignoring aerodynamic resistance, if the driver wants the truck to be able to achieve a minimum theoretical stopping distance of 90 m from the point of brake application, what is the maximum amount of cargo (in newtons) that can be carried?

2.22 Consider the conditions in Example 2.8. The car has $C_D = 0.5$, $A_f = 2.3$ m², $W = 15.6$ kN, $\rho = 1.2256$ kg/m³, and a coefficient of rolling friction that is approximated as 0.018 for all speed conditions. If aerodynamic resistance is considered in stopping, estimate how fast the car will be going when it strikes the object on a level and a +5% grade [all other conditions (speed, etc.) as described in Example 2.8].

2.23 A car is traveling at 120 km/h down a 3% grade on poor, wet pavement. The car's braking efficiency is 90%. The brakes were applied 90 m before impacting an object. The car had an antilock braking system, but the system failed 60 m after the brakes had been applied (wheels locked). What speed was the car traveling at just before it impacted the object? (Assume theoretical stopping distance, ignore air resistance, and let $f_{rl} = 0.015$.)

2.24 A driver traveling down a 4% grade collides with a roadside object in rainy conditions, and is issued a ticket for driving too fast for conditions. The posted speed limit is 105 km/h. The accident investigation team determined the following: The vehicle was traveling 65 km/h when it struck the object, braking skid marks started 61 m before the struck object, the pavement is in good condition, and the braking efficiency of the vehicle was 0.95. Using theoretical stopping distance, assuming aerodynamic resistance is negligible, and with the coefficient rolling resistance approximated as 0.015, should the driver appeal the ticket? Why or why not?

2.25 A driver is traveling 110 km/h on a road with a negative 3% grade. There is a stalled car on the road 305 m ahead of the driver. The driver's vehicle has a braking efficiency of 90%, and it has antilock brakes. The road is in good condition and is initially dry, but it becomes wet 45 m before the stalled car (and stays wet until the car is reached). What is the minimum distance from the stalled car at which the driver could apply the brakes and still stop before hitting it? (Assume theoretical stopping distance, ignore air resistance, and let $f_{rl} = 0.013$.)

2.26 Two cars are traveling on level terrain at 100 km/h on a road with a coefficient of adhesion of 0.8. The driver of car 1 has a 2.5-s perception/reaction time and the driver of car 2 has a 2.0-s perception/reaction time. Both cars are traveling side by side, and the drivers are able to stop their respective cars in the same amount of distance from first seeing a roadway obstacle (perception/reaction plus vehicle stopping distance). If the braking efficiency of car 2 is 0.75, determine the braking efficiency of car 1. (Assume minimum theoretical stopping distance and ignore aerodynamic resistance.)

2.27 An engineering student is driving on a level roadway and sees a construction sign 185 m ahead in the middle of the roadway. The student strikes the sign at a speed of 55 km/h. If the student was traveling at 90 km/h when the sign was first spotted, what was the student's associated perception/reaction time (use practical stopping distance)?

2.28 An engineering student claims that a country road can be safely negotiated at 110 km/h in rainy weather. Because of the winding nature of the road, one stretch of level pavement has a sight distance of only 180 m. Assuming practical stopping distance, comment on the student's claim.

2.29 A driver is traveling at 90 km/h on a wet road. An object is spotted on the road 140 m ahead and the driver is able to come to a stop just before hitting the object. Assuming standard perception/reaction time and practical stopping distance, determine the grade of the road.

2.30 A test of a driver's perception/reaction time is being conducted on a special testing track with wet pavement and a driving speed of 90 km/h. When the driver is sober, a stop can be made just in time to avoid hitting an object that is first visible 160 m ahead. After a few drinks under the exact same conditions, the driver fails to stop in time and strikes the object at a speed of 55 km/h. Determine the driver's perception/reaction time before and after drinking. (Assume practical stopping distance.)

CHAPTER 3

3.1 A 490-m–long sag vertical curve (equal tangent) has a PVC at station 3 + 700 and elevation 460 m. The initial grade is −3.5% and the final grade is +6.5%. Determine the elevation and stationing of the low point, PVI, and PVT.

3.2 A 150-m–long equal-tangent crest vertical curve connects tangents that intersect at station 10 + 360 and elevation 400 m. The initial grade is +4.0% and the final grade is −2.5%. Determine the elevation and stationing of the high point, PVC, and PVT.

3.3 Consider Example 3.4. Solve this problem with the parabolic equation (Eq. 3.1) rather than by using offsets.

3.4 Again consider Example 3.4. Does this curve provide sufficient stopping sight distance for a speed of 100 km/h?

3.5 An equal-tangent sag vertical curve is designed with the PVC at station 3 + 320 and elevation 290 m, the PVI at station 3 + 375 and elevation 288.74 m, and the low point at station 3 + 365. Determine the design speed of the curve.

3.6 An equal-tangent vertical curve was designed in 2002 (to 2001 AASHTO guidelines) for a design speed of 110 km/h to connect grades $G_1 = +1.0\%$ and $G_2 = -2.0\%$. The curve is to be redesigned for a 110-km/h design speed in the year 2025. Vehicle braking technology has advanced such that the recommended design deceleration rate is 25% greater than its 2001 value used to develop Table 3.1, but due to the higher percentage of older persons in the driving population, design reaction times have increased by 20%. Also, vehicles have become smaller such that the driver's eye height is assumed to be 900 mm above the pavement, and roadway objects are assumed to be 300 mm above the pavement surface. Compute the difference in design curve lengths for the 2002 and 2025 designs.

3.7 A 365-m equal-tangent crest vertical curve is currently designed for 80 km/h. A civil engineering student contends that 100 km/h is safe in a van because of the higher driver's eye height. If all other design inputs are standard, what must the driver's eye height (in the van) be for the student's claim to be valid?

3.8 A highway reconstruction project is being undertaken to reduce crash rates. The reconstruction involves a major realignment of the highway such that a 100-km/h design speed is attained. At one point on the highway, a 245-m equal-tangent crest vertical curve exists. Measurements show that at 0 + 107.3 stations from the PVC, the vertical curve offset is 1 m. Assess the adequacy of this existing curve in light of the reconstruction design speed of 100 km/h and, if the existing curve is inadequate, compute a satisfactory curve length.

3.9 Two level sections of an east-west highway ($G = 0$) are to be connected. Currently, the two sections of highway are separated by a 1200-m (horizontal distance), 2% grade. The westernmost section of highway is the higher of the two and is at elevation 30 m. If the highway has a 100-km/h design speed, determine, for the crest and sag vertical curves required, the stationing and elevation of the PVCs and PVTs given that the PVC of the crest curve (on the westernmost level highway section) is at station 0 + 000 and elevation 30 m. In solving this problem, assume that the curve PVIs are at the intersection of $G = 0$ and the 2% grade, that is, $A = 2$.

3.10 Consider Problem 3.9. Suppose it is necessary to keep the entire alignment within the 1200 m that currently separate the two level sections. It is determined that the crest and sag curves should be connected (the PVT of the crest and PVC of the sag) with a constant-grade section that has the lowest grade possible. Again using a 100-km/h design speed, determine, for the crest and sag vertical curves, the stationing and elevation of the PVCs and PVTs given that the westernmost level section ends at station 0 + 000 and elevation 30 m. (Note that A must now be determined and will not be equal to 2.)

3.11 An equal-tangent crest vertical curve is designed for 100 km/h. The initial grade is +4.0% and the final grade is negative. What is the elevation difference between the PVC and the high point of the curve?

3.12 An equal-tangent crest vertical curve has an 80-km/h design speed. The initial grade is +3%. The high point is at station 1 + 017.2 and the PVT is at station 1 + 133.3. What is the elevation difference between the high point and the PVT?

3.13 A vertical curve is designed for 90 km/h, and it has an initial grade of +2.5% and a final grade of −1.0%. The PVT is at station 3 + 480. It is known that a point on the curve at station 3 + 440 is at elevation 75 m. What is the stationing and elevation of the PVC? What is the stationing and elevation of the high point on the curve?

3.14 An equal-tangent crest curve connects a +1.0% and a −0.5% grade. The PVC is at station 1 + 670 and the PVI is at station 1 + 750. Is this curve long enough to provide passing sight distance for a 90-km/h design speed?

3.15 Due to crashes at a railroad crossing, an overpass (with a roadway surface 7.5 m above the existing road) is to be constructed on an existing level highway. The existing highway has a design speed of 80 km/h. The overpass structure is to be level, centered above the railroad, and 60 m long. What length of the existing level highway must be reconstructed to provide an appropriate vertical alignment?

3.16 A section of a freeway ramp has a +4.0% grade and ends at station 3 + 870 and elevation 42 m. It must be connected to another section of the ramp (which has a 0.0% grade) that is at station 4 + 940 and elevation 30 m. It is determined that the crest and sag curves required to connect the ramp should be connected (the *PVT* of the crest and *PVC* of the sag) with a constant-grade section that has the lowest grade possible. Design a vertical alignment to connect between these two stations using an 80-km/h design speed. Provide the lengths of the curves and constant-grade section.

3.17 A tangent section of highway has a −1.0% grade and ends at station 0 + 145 and elevation 25 m. It must be connected to another section of highway that has a −1.0% grade and that begins at station 1 + 345 and elevation 40 m. The connecting alignment should consist of a sag curve, constant-grade section, and crest curve, and be designed for a speed of 80 km/h. What is the lowest grade possible for the constant-grade section that will complete this alignment?

3.18 A roadway has a design speed of 80 km/h, and at station 3 + 045 a +3.0% grade roadway section ends and at station 3 + 960 a +2.0% grade roadway section begins. The +3.0% grade section of highway (at station 3 + 045) is at a higher elevation than the +2.0% grade section of highway (at station 3 + 960). If a −5% constant-grade section is used to connect the crest and sag vertical curves that are needed to link the +3.0 and +2.0% grade sections, what is the elevation difference between stations 3 + 045 and 3 + 960? (The entire alignment, crest and sag curves, and constant-grade section must fit between stations 3 + 045 and 3 + 960.)

3.19 A sag curve and crest curve connect a −3.0% tangent section of highway (to the west) with a +2.0% tangent section of highway (to the east). The +2.0% tangent section is at a higher elevation than the −3.0% tangent section. The two tangent sections are separated by 390 m of horizontal distance. If the design speed of the curves is 80 km/h, what is the common grade between the sag and crest curves (G_2 of sag and G_1 of crest, from west to east), and what is the elevation difference between PVC_s and PVT_c?

3.20 An overpass is being built over the *PVI* of an existing equal-tangent sag curve. The sag curve has a 110-km/h design speed, and $G_1 = -6\%$, $G_2 = +3\%$. Determine the minimum necessary clearance height of the overpass and the resultant elevation of the bottom of the overpass over the *PVI*. (Ignore the cross-sectional width of the overpass.)

3.21 An equal-tangent sag curve has its *PVI* at station 0 + 305 and elevation at 42 m. Directly above the *PVI*, the bottom of an overpass structure is at elevation 50 m.

The *PVC* is at station 0 + 120. If the initial grade is −4%, what is the highest possible value of the final grade given that a 110-km/h design speed is to be provided in daytime conditions? What is the highest possible final grade in nighttime conditions? (*Note:* Be careful of units of A, and ignore the cross-sectional width of the overpass.)

3.22 An existing highway-railway at-grade crossing is being redesigned as grade separated to improve traffic operations. The railway must remain at the same elevation. The highway is being reconstructed to travel under the railway. The underpass will be a sag curve that connects to 2% tangent sections on both ends, and the *PVI* will be centered under the railway (a symmetrical alignment). The sag curve design speed is 70 km/h. How many meters below the railway should the curve *PVI* be located?

3.23 You are asked to design a horizontal curve for a two-lane road. The road has 3.6-m lanes. Due to expensive excavation, it is determined that a maximum of 10 m can be cleared from the road's centerline toward the inside lane to provide for stopping sight distance. Also, local guidelines dictate a maximum superelevation of 0.08 m/m. What is the highest possible design speed for this curve?

3.24 A horizontal curve on a single-lane highway has its *PC* at station 3 + 780 and its *PI* at station 4 + 000. The curve has a superelevation of 0.06 m/m and is designed for 110 km/h. What is the station of the *PT*?

3.25 A horizontal curve is being designed through mountainous terrain for a four-lane road with lanes that are 3 m wide. The central angle (Δ) is known to be 40 degrees, the tangent distance is 155 m, and the stationing of the tangent intersection (*PI*) is 82 + 300. Under specified conditions and vehicle speed, the roadway surface is determined to have a coefficient of side friction of 0.08, and the curve's superelevation is 0.09 m/m. What is the stationing of the *PC* and *PT* and what is the safe vehicle speed?

3.26 A new interstate highway is being built with a design speed of 110 km/h. For one of the horizontal curves, the radius (measured to the innermost vehicle path) is tentatively planned as 275 m. What rate of superelevation is required for this curve?

3.27 A developer is having a single-lane raceway constructed with a 160-km/h design speed. A curve on the raceway has a radius of 305 m, a central angle of 30 degrees, and *PI* stationing at 34 + 300. If the design coefficient of side friction is 0.20, determine the superelevation required at the design speed (do not ignore the normal component of the centripetal force). Also, compute the degree of curve, length of curve, and stationing of the *PC* and *PT*.

3.28 A horizontal curve is being designed for a new two-lane highway (3.6-m lanes). The *PI* is at station 7 + 640, the design speed is 105 km/h, and a maximum superelevation of 0.08 m/m is to be used. If the central angle of the curve is 35 degrees, design a curve for the highway by computing the radius and stationing of the *PC* and *PT*.

3.29 You are asked to design a horizontal curve with a 40-degree central angle ($\Delta = 40$) for a two-lane road with 3-m lanes. The design speed is 110 km/h and superelevation is limited to 0.06 m/m. Give the radius, degree of curvature, and length of curve that you would recommend.

3.30 A horizontal curve on a single-lane freeway ramp is 120 m long, and the design speed of the ramp is 70 km/h. If the superelevation is 10% and the station of the *PC* is 0 + 530, what is the station of the *PI* and how much distance must be cleared from the center of the lane to provide adequate stopping sight distance?

3.31 A freeway exit ramp has a single lane and consists entirely of a horizontal curve with a central angle of 90 degrees and a length of 190 m. If the distance cleared from the centerline for sight distance is 5.9 m, what design speed was used?

3.32 A horizontal curve on a two-lane highway (3.6-m lanes) has *PC* at station 3 + 765 and *PT* at station 3 + 940. The central angle is 34 degrees, the superelevation is 0.08, and 6 m are cleared (for sight distance) from the inside edge of the innermost lane. Determine a maximum safe speed (assuming current design standards) to the nearest 10 km/h.

3.33 For the horizontal curve in Problem 3.29, what distance must be cleared from the inside edge of the inside lane to provide adequate stopping sight distance?

3.34 A horizontal curve was designed for a four-lane highway for adequate SSD. Lane widths are 3.6 m, and the superelevation is 0.06 and was set assuming maximum f_s. If the necessary sight distance required 16 m of lateral clearance from the roadway centerline, what design speed was used for the curve?

3.35 A section of highway has vertical and horizontal curves with the same design speed. A vertical curve on this highway connects a +1% and a +3% grade and is 124 m long. If a horizontal curve on this highway is on a two-lane section with 3.6-m lanes, has a central angle of 37 degrees, and has a superelevation of 6%, what is the length of the horizontal curve?

3.36 A section of a two-lane highway (3.6-m lanes) is designed for 120 km/h. At one point a vertical curve connects a −2.5% and a +1.5% grade. The *PVT* of this curve is at station 0 + 765. It is known that a horizontal curve starts (has *PC*) 90 m before the vertical curve's *PVC*. If the superelevation of the horizontal curve is 0.08 and the central angle is 38 degrees, what is the station of the *PT*?

3.37 Two straight sections of freeway cross at a right angle. At the point of crossing, the east-west highway is at elevation 45 m and has a constant +5.0% grade (upgrade in the east direction), and the north-south highway is at elevation 38 m and has a constant −3.0% grade (downgrade in the north direction). Design a 90-degree ramp that connects the northbound direction of travel to the eastbound direction of travel. Design the ramp for the highest design speed (to the nearest 10 km/h) with the constraint that the minimum allowable value of D is 8.0. (Assume that the *PC* of the horizontal curve is at station 0 + 460, and the vertical curve *PVI*s are at the *PC* and *PT*.) Give the stationing and elevations of the *PC*, *PT*, *PVC*s, and *PVT*s.

CHAPTER 4

4.1 A tire carries a 22.2-kN load and has a pressure of 690 kPa. The pavement that the tire is on is constructed with a modulus of elasticity of 300,000 kPa. A deflection of 0.41 mm is observed at a point at the pavement surface 20.32 mm from the center of the tire load. Using the Ahlvin and Ulery equations, what is the radial-horizontal stress at this point?

4.2 A wheel carrying a 29.8-kN load generates a deflection of 0.889 mm, 50.8 mm below the center of the tire load. The contact area is measured at 0.052 m² and the Poisson ratio is 0.5. Using the Ahlvin and Ulery equations, what is the pavement's modulus of elasticity?

4.3 A wheel load produces a circular contact radius of 88.9 mm and the pavement material has a modulus of elasticity of 300,000 kPa. At a point on the pavement surface under the center of the tire load, the radial-horizontal stress is 600 kPa and the deflection is

0.42 mm. Using the Ahlvin and Ulery equations, what is the load applied to the wheel?

4.4 A pavement is 635 mm thick and has a modulus of elasticity of 250,000 kPa with a Poisson ratio of 0.40. A wheel load is applied 1.27 m from the edge of the pavement. The wheel's tire has a pressure of 700 kPa and a circular contact radius of 50 mm. Using the Ahlvin and Ulery equations, determine the vertical stress, radial-horizontal stress, and deflection at a point at the bottom of the pavement, at the pavement's edge.

4.5 Truck A has two single axles. One axle weighs 53.38 kN and the other weighs 102.3 kN. Truck B has a 35.6-kN single axle and a 191.3-kN tandem axle. On a flexible pavement with a 76.2-mm hot-mix asphalt wearing surface, a 152.4-mm soil-cement base, and a 203.2-mm crushed stone subbase, which truck will cause more pavement damage? (Assume drainage coefficients are 1.0.)

4.6–4.32 The applicable design equations are based only on U.S. Customary units, so a metric problem statement is not given here.

CHAPTER 5

5.1 On a specific westbound section of highway, studies show that the speed-density relationship is

$$u = u_f \left[1 - \left(\frac{k}{k_j} \right)^{3.5} \right]$$

It is known that the capacity is 3800 veh/h and the jam density is 140 veh/km. What is the space-mean speed of the traffic at capacity, and what is the free-flow speed?

5.2 A section of highway has a speed-flow relationship of the form

$$q = au^2 + bu$$

It is known that at capacity (which is 2900 veh/h) the space-mean speed of traffic is 50 km/h. Determine the speed when the flow is 1400 veh/h and the free-flow speed.

5.3 A section of highway has the following flow-density relationship:

$$q = 80k - 0.4k^2$$

What is the capacity of the highway section, the speed at capacity, and the density when the highway is at one-quarter of its capacity?

5.4 Assume you are observing traffic in a single lane of a highway at a specific location. You measure the average headway and average spacing of passing vehicles as 3 seconds and 45 m, respectively. Calculate the flow, average speed, and density of the traffic stream in this lane.

5.5 Assume you are an observer standing at a point along a three-lane roadway. All vehicles in lane 1 are traveling at 50 km/h, all vehicles in lane 2 are traveling at 75 km/h, and all vehicles in lane 3 are traveling at 100 km/h. There is also a constant spacing of 0.8 km between vehicles. If you collect spot speed data for all vehicles as they cross your observation point, for 30 minutes, what will be the time-mean speed and space-mean speed for this traffic stream?

5.6 Four race cars are traveling on a 4.0-km tri-oval track. The four cars are traveling at constant speeds of 315 km/h, 305 km/h, 300 km/h, and 290 km/h, respectively. Assume you are an observer standing at a point on the track for a period of 30 minutes and are recording the instantaneous speed of each vehicle as it crosses your point. What is the time-mean speed and space-mean speed for these vehicles for this time period? (*Note:* Be careful with rounding.)

5.7 For Problem 5.6, calculate the space-mean speed assuming you were provided with only an aerial photo of the circling race cars and the constant travel speed of each of the vehicles.

5.8–5.35 Same as in Chapter 5.

CHAPTER 6

6.1 A six-lane rural freeway has regular weekday users and currently operates at maximum LOS C conditions. The base free-flow speed is 105 km/h, lanes are 3.4 m wide, the right-side shoulder is 1.2 m wide, and the interchange density is 0.4 per kilometer. The highway is on rolling terrain with 10% large trucks and buses (no recreational vehicles), and the peak hour factor is 0.90. Determine the hourly volume for these conditions.

6.2 Consider the freeway in Problem 6.1. At one point along this freeway there is a 4% upgrade with a directional hourly traffic volume of 5435 vehicles. If all other conditions are as described in Problem 6.1, how long can this grade be without the freeway LOS dropping to F?

Appendix B Metric End-of-Chapter Problems

6.3 A four-lane urban freeway is located on rolling terrain and has 3.6-m lanes, no lateral obstructions within 1.8 m of the pavement edges, and an interchange every 3.2 kilometers. The traffic stream consists of cars, buses, and large trucks (no recreational vehicles). A weekday directional peak-hour volume of 1800 vehicles (familiar users) is observed, with 700 arriving in the most congested 15-min period. If a level of service no worse than C is desired, determine the maximum number of large trucks and buses that can be present in the peak-hour traffic stream.

6.4 Consider the freeway and traffic conditions described in Problem 6.3. If 180 of the 1800 vehicles observed in the peak hour were large trucks and buses, what would the level of service of this freeway be on an 8-km, 6% downgrade?

6.5 A six-lane freeway in a scenic area has a measured free-flow speed of 90 km/h. The peak-hour factor is 0.80, and there are 8% large trucks and buses and 6% recreational vehicles in the traffic stream. One upgrade is 5% and 0.8 km long. An analyst has determined that the freeway is operating at capacity on this upgrade during the peak hour. If the peak-hour traffic volume is 3900 vehicles, what value for the driver population factor was used?

6.6 A four-lane freeway is on mountainous terrain with 3.4-m lanes, a 1.5-m right-side shoulder, interchange spacing of one every 16 kilometers, and a 100-km/h base free-flow speed. During the peak hour there are 12% large trucks and buses and 6% recreational vehicles. PHF is 0.88 and the driver population adjustment is determined to be 0.90. The freeway currently operates at capacity during the peak hour, in the direction in question. If an additional 3.4-m lane is added (each direction) and all other factors stay the same, what will be the new level of service?

6.7 A segment of urban four-lane freeway has a 3% upgrade that is 455 m long followed by a 305-m 4% upgrade. It has 3.6-m lanes and 0.9-m shoulders. The base free-flow speed is 105 km/h, and the directional hourly traffic flow is 2000 vehicles with 5% large trucks and buses (no recreational vehicles). There are no interchanges in the vicinity of this freeway segment. If the peak-hour factor is 0.90 and all of the drivers are regular users, what is the level of service of this compound-grade freeway segment?

6.8 Consider Example 6.2, in which it was determined that 1678 vehicles could be added to the peak hour before capacity is reached. Assuming rolling terrain as in Example 6.1, how many passenger cars could be added to the original traffic mix before peak-hour capacity is reached? (Assume only passenger cars are added and that the number of large trucks and buses originally in the traffic stream remains constant.)

6.9 An eight-lane urban freeway is on rolling terrain and has 3.4-m lanes with a 1.2-m right-side shoulder. The interchange density is 0.8 per kilometer. The base free-flow speed is 110 km/h. The directional peak-hour traffic volume is 5400 vehicles with 6% large trucks and 5% buses (no recreational vehicles). The traffic stream consists of regular users and the peak-hour factor is 0.95. It has been decided that large trucks will be banned from the freeway during the peak hour. What will the freeway's density and level of service be before and after the ban? (Assume that the trucks are removed and all other traffic attributes are unchanged.)

6.10 A 5% upgrade on a six-lane freeway is 2.0 km long. On this segment of freeway, the directional peak-hour volume is 3800 vehicles with 2% large trucks and 4% buses (no recreational vehicles), the peak-hour factor is 0.90, and all drivers are regular users. The base free-flow speed is 105 km/h, the lanes are 3.6 m wide, there are no lateral obstructions within 3 m of the roadway, and the interchange density is 0.9 per kilometer. A bus strike will eliminate all bus traffic, but it is estimated that for each bus removed from the roadway, six additional passenger cars will be added as travelers seek other means of travel. What are the density, volume-to-capacity ratio, and level of service of the upgrade segment before and after the bus strike?

6.11 A multilane highway (two lanes in each direction) is on level terrain. The free-flow speed has been measured at 70 km/h. The peak-hour directional traffic flow is 1300 vehicles with 6% large trucks and buses and 2% recreational vehicles ($f_p = 0.95$). If the peak-hour factor is 0.85, determine the highway's level of service.

6.12 Consider the multilane highway in Problem 6.11. If the proportion of vehicle types and peak-hour factor remain constant, how many vehicles can be added to the directional traffic flow before capacity is reached?

6.13 A six-lane multilane highway has a peak-hour factor of 0.90, 3.4-m lanes with a 1.2-m right-side shoulder, and a two-way left-turn lane in the median. The directional peak-hour traffic flow is 4000 vehicles with 8% large trucks and buses and 2% recreational vehicles. The driver population factor has been estimated at 0.95. What will the level of service of this highway be on a 4% upgrade that is 2.4 kilometers long if the speed limit is 90 km/h and there are 9 access points per kilometer?

6.14 A divided multilane highway in a recreational area ($f_p = 0.90$) has four lanes and is on rolling terrain. The highway has 3-m lanes with a 1.8-m right-side shoulder and a 0.9-m left-side shoulder. The posted speed is 80 km/h. Formerly there were 2.5 access points per kilometer, but recent development has increased the number of access points to 7.5 per kilometer. Before the development, the peak-hour factor was 0.95 and the directional hourly volume was 2300 vehicles with 10% large trucks and buses and 3% recreational vehicles. After the development, the peak-hour directional flow is 2700 vehicles with the same vehicle percentages and peak-hour factor. What is the level of service before and after the development?

6.15 A multilane highway has four lanes and a measured free-flow speed of 90 km/h. One upgrade is 5% and is 1 km long. Currently trucks are not permitted on the highway, but there are 2% buses (no recreational vehicles) in the directional peak-hour volume of 1900 vehicles (the peak-hour factor is 0.80). Local authorities are considering allowing trucks on this upgrade. If this is done, they estimate that 150 large trucks will use the highway during the peak hour. What would be the level of service before and after the trucks are allowed (assuming the driver population adjustment to be 1.0 before and 0.97 after)?

6.16 A four-lane undivided multilane highway has 3.4-m lanes, 1.2-m shoulders on both sides, and 6 access points per kilometer. It is determined that the roadway currently operates at capacity with $PHF = 0.80$ and a driver population adjustment of 0.9. If the highway is on level terrain with 8% trucks and buses (no recreational vehicles) and the speed limit is 90 km/h, what is the directional hourly volume?

6.17 A new four-lane divided multilane highway is being planned with 3.6-m lanes, 1.8-m shoulders on both sides, and a 80-km/h speed limit. One 3% downgrade is 7.2 km long, and there will be 2.5 access points per kilometer. The peak-hour directional volume along this grade is estimated to consist of 1800 passenger cars, 140 large trucks, 40 buses, and 10 recreational vehicles. If the peak-hour factor is estimated to be 0.85, what level of service will this segment of highway operate under?

6.18 A six-lane divided multilane highway has a measured free-flow speed of 80 km/h. It is on mountainous terrain with a traffic stream consisting of 6% large trucks and buses and 2% recreational vehicles. The driver population adjustment is 0.92. One direction of the highway currently operates at maximum LOS C conditions, and it is known that the highway has $PHF = 0.90$. How many vehicles can be added to this highway before capacity is reached, assuming the proportion of vehicle types remains the same but the peak-hour factor increases to 0.95?

6.19 A four-lane undivided multilane highway has 3.4-m lanes and 1.5-m shoulders. At one point along the highway there is a 4% upgrade that is 1.0 km long. There are 15 access points along this grade. The peak-hour traffic volume has 2500 passenger cars and 200 trucks and buses (no recreational vehicles), and 720 of these vehicles arrive in the most congested 15-min period. This traffic stream is primarily commuters. The measured free-flow speed is 90 km/h. To improve the level of service, the local transportation agency is considering reducing the number of access points by blocking some driveways and rerouting their traffic. How many of the 15 access points must be blocked to achieve LOS C?

6.20 A Class I two-lane highway is on level terrain, has a measured free-flow speed of 105 km/h, and has 50% no-passing zones. The peak-hour traffic volume is 180 vehicles, with $PHF = 0.90$. There are 15% trucks and buses (no RVs) and the directional split is 70/30. Determine the level of service.

6.21 A Class I two-lane highway carries a peak-hour volume of 540 vehicles and has a peak-hour factor of 0.87. The traffic stream has a 60/40 directional split with 5% large trucks, 10% recreational vehicles, no buses, and 85% passenger cars. The highway is on rolling terrain and has 80% no-passing zones. The free-flow speed was measured at 92 km/h, but this was during a flow rate of 275 pc/h. Determine the level of service.

6.22 A Class I two-lane highway is on level terrain with passing permitted throughout. The highway has 3.4-m lanes with 1.2-m shoulders. There are 10 access points per kilometer. The base FFS is 100 km/h. The highway is oriented north and south, and during the peak hour, 440 vehicles are going northbound and 360 vehicles are going southbound. If the *PHF* is 0.87 and there are 4% large trucks, 3% buses, and 1% recreational vehicles, what is the level of service?

6.23 A two-lane highway is currently operating at its two-way capacity in rolling terrain. The traffic stream consists of cars and trucks only. A recent traffic count revealed 720 vehicles (total, both directions) arriving in the most congested 15-min interval. What is the percentage of trucks in the traffic stream based on the *ATS* service measure?

6.24 A Class II two-lane highway needs to be redesigned for an area with rolling terrain. The peak-hour traffic volume is 500 vehicles with a directional split of 60/40 and a *PHF* of 0.85. The traffic stream includes 8% large trucks, 2% buses, and 5% recreational vehicles. What is the maximum percentage of no-passing zones that can be built into the design with LOS B maintained?

6.25 A four-lane freeway segment consists of passenger cars only, a driving population of regular users, a peak-hour directional distribution of 0.70, a peak-hour factor of 0.80, and a measured free-flow speed of 110 km/h. Assuming Fig. 6.7 applies, if the AADT is 30,000 veh/day, determine the level of service for the 10th, 50th, and 100th highest annual hourly volumes.

6.26 A four-lane urban freeway operates at capacity during the peak hour. It has 3.4-m lanes and 1.2-m shoulders. The freeway has only regular users and there are 8% large trucks and buses (no recreational vehicles), and the freeway is on rolling terrain with a peak-hour factor of 0.85. There is one interchange every 3.2 km. It is known that 12% of the AADT occurs in the peak hour and that the directional factor is 0.6. What is the freeway's AADT?

6.27 A six-lane highway has regular weekday users and currently operates at maximum LOS C conditions. The measured free-flow speed is 90 km/h. The highway is on rolling terrain with 10% large trucks and buses and 5% recreational vehicles, and the peak-hour factor is 0.94. If 20% of all directional traffic occurs during the peak hour, determine the total directional traffic volume.

CHAPTER 7

7.1–7.36 Same as in Chapter 7.

7.37 Consider the intersection of Vine and Maple streets as shown in Fig. 7.9. Suppose Vine Street's northbound and southbound approaches are both on an 8% upgrade, and the assumed vehicle approach speed is 50 km/h. What should the yellow and all-red times be?

7.38 Consider Example 7.8. Two additional 3.6-m through lanes are added to Vine Street (the street in the intersection shown in Fig. 7.9), one lane in each direction. If the peak-hour traffic volumes are unchanged, but the Vine Street left-turn saturation flow rates increase by 100 veh/h because of the added through lanes, what would the revised effective green time, yellow time, and all-red time be for each phase?

7.39 Same as in Chapter 7.

7.40 Consider Problem 7.38. Calculate the new minimum pedestrian green time, assuming the effective crosswalk width is 2 m and the maximum number of crossing pedestrians in any phase is 20.

7.41–7.42 Same as in Chapter 7.

CHAPTER 8

8.1–8.7 Same as in Chapter 8.

8.8 It is known that 4000 automobile trips are generated in a large residential area from noon to 1:00 P.M. on Saturdays for shopping purposes. Four major shopping centers have the following characteristics:

Shopping center	Distance from residential area (km)	Commercial floor space (thousands of m^2)
1	4	20
2	9	15
3	8	30
4	14	60

If a logit model is estimated with coefficients of -0.283 for distance and 0.172 for commercial space (in thousands of m^2), how many shopping trips will be made to each of the four shopping centers?

8.9 Consider the shopping trip situation described in Problem 8.8. Suppose that shopping center 3 goes out of business and shopping center 2 is expanded to 50,000 m^2 of commercial space. What would be the new distribution of the 4000 Saturday afternoon shopping trips?

8.10 If shopping center 3 is closed (see Problem 8.9), how much commercial floor space is needed in shopping centers 1 and 2 to ensure that each of them has the same probability of being selected as shopping center 4?

8.11 Consider the situation described in Example 8.7. If the construction of a new freeway lowers auto and transit travel times to shopping center 2 by 20%, determine the new distribution of shopping trips by destination and mode.

8.12 Consider the conditions described in Example 8.7. Heavily congested highways have caused travel times to shopping center 2 to increase by 4 min for both auto and transit modes (travel times to shopping center 1 are not affected). In order to attract as many total trips (auto and transit) as it did before the congestion, how much commercial floor space must be added to shopping center 2 (given that the total number of departing shopping trips remains at 900)?

8.13 A total of 700 auto-mode social/recreational trips are made from an origin (residential area) during the peak hour. A logit model estimation is made, and three factors were found to influence the destination choice: (1) population at the destination, in thousands (coefficient = 0.2); (2) distance from origin to destination, in kilometers (coefficient = −0.15); and (3) square meters of amusement floor space (movie theaters, video game centers, etc.), in thousands (coefficient = 0.9). Four possible destinations have the following characteristics:

	Population (thousands)	Distance from origin (km)	Amusement space (thousands of m^2)
Destination 1	15.5	12	0.5
Destination 2	6.0	8	1.0
Destination 3	0.8	3	0.8
Destination 4	5.0	11	1.5

Determine the distribution of trips among possible destinations.

8.14 Consider the situation described in Problem 8.13. If a new 1500-m^2 arcade center is built at destination 3, determine the distribution of the 700 peak-hour social/recreational trips.

8.15 Note that with the situation described in Example 8.8, 26.6% [(110 + 23)/500] of all social/recreational trips are to destination 1. If the total number of trips remains constant, how much additional amusement floor space would have to be added to destination 1 to have it capture 40.0% of the total social/recreational trips?

8.16 Consider the situation described in Problem 8.13. Destination 2 currently attracts 146 of the 700 social/recreational trips. If the total number of trips remains constant, determine the amount of amusement floor space that must be added to destination 2 to attract a total of 250 social/recreational trips.

8.17–8.36 Same as in Chapter 8.

Appendix C

Unit Conversions

LENGTH

1 inch (in) = 25.40 mm
 2.540 cm
 2.540×10^{-2} m
 2.540×10^{-5} km
 8.333×10^{-2} ft
 1.578×10^{-5} mi

1 foot (ft) = 304.8 mm
 30.48 cm
 0.3048 m
 3.048×10^{-4} km
 12 in
 1.894×10^{-4} mi

1 mile (mi) = 1.609×10^{6} mm
 1.609×10^{5} cm
 1609 m
 1.609 km
 6.336×10^{4} in
 5280 ft

1 millimeter (mm) = 1.0×10^{-1} cm
 1.0×10^{-3} m
 1.0×10^{-6} km
 0.03937 in
 3.281×10^{-3} ft
 6.214×10^{-7} mi

1 centimeter (cm) = 10 mm
 1.0×10^{-2} m
 1.0×10^{-5} km
 0.3937 in
 3.281×10^{-2} ft
 6.214×10^{-6} mi

1 meter (m) = 1000 mm
 100 cm
 10^{-3} km
 39.37 in
 3.281 ft
 6.214×10^{-4} mi

1 kilometer (km) = 1.0×10^{6} mm
 1.0×10^{5} cm
 1000 m
 3.937×10^{4} in
 3281 ft
 0.6214 mi

AREA

1 square inch (in^2) = 645.2 mm^2
6.452 cm^2
6.452 × 10^{-4} m^2
6.944 × 10^{-3} ft^2

1 square foot (ft^2) = 9.290 × 10^4 mm^2
9.290 × 10^2 cm^2
9.290 × 10^{-2} m^2
144 in^2

1 square millimeter (mm^2) = 1.0 × 10^{-2} cm^2
1.0 × 10^{-6} m^2
1.55 × 10^{-3} in^2
1.076 × 10^{-5} ft^2

1 square centimeter (cm^2) = 100 mm^2
1.0 × 10^{-4} m^2
0.1550 in^2
1.076 × 10^{-3} ft^2

1 square meter (m^2) = 1.0 × 10^6 mm^2
1.0 × 10^4 cm^2
1550 in^2
10.76 ft^2

SPEED

1 foot per second (ft/s) = 1.097 km/h
0.3048 m/s
0.6818 mi/h

1 mile per hour (mi/h) = 1.609 km/h
0.4470 m/s
1.467 ft/s

1 meter per second (m/s) = 3.6 km/h
3.281 ft/s
2.237 mi/h

1 kilometer per hour (km/h) = 0.2778 m/s
0.9113 ft/s
0.6214 mi/h

MASS

1 slug = 1.459 × 10^4 g
14.59 kg

1 gram (g) = 1.0 × 10^{-3} kg
6.852 × 10^{-5} slug

1 kilogram (kg) = 1000 g
6.852 × 10^{-2} slug

DENSITY

1 slug per cubic foot (slug/ft^3) = 515.4 kg/m^3

1 kilogram per cubic meter (kg/m^3) = 1.940 × 10^{-3} slug/ft^3

FORCE

1 pound (lb) = 4.448 N
           0.001 kip

1 kip = 4.448 × 10³ N
        1000 lb

1 newton (N) = 0.2248 lb
              2.248 × 10⁻⁴ kip

PRESSURE

1 pound per square inch (lb/in²) = 6.895 kPa
                                   6.895 × 10³ N/m²
                                   6.895 × 10⁻³ N/m²

1 kilopascal (kPa) = 0.145 lb/in²
                    1.0 × 10³ N/m²
                    1.0 × 10⁻³ N/mm²

WORK

1 foot-pound (ft-lb) = 1.356 N-m

1 newton-meter (N-m) = 0.7376 ft-lb

POWER

1 foot-pound per second (ft-lb/s) = 1.356 × 10⁻³ kW
                                    1.818 × 10⁻³ hp

1 horsepower (hp) = 550 ft-lb/s
                    0.7457 kW

1 kilowatt (kW) = 737.6 ft-lb/s
                 1.341 hp

Index

A

AADT (Annual Average Daily Traffic), 213–215
AASHO Road Test, 103–104, 107, 119
AASHTO:
    flexible pavement design procedure, 103–115
    geometric design guidelines, 59, 61–62, 64–65, 70–71, 73–74
    recommended deceleration rate, 36–37
    rigid pavement design procedure, 119–129
Access point frequency adjustment:
    for multilane highways, 194, 196
    for two-lane highways, 201–203
Aerodynamic resistance, 7–12, 16, 27, 34, 37
Aggregates, *see* Pavement
Aggregation of decision-making units, 275
Ahlvin and Ulery equations, 96, 103
Air density, 9–10
All-red time, 223, 228, 229, 256–258
Analysis flow rate:
    for freeways, 182
    for multilane highways, 197
    for signalized intersections, 238, 249, 261
    for two-lane highways, 203, 205–206, 208
Antilock braking systems, 28, 31, 32
Arrivals:
    deterministic, 146, 152, 153, 156–157
    exponential, 153, 156, 158
    Poisson, 146, 149, 151, 152, 156, 239
    random, 146, 157, 237–239
    time-varying, 157, 235
    uniform, 146, 152–153, 156, 157, 236–238
Asphaltic cement, 91, 92
Asphaltic concrete, 92, 93, 112
Available tractive effort, 8, 15–17, 21
Average length of queue, 157, 159, 161
Average passenger car speed, 178, 188, 193, 197
Average time waiting in queue, 157–159, 161
Average time waiting in system, 157–159, 161–162
Average travel speed (ATS):
    adjustment for the percentage of no-passing zones, 205–206
    defined, 201
    formula for, 205

B

Base conditions (for highway capacity computations):
    defined, 174
    for freeways, 176, 183, 187
    for multilane highways, 193–195
    for two-lane highways, 201–202
Base layer, 92, 93, 94, 106, 112, 116
Basic freeway segment:
    analysis of, 175–188
    defined, 175
Beam action, 115, 116
Bottlenecks, analysis of, 163–164
Boussinesq theory, 94–96, 103
Braking:
    antilock systems, 28, 31, 32
    distance, 36, 39
    efficiency, 31–34, 37
    force ratio, 28–31
    forces, 26–28, 30, 32
Bus PCE factors, 183–187, 204, 205

C

California bearing ratio (CBR), 111, 116, 123, 126
Capacity:
    defined, 143, 144, 170–172, 174, 289, 290
    for freeways, 176–177, 179
    for multilane highways, 192–195
    for signalized intersections, 229–231, 233, 238
    for two-lane highways, 201, 208
Central angle, 81, 83, 84
Centripetal force, 76, 77
Change interval, 223, 256, 258
Circular curve, 80
Circular load, 96, 117
Clearance interval, 223, 256, 258
Coefficient:
    of drag, 9–11, 25
    of road adhesion, 17, 25, 27, 28, 31, 32
    of rolling resistance, 7, 8, 12, 13, 16, 25, 27, 33
    of side friction, 76
Compound curve, 80
Concrete modulus of rupture, 120, 121
Contraction joint, 93, 116
Crest vertical curve:
    minimum length of, 58–61
    passing sight distance and design of, 70–72
    sight distance on, 58–59
    stopping sight distance and design of, 58–63
Critical flow ratio, 250, 253, 255, 265
Curve radius 76, 78–83

Curves, *see* Horizontal curve, Vertical curve
Cycle length:
    defined, 223, 253
    minimum, 253–254
    optimal, 253–254

D

Deceleration:
    for stopping sight distance, 28, 30–31, 35–37
    for yellow time calculation, 257
Degree of curve, 80
Delay:
    aggregation, 237, 260–263
    average, 231
    initial queue, 237, 239
    maximum, 231
    random, 238
    total, 155, 231
    uniform, 237
Demographic trends, 5
Density (of traffic):
    calculation of, 139–140
    defined, 136, 139
    relationship to flow, 143–144
    relationship to speed, 141–143
Departure channels, 152, 160
Departures:
    deterministic, 146, 152, 153, 156–157
    exponential, 153, 156, 158
    random, 146, 157, 237–239
    time-varying, 163–164
    uniform, 146, 152, 153, 156, 157
Design hour volume, 213–214
Design-lane axle loads, 128
Design speed, 57
Design traffic volumes, 211–214
Deterministic arrivals, 146, 152, 153, 156–157
Deterministic departures, 146, 152, 153, 156–157
Dilemma zone, 257–258
Directional design hour volume, 214
Directional distribution of traffic, 206–207, 214
Drainage coefficient, 112, 120, 121
Driveline efficiency, 20–21
Driver population adjustment, 182, 187
Driver's eye height, *see* Height of driver's eye
Dual-ring signal control, 224–226, 250

E

Effective green time, 229
Effective red time, 229

Engine-generated tractive effort, 18–21
Equilibration, 272–274, 279
Equilibrium problem, 289
Equivalent single-axle load (ESAL), 106, 107, 120, 122, 128
Exponential arrivals, 153, 156
Exponential departures, 153, 158
Exponential distribution, 150–153, 160
External distance, 81

F

Fatalities, 1, 2
First-in-first-out (FIFO), 153
Flexible pavement:
    defined, 92, 93
    deflections, 96
    design equation, 104–111
    stresses, 94, 96
    see also Pavement, Rigid pavement
Flow, of traffic:
    defined, 136–137
    relationship to density, 143–144
    relationship to speed, 144–145
    see also Analysis flow rate
Flow-density model, 143–144
Forecasting vehicle traffic, 270–275
Free-flow speed (FFS):
    defined, 141–142, 174–175
    for freeways, 179–180
    for multilane highways, 194–195
    for two-lane highways, 202–203
Freeway, *see* Basic freeway segment
Fuel efficiency, 25
Fully actuated signal control, 224

G

Gear reductions, 20–21
Grade:
    average, 184
    final, 48
    initial, 48
Grade adjustment (for two-lane highways), 203–204
Grade resistance, 7–8, 14
Gravity model, 301
Green time, 223, 229, 255

H

Harmonic mean of speed, 138
Headlight angle, 63–64
Headlight height, 63–64

Headway, 136–137
Heavy-vehicle adjustment factor, 182–184, 187, 202–204
Height of driver's eye, 59, 70, 73
Height of roadway object, 59, 70, 73
Highway:
  capital investment, 3
  fatalities, 1–2
  performance functions, 289–290
  safety, 1
Horizontal curve:
  alignment, 75
  centripetal force on, 76, 77
  degree of curve, 80
  external distance, 81
  length, 81–83
  middle ordinate, 81
  PC, 81
  PI, 81
  PT, 81
  radius, 76, 78–83
  side frictional force on, 76
  sight distance on, 82–83
  stopping sight distance and design of, 82–86
  superelevation on, 77, 78–79
  tangent length, 81
  types of, 80
Hourly volume, 175

I
Infrastructure, 1–4, 6
Intelligent transportation systems (ITS), 4
Interchange density adjustment (for freeways), 179–180, 182
Interstate highway system, 1, 3

J
Jam density, 142–143

K
K-factor, 214
K-value, 54, 61–62, 64–65, 71

L
Lane groups, 247–249
Lane width adjustment:
  for freeways, 179–180
  for multilane highways, 194–195
  for two-lane highways, 202–203
Last-in-first-out (LIFO), 152

Lateral clearance adjustment:
  for freeways, 179–181
  for multilane highways, 194–196
Layer coefficients, 112
Least squares regression, 302–304
Length:
  of horizontal curve, 81–83
  of vertical curve, 48
Level of service:
  defined, 171–173
  for freeways, 188
  for multilane highways, 197
  for signalized intersections, 260–261
  for two-lane highways, 208
Level terrain:
  defined, 184
  PCE values for, 184, 204–205
Linear performance function (for travel time), 289–290
Load transfer coefficient, 120, 121, 123
Logit model:
  applications, 283–289
  methodological approach, 281–283
LOS, see Level of service
Lost time, 227–228

M
Mass factor:
  acceleration, 22
  braking, 32, 37
Mathematical programming, 296–297, 303
Maximum braking force, 27–28
Maximum flow rate, 143, 174, 176, 227
Maximum likelihood estimation, 280, 283, 304–306
Maximum service flow rate, 176
Maximum tractive effort, 15–17, 21
Median type adjustment (for multilane highways), 194, 196
Middle ordinate, 81
Mode and destination choice, 272, 281–288
Modulus of elasticity, 96, 115, 117, 123
Modulus of subgrade reaction, 116, 117, 120, 121, 123, 126
Mountainous terrain:
  defined, 184
  PCE values for, 184
Multilane highway:
  analysis of, 191–197
  defined, 191

N

Negative exponential distribution, 150
Nonlinear performance function, 289–290
Number-of-lanes adjustment (for freeways), 179–181

O

Object height, *see* Height of roadway object
Offsets, 52–54
Optimal brake force proportioning, 28, 30
Optimal traffic signal timing, 241
Overall standard deviation, 106, 111, 120

P

Passenger-car equivalency (PCE) factors:
    buses, 183–187, 204, 205
    recreational vehicles, 183–184, 186–187, 204, 205
    trucks, 183–187, 204, 205
Passing sight distance:
    for crest vertical curve, 70–71
    for horizontal curve, 84
Pavement:
    aggregates, 92–93
    failure, 92, 103, 104, 119
    performance, 104, 105
    reliability, 106–108, 111, 112, 120, 122
    serviceability, 104, 106, 107, 108, 111, 120–122
    see also Flexible pavement, Rigid pavement
Peak-hour factor, 182–183, 203
Pedestrian green/crossing time, 259
Percent time spent following (PTSF):
    adjustment for combined effect of directional distribution of traffic and the percentage of no-passing zones, 206–207
    defined, 201
    formula for, 206
Perception/reaction time:
    for stopping sight distance, 39, 57
    for yellow time calculation, 257
Performance functions (for travel time), 289–290
Permitted movement, 223
Plan view, 46–47
Point of curve (PC), 81
Point of intersection (PI), 81
Point of tangent (PT), 81
Point of vertical curve (PVC), 48
Point of vertical intersection (PVI), 48
Point of vertical tangent (PVT), 48
Poisson arrivals, 146, 149, 151–152, 156, 239
Poisson distribution, 146, 149, 151–152, 239
Poisson models, 279–280, 305–306
Poisson ratio (for pavement design), 96, 117
Poisson regression, 279–280, 305–306
Portland cement concrete (PCC), 91, 93, 115, 123
Present serviceability index (PSI), 104
Pretimed signal control, 224
Private vehicles, 4–6
Probabilistic choice model, 282
Profile view, 46–47
Progression adjustment factor (PF), 237
Protected movement, 223

Q

Queuing:
    arrival pattern, 152–153
    departure channels, 152–153, 160–161
    departure/service pattern, 152–153
Queuing models:
    $D/D/1$, 153–156
    $M/D/1$, 156–158
    $M/M/1$, 158–160
    $M/M/N$, 160–163

R

Radial-horizontal stress, 96
Radius (of horizontal curve), 76–77, 78–79, 80–81
Random arrivals, 146, 157, 237–239
Random departures, 146, 157, 237–239
Recreational vehicle PCE factors, 183–184, 186–187, 204–205
Red time, 223, 229
Reliability, *see* Pavement
Resistance:
    aerodynamic, 7–12, 16, 27, 34, 37
    grade, 7–8, 14
    rolling, 7–8, 12–13
Reverse curve, 80
Rigid pavement:
    beam action, 115, 116
    contraction joint, 93, 116
    defined, 93
    deflections, 93, 116, 117
    design equation, 120–126
    stresses, 93, 116, 117
    see also Flexible pavement, Pavement
Rolling resistance, 7–8, 12–13
Rolling terrain:
    defined, 184
    PCE values for, 184, 204–205
Route choice, 272, 289–296

S

Sag vertical curve:
  minimum length of, 64
  sight distance on, 63–64
  stopping sight distance and design of, 63–69
  underpass sight distance and design of, 72–75
Saturation flow rate:
  adjusted, 227
  base, 227
  defined, 226
  maximum, 227
Segmentation of trips by type, 275
Semi-actuated signal control, 224
Serviceability, see Pavement
Service flow rate, 176
Service measure:
  defined, 171
  for freeways, 179, 188
  for multilane highways, 194, 197
  for signalized intersections, 236, 260
  for two-lane highways, 201, 205–207
Shoulder width adjustment (for two-lane highways), 202–203
Side frictional force, 76
Sight distance, see Crest vertical curve, Horizontal curve, Sag vertical curve
Signalized intersection analysis:
  average delay, 231
  $D/D/1$ queuing, 230–236
  filtering/metering adjustment, 239
  lane groups, 247–249
  progression adjustment factor, 237
  random delay, 238
  signal controller mode delay adjustment, 238
  total delay, 155, 231
  uniform delay, 237
Signalized intersection control:
  dual-ring signal control, 224–226, 250
  fully actuated, 224
  permitted movement, 223
  phasing, 223
  pretimed, 224
  protected movement, 223
  semi-actuated, 224
Signalized intersection timing:
  all-red time, 228, 229, 256–258
  cycle length, see Cycle length
  effective green time, 229
  effective red time, 229
  green time, 223, 229, 255
  lost time, 227–228
  optimal timing, 241
  pedestrian green/crossing time, 259
  red time, 223, 229
  yellow time, 223, 229, 252, 256–258
Soil-bearing capacity, 92
Soil resilient modulus, 106, 107, 111, 112, 116
Space-mean speed, 137–140, 142
Spacing, see Vehicle
Speed:
  average passenger car, 178, 188, 193, 197
  at capacity, 143–144
  design, 57
  free flow, see Free-flow speed (FFS)
  harmonic mean, 138
  relationship to density, 141–143
  relationship to flow, 144–145
  space-mean, 137–140, 142
  time-mean, 137
Speed-density model, 141–143
Speed-flow model, 144–145
Spiral curve, 80
Split phasing, defined, 243
Stationing (for highways), 46–47
Stopping distance:
  practical, 35–36
  theoretical, 32–34, 37
Stopping sight distance:
  for crest vertical curves, see Crest vertical curve
  defined, 57–58
  for horizontal curves, see Horizontal curve
  for sag vertical curves, see Sag vertical curve
Structural-layer coefficients, 112
Structural number, 104, 106–112, 120
Subbase, 92, 94, 106, 112
Subgrade, 92–94, 106, 111, 116, 117, 120, 121, 123, 126
Superelevation, 77, 78–79
System optimal route choice, 297

T

Tangent length, 81
Temporal aggregation, 276
Temporal decisions, 271–272
Temporal distribution of traffic, 171, 175, 211–213
Terminal serviceability index (TSI), 104, 107, 120–121
Terrain, see Level terrain, Rolling terrain, Mountainous terrain
Thickness index, 104–106
Time-mean speed, 137

Tire:
    footprint, 95, 96, 117, 118
    load, 92, 95, 96, 118
    pressure, 92, 93, 95, 103, 118
Torque, 18–21
Tractive effort:
    available, 8, 15–17, 21
    defined, 7
    engine-generated, 18–21
    maximum, 15–17, 21
Traffic:
    control, *see* Signalized intersection control
    density, *see* Density
    design volumes, 211–214
    equilibrium, 289
    flow, *see* Flow
    forecasting, 270–275
    forecasting methodology in practice, 301–302
    intensity, 156–157, 160, 231, 250
    speed, *see* Speed
    volume, *see* Volume
Traveler decisions, 271–274
Travel time:
    free-flow, 289–290
    route, 289–291
Trip generation, 275
Trip-generation models:
    linear, 276, 279
    Poisson, 279–280
Truck PCE factors, 183–187, 204, 205
Two-lane highway:
    analysis of, 200–208
    defined, 200

U

Underpass sight distance, 72–74
Uniform arrivals, 146, 152–153, 156–157, 236–238
Uniform departures, 146, 152, 153, 156, 157
User equilibrium route choice:
    mathematical program, 296–297
    theory of, 290–291
Utility:
    function, 281–282
    maximization, 281

V

Vehicle:
    acceleration, 7–8, 21–23
    arrival rate, 146
    braking, *see* Braking
    cornering, 7, 75–76
    deceleration, *see* Deceleration
    departure/service rate, 152
    drag coefficient, 9–11, 25
    frontal area, 9
    fuel efficiency, 25
    headway, 136–137
    occupancy, 5
    power, 11–13, 18–19
    spacing, 139–140
    technologies, 2–4
    torque, 18–21
    tractive effort, *see* Tractive effort
Vertical curve:
    alignment, 47–48
    elevation (parabolic) formula, 48–49
    high point, formula for, 54
    low point, formula for, 54
    measurement of length on, 46, 48
    offsets, 52–54
    PVC, 48
    PVI, 48
    PVT, 48
    rate of change of slope, 48–49
    see also Crest vertical curve, Sag vertical curve
Vertical stress, 96
Volume:
    design hour, 213–214
    directional design hour, 214
    highest annual hourly, 213–214
    peak hour, 175, 183
Volume/capacity ratio, 176, 191, 237, 253

W

Wardrop's rule, 291
Warrants (for signal installation), 220
Webster's optimal cycle length formula, *see* Cycle length
Westergaard equations, 116, 117

Y

Yellow time, 223, 229, 252, 256–258
Young's modulus, 96, 123

Z

z-statistic, 106, 108, 120